Additional praise for Christine Ogren's *The American State Normal School*:

"A significant contribution to the literature in the history of American education . . . this book is amazingly well researched . . . a first rate addition to any history of education course." —James W. Fraser, Professor of History and Education, Northeastern University

"[*The American State Normal School*] is a terrific book, one that will have a major impact on many fields. This is an important contribution to the history of higher education, the history of teacher education, the history of teaching, and the history of gender in education. There is simply nothing else like it. . . . Christine Ogren carefully locates the normal school period in light of what came before and after, but she wisely keeps her attention on the period when it was anything but generic and when its identity was clear. . . . Ogren reconstructs the normal school as a vibrant, purposive, engaging, intellectually alive, and socially relevant institution, one that provided a rich experience of education and professional preparation for a large number of working and lower-middle class students, launching them on interesting trajectories of professional accomplishment and social mobility . . . Her effort to resurrect the valuable role that the normal school played in the lives of its students is an important part of what makes this book so effective, and this approach will serve to sharpen its impact on the literature. Like many people in my field, I will be assigning this book in my history of education classes as soon as it appears in print." —David Labaree, Professor, School of Education, Stanford University, author of *The Trouble with Ed Schools*

The American State Normal School
"An Instrument of Great Good"

Christine A. Ogren

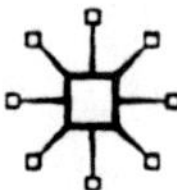

THE AMERICAN STATE NORMAL SCHOOL

First published in 2005 by
PALGRAVE MACMILLAN™
175 Fifth Avenue, New York, N.Y. 10010 and
Houndmills, Basingstoke, Hampshire, England RG21 6XS
Companies and representatives throughout the world.

PALGRAVE MACMILLAN is the global academic imprint of the Palgrave Macmillan division of St. Martin's Press, LLC and of Palgrave Macmillan Ltd. Macmillan® is a registered trademark in the United States, United Kingdom and other countries. Palgrave is a registered trademark in the European Union and other countries.

ISBN: 978-1-4039-6838-8

Library of Congress Cataloging-in-Publication Data

Ogren, Christine A.
The American state normal school : an instrument of great good / by Christine A. Ogren.
p. cm.
Includes bibliographical references and index.
ISBN 1–4039–6837–3—ISBN 1–4039–6838–1 (pbk.)
1. Teachers colleges—United States—History. 2. Educational change—United States—History. I. Title.

LB1811.047 2005
370′.71′173—dc22 2004062149

A catalogue record for this book is available from the British Library.

Design by Newgen Imaging Systems (P) Ltd., Chennai, India.

First edition: May 2005.

10 9 8 7 6 5 4 3 2 1

Printed in the United States of America.

Transferred to Digital Printing 2011

To my teachers

Table of Contents

Illustrations

Figures

Tables

Acknowledgments

This book is dedicated to my teachers. Beginning at Indian Landing Elementary School, I was fortunate to work with teachers who both challenged and encouraged me. Mr. Shay, Mr. Stewart, and the other social studies teachers at Penfield High School fostered my love of American history, and Clifford Clark and the other history professors at Carleton College taught me to think like a historian. Carl Kaestle, Bill Reese, Jurgen Herbst, and other members of the Educational Policy Studies Department at the University of Wisconsin helped me to become a historian of education. My doctoral advisor, Jurgen Herbst, introduced me to state normal schools and encouraged me to pursue my interest in women's experiences as normalites. Thanks also to the other members of my dissertation committee: Bill Reese, Clifton Conrad, Herbert Kliebard, and the late Sterling Fishman.

I am very grateful for the thoughtful reflections and suggestions, as well as good humor of many members of the History of Education Society, Division F (History and Historiography) of the American Educational Research Association, and the Association for the Study of Higher Education. When I presented various strands of my research at annual meetings, I received invaluable feedback from session chairs, discussants, and fellow panelists, including Sarah Barnes, Barbara Beatty, Amy Bix, Mary Ann Dzuback, Linda Eisenmann, Ruby Heap, David Labaree, Robert Levin, Yvonna Lincoln, Patricia Palmieri, Linda Perkins, Kristin Renn, Julie Reuben, John Rury, Susan Talburt, Laura D. Thornburg, and Amy Wells. Outside of formal sessions, many other members of these societies provided very helpful advice on my research, publishing, and career. I could not possibly list them all, but am especially grateful to the women's dinner group (particularly for helping finalize my title during the 2004 AERA meeting), the historians in ASHE ("HASHE"), and my perennial mentor as well as conference roommate, Linda Eisenmann. I always returned from professional meetings with renewed enthusiasm for my normal-school project.

Many archivists and librarians provided crucial assistance for my research. In the mid-1990s, the staffs at San Jose State University Archives, Castleton State College Archives, the Milne Library College Archives at SUNY Geneseo, Southwest Texas State University Archives, Collier Library Archives at the University of North Alabama, and the University of Wisconsin-Oshkosh Archives not only made their collections available to me, but also expressed excitement about my work and occasionally provided photocopying at a reduced price. At the University of Arkansas at Pine Bluff, art professor and de facto college historian Henri Linton let me loose in an exhibit on the history of the school and even temporarily dismantled displays in order to photocopy documents for me. In the early 2000s, archivists and librarians at former normal schools throughout the country answered e-mail inquiries about documents and photographs; their long-distance help is much appreciated.

I would also like to acknowledge financial support from the Spencer Foundation and the University of Iowa: a postdoctoral fellowship administered by the National Academy of Education and financed by the Spencer Foundation enabled me to work full-time on this project for the 1998–1999 academic year, and an Old Gold Summer Fellowship and an Arts and Humanities Initiative grant from the University of Iowa financed my research and writing during the summers of 2000 and 2001. Thanks to the Department of Psychological and Social Foundations of Education at the University of South Florida and the Educational Policy and Leadership Studies Department at the University of Iowa for providing me with excellent research assistants; at Iowa, Richard Breaux, Angie Hetler, and especially Christy Wolfe went far beyond their official duties to provide valuable help. In addition, I am grateful for the support and resourcefulness of Jan Latta, Karen Bixby and my faculty colleagues in the EPLS department at Iowa, as well as Amanda Johnson and the rest of the editorial staff at Palgrave.

Finally, I would like to thank my friends and family for their encouragement and patience. I am sorry that my grandmother, Nancy B. Foxton will not be here to see the book; she always supported my scholarship and my career. Thanks to Michael Rhine for helping me drive to several archives and many former normal schools throughout the country, DJ Chandler for hours of encouragement by telephone, and Bruce Hostager, whose support and sense of humor helped me to clear the final hurdles.

Christine A. Ogren
Iowa City, Iowa
August 2004

Introduction: "It Wasn't Much of a College"

Nearly five decades ago, David Riesman lamented "institutional homogenization" in higher education in the United States. He presented a "concededly oversimplified picture" of "a snake-like procession—the head of which is often turning back on itself . . . while the middle part seeks to catch up with where the head once was." He explained, "The assumption is that every decent university will offer courses in archeology, in Tudor history, or in the sociology of small groups, whether or not there exist topflight people to fill these lines, and even if to get them filled means sacrificing the possibility of building up a uniquely exhilarating department out of offerings not currently regarded as among the blue chips of academia." In the tail of the snake, Riesman located some denominational colleges, technical schools, and teachers colleges, which he called "colleges only by grace of semantic generosity."[1]

Although Riesman did not acknowledge them, gender, race, and social-class assumptions helped shape and direct the academic procession. Ivy League colleges and research universities made up the snake's head partly because they catered to white, male social elites, while the institutions in the snake's middle had fewer, and those in the tail had very few, such students. As less prestigious institutions increasingly emulated the universities in the snake's head, they marginalized women, racial minorities, and working-class students. In addition to institutionalizing gender, race, and class bias, the academic procession has implicitly shaped the historiography of higher education. As this field has grown in the decades following Riesman's observations, historians have assumed that the story of elite institutions captures *the* history of higher education. By focusing on one type of college "only by grace of semantic generosity," this book illustrates the fallacy of this assumption as well as the damaging effects of the academic procession.

State normal schools grew out of the common school revival of the early to middle nineteenth century. Responding to an increased need for trained teachers, education reformers adapted the German teacher seminary and the French *école normale* to serve the growing system of American common schools. Massachusetts established the first state normal schools, in 1839. Within a decade, Connecticut and New York followed suit. By 1870, 18 (of 37) states had at least 1, and a total of 39 state normal schools were located in New England, the mid-Atlantic states, the Midwest, and California. Twenty years later, state normal schools numbered 103, and were located in

35 (of 44) states, as well as Arizona Territory. By 1910, there were 180 normal schools in states north, south, east, and west; 42 (of 46) states, as well as 3 territories, had state normals. A few additional normals opened during the 1910s and 1920s; only four states would never establish normal schools.[2] These 180-plus institutions did not offer bachelor's degrees and their official purpose was to prepare students for a low-status profession, which colleges and universities had little interest in doing. Normal schools had so little prestige that they were beyond the bounds of the academic procession, yet the great snake would have a profound impact upon them.

By the end of the nineteenth century, the spread of public high schools meant that increasing numbers of normal-school matriculants were high-school graduates, and that many states began to look to the normals to prepare teachers for the burgeoning number of positions on high-school faculties. As a result, a few institutions took steps forward in the procession. Before the turn of the century, the schools in Albany, New York and Ypsilanti, Michigan adopted the name "normal college." Beginning with Ypsilanti in 1903, several normals began to offer four years of college work and to grant bachelor's degrees. Replacing the title "normal school" with "teachers college" generally indicated that an institution required high-school graduation for admission and granted college degrees. The majority of state normal schools became teachers colleges during the 1920s and 1930s. But their quest to advance in the hierarchical procession did not stop there; beginning in the 1940s, they dropped teacher education as their organizing purpose. The flood of World War II veterans seeking higher education accelerated the move away from teacher education because all-purpose state colleges were better able to meet their varying needs. The 1940s, 1950s, and 1960s witnessed another flurry of name changes as the former normals became state colleges, first in the West, Midwest, and South, and later in the East. And by the end of the century, many would become state universities. In fact, institutions that began as normal schools formed the nucleus of state university systems in California, Maine, Maryland, Minnesota, Pennsylvania, New York, and Wisconsin. Western, Central, Southern, and Eastern Connecticut State Universities, like the "directional" universities in Colorado, Illinois, Kentucky, Missouri, Oklahoma, Oregon, Tennessee, Texas, and Washington, also began as normal schools.

Although institutions whose roots are in nineteenth-century normal schools play a central role in mass higher education today, their story is not well known. Looking only forward to the head of the snake, former normals have generally buried their roots as deep as possible. As education scholar Paul Woodring explained, "A speaker before an academic group (or a radio announcer of football scores) could always get a laugh by mentioning 'Slippery Rock State Normal School.' Students and faculty members in such institutions were sensitive to their lack of status and hence eager to transform the normal schools into colleges."[3] With each advance in the procession, these institutions breathed a sigh of relief. For example, a 1940s in-house history of the New York State College for Teachers at Albany (as it was then called) reported that, in 1908, "the last two-year class was graduated. Rid at last of this final incubus of its normal school days, the institution in all its departments reached full collegiate stature."[4] Other teachers colleges sandblasted buildings to remove the ignominious word "Normal," and saw to it that their town's Normal Avenue became College Avenue, and then University Avenue.[5] In his study of the history of teacher education, John Goodlad found that in the late twentieth century, such schools paid little attention to their past in teacher education. He reported,

> The bundle of catalogues, recruitment documents, and the like forwarded to us prior to our arrival to one such campus, known by us to be one of the earliest and most respected normal schools, included no mention of this august past. At another university, the one-room schoolhouse recently transported to the campus appeared to be less a nostalgic symbol of worthy services rendered than a monument to an impoverished past thankfully left behind.[6]

Able to view themselves only according to the model of the academic procession, these institutions were quick to discard their past.

Historians of education have tended, with only a few exceptions, to be co-conspirators in the former normals' efforts to bury their roots.[7] With little respect for nineteenth-century approaches, historians of teacher education have painted a disparaging picture of these institutions. And, interested only in the head of the snake, historians of higher education have simply omitted them. Their silence has further encouraged society and the schools themselves to measure their worth only according to the great snake. Thus, biographer Robert Caro's description of the institution that Lyndon B. Johnson entered in 1927 encapsulates the dominant condescending view of state normals and infant teachers colleges:

> In all the 24,000 square miles of the Hill Country, there was only one college.
>
> It wasn't much of a college. Its Main Building—surmounted by four spires and by layers of arches, gables, pinnacles and parapets—had been built to impress, and had been placed on the highest hill in the San Marcos area, so that its red spires, trimmed with gold paint, glittered for miles across the hills as if Camelot had been set down in dog-run country. But "Old Main," as it was known, and three other buildings lined up on the steep stairstep campus—a library so rickety that when, the year before, it had enlarged its reference department on the second floor, that floor had begun to cave in and all encyclopedias had had to be hastily moved downstairs; a rough, wooden, barn-like "gymnasium"; and a squat, unadorned classroom structure—were, except for a few frame houses, converted to classrooms, the extent of the campus of Southwest Texas State Teachers College at San Marcos.[8]

The normal school/teachers college certainly fell far short of elite institutions of higher education.

Casting aside the paradigm of the academic procession reveals, however, a very different image of this institution. After discussing their early years, this book presents a wide-ranging look at state normals during their heyday from 1870 through the 1900s, based upon archival research at seven former normal schools (located in Castleton, Vermont; Geneseo, New York; Florence, Alabama; Pine Bluff, Arkansas; San Marcos, Texas; Oshkosh, Wisconsin; and San Jose, California) and a review of various sources on close to one hundred other former state normals. These sources include some additional archival materials and many institutional histories, which I have used selectively—like Frederick Rudolph, I have "carefully culled episodes and illustrations" to use as primary material for my own analysis.[9] Examining the many dimensions of curricular and extracurricular life reveals that, although it "wasn't much of a college," the state normal school was a revolutionary institution in the field of higher education.

In their early years during the 1840s, 1850s, and 1860s, state normal schools had the potential to be revolutionary, but their future was far from certain. Part I describes the successful and unsuccessful struggles of educators in various states to establish normals, as well as the experiences of early normal-school students. In many ways, early state normals deserved the criticism they received at the time, and have since received from historians. Chapter 1 explains that they were bare-bones institutions, most of whose students were barely educated beyond the elementary level. The academic curriculum was necessarily low-level, and the teacher-education curriculum was immature. At the same time, however, early normal schools broke new ground in even offering teacher education. They served women, who were not welcome at most higher-education institutions, and children of struggling small farmers, who had limited access to any other type of higher education. And many of these students seized their opportunity, however limited it was. Chapter 1 also documents what early students made of their normal-school experiences, and how attending normal school helped to shape the course of their lives.

The revolutionary spark of the early normal schools ignited during the institutions' heyday between the 1870s and the 1900s. Part II focuses on this important period in educational history. Chapter 2 describes the growth and expansion of state normal schools during these decades and their accessibility to students who were new to higher education. Normal schools welcomed students who were female, older than typical college students, had work experience (usually in teaching), were not well-off financially, and were from provincial, educationally unsophisticated backgrounds. Also during this period, normal schools began to educate members of racial minority groups, both in separate institutions and in majority-white schools. Not only did normal schools serve such "nontraditional" students, but many of their policies also met these students' needs for guidance and financial assistance.

The following chapters describe the many dimensions of normal-school life during the schools' heyday. Chapter 3 focuses on academic studies and intellectual life. Although still somewhat restricted, the curriculum introduced students to the life of the mind. Required classes in mathematics, the sciences, history, and English language arts developed students' reasoning and analytical skills as well as their abilities to express their ideas, while optional advanced studies presented the opportunity to acquire prestigious cultural knowledge. Student-organized extracurricular activities, especially the literary societies, enabled students to gain further intellectual capital, or the cultural knowledge of the middle and upper classes. At the same time, as chapter 4 explains, the teacher-education program fostered a sense of professionalism. As students studied teaching, observed and practiced in "model" elementary schools, and delved into educational topics in their societies, they viewed themselves as dedicated educators. Other aspects of public life at state normal schools, described in chapter 5, also prepared students for middle-class life, and sparked in them a sense of possibility, especially for women. During a time when women struggled for acceptance in the public sphere, female students served as class officers, debated current issues, and dominated the basketball court. Not surprisingly, many graduates went on to lead lives that reached far beyond their humble origins, and in the case of female graduates, their gender. Among graduates, long teaching careers were common, even for some women who married. Teaching and missionary work sometimes took them far from home. Numerous normal-school graduates went further educationally, and a few pursued careers in law and medicine.

When Massachusetts opened its second normal school, at Barre in 1839, Governor Edward Everett congratulated his "fellow citizens and friends . . . on the establishment, in this community, of an institution, destined, we trust, to be an instrument of great good."[10] After a few decades of uncertainty, normal schools between the 1870s and the 1900s lived up to Everett's prediction in ways he probably did not foresee. Then, as the state normal school reached its peak in the second decade of the twentieth century, the institution changed in significant respects. The epilogue to this book describes how normals began by the 1910s to emulate collegiate institutions in the curriculum and extracurriculum, in the process of transforming themselves into state teachers colleges. While the model of the academic procession might suggest that such an increase in stature was nothing but positive, this book suggests otherwise. To dismiss the state normal school as not "much of a college" is to sell it short, for it was not only "an instrument of great good," but also a revolutionary institution of higher education.

Part I

Early Normal Schools, 1840s–1860s

1

"To Awaken the Conscience": Establishing Teacher Education and State Normal Schools

Marshall Conant was principal in the 1850s of the Bridgewater State Normal School in Massachusetts, one of only a handful of such institutions. A former district-school teacher and head of a private school, Conant had also run the topographical department of the Boston Water Works and served as a consulting engineer for both a railroad and cotton gin company. While his route to the principalship had been circuitous, Conant's sense of the normal-school mission was straightforward; he explained, "I have sought to awaken the conscience to feel the responsibilities and duties that devolve upon the teacher . . ."[1] In this statement, Conant captured the spirit of the preceding three decades of advocacy for teacher education. Education reformers of the early to middle nineteenth century sought to awaken the conscience of the public and state legislators to the importance of teaching and teacher training, and to establish state normal schools as the primary vehicle for shaping a professional teaching force. For more than a quarter-century following the establishment of the first one in 1839, however, state normal schools did little more than "awaken the conscience." They remained an unpopular option among many for the education of teachers, and their methods of teacher training lacked substance. Early state normal schools did succeed in instilling future teachers with the sense that they were undertaking a consecrated mission and, in the process, they also awakened students' consciousness of the wider world.

Education Reformers Make a Cause of Teacher Education

Teacher education was one plank of the antebellum common school reform movement, which sought to improve society by instilling uniform values in standardized elementary-level schools. Horace Mann explained, "Above all others, must the children of a Republic be fitted for society, as well as for themselves. . . . however loftily the intellect of man may have been gifted, however skillfully it may have been trained, if it be not guided by a sense of justice, a love of mankind and a devotion to duty, its possessor is only a more splendid, as he is a more dangerous barbarian." During the decades following the American Revolution, rudimentary schooling had gradually become widespread while remaining remarkably unsystematic. In rural areas, locally controlled district schools had short, rather chaotic sessions and were subject to the whims of the community. Urban schools included a variety of independent pay schools tailored to different parental interests and income levels, and charity schools to acculturate the poor. Rapid social change by the 1830s challenged this laissez-faire approach to schooling. The extension of voting rights among white males, the growth of manufacturing and factory production and the corresponding move away from a predominantly agricultural economy, increasing immigration of Roman-Catholic Europeans, and rapid urbanization, ignited leaders' fears of social and political instability. As a state senator in Massachusetts, Mann proposed a solution: state-controlled education to instill republican virtue, Protestant morality, and capitalist sensibilities.[2]

Henry Barnard in Connecticut, Horace Eaton in Vermont, John Pierce in Michigan, Calvin Stowe in Ohio, Calvin Wiley in North Carolina, and many others, joined Mann in a self-styled crusade for common schools. Primarily members of the Whig party, middle-class, and Protestant, these reformers adhered the "cosmopolitan" belief in government intervention and institution building to shape the country's economic and social growth. Like state-run canals, prisons, insane asylums, and poor houses, a system of public schools would foster the nation's orderly advancement. The school activists admired the development of state-regulated schooling in Prussia, apparently overlooking its monarchial government structure. Despite fears of "Prussianization," reform campaigns were most successful in Massachusetts—where Mann was appointed state superintendent in 1837—and in other northeastern states. The Middle Atlantic and Middle West also saw many reforms before the Civil War, but it would take longer for substantial systematization to reach the South and Far West. Everywhere the reformers called for bureaucratic changes in the interest of creating centralized, uniform systems. The top of the bureaucracy would be a state department of education, headed by a superintendent or secretary of education. Throughout the state, city or county superintendents would answer to the state superintendent and disseminate state-department policies. Education journals and reports would help spread information about policies and effective practices. Tax money would ensure free schooling for all citizens. In urban areas, pay and charity schools would come together to form public systems, while rural districts would consolidate into larger, more efficient systems. Wherever possible, principals would oversee age-graded schools. Longer school terms, increased daily attendance, and uniform textbooks would enable all pupils to absorb the morals and virtues of a Protestant, capitalist republic—as long as they had effective teachers. Thus, teachers, the foot soldiers in the crusade, received much of the reformers' attention.[3]

The common school reformers' mission was to reverse two centuries of high turnover and ignominy in the teaching ranks. Before the antebellum period, teachers were most often young, white males who were preparing for other professions, especially the ministry. Short school sessions enabled them to teach during college vacations or while they awaited a church appointment. Longer-term teachers tended to be men who farmed or ran other businesses on the side, or whose handicaps made them unsuitable for more physically taxing occupations. The few women who taught ran rudimentary "dame schools" in their homes or perhaps presided over the local school's summer session. Historian Willard Elsbree's tongue-in-cheek description of the character of early schoolmasters highlighted their unfavorable reputation: while most were "sober, upright, virtuous, and God-fearing," Elsbree wrote, the "schoolteacher who was a rogue, scoundrel, defamer, souse, or knave" dominated popular memory. Whether virtuous or scoundrels, teachers before the antebellum period had no specialized training. They were usually hired by town elders or some sort of community group, who attempted to test applicants' subject matter and pedagogical knowledge, as well as character and religion. In some towns, the ignorance of the hiring committee or lack of applicants made the interview process a bit of a farce; other committees forewent the interview to hire their relatives. Disturbed by such trends, school reformers called for a wider pool of better-prepared applicants as a step toward the professionalization of teaching.[4]

At the same time that reformers wanted to select teachers more carefully, their agenda called for more oversight of teachers and for teaching to occupy more of the year, which made it less attractive to men looking for independence and temporary work. Furthermore, as the agricultural economy declined, so too did the family unit of production. Middle-class gender ideology told men to make their way in the public world, and the growth of manufacturing and trade created changes in the labor market that presented them with all sorts of new options. Teaching, especially of small children, was outside the public, male sphere because the school was seen as an extension of the family. Catharine Beecher lamented in 1846, "Thus it is that two millions of American children are left without any teachers at all, while, of those who go to school, a large portion of the youngest and tenderest are turned over to coarse, hard, unfeeling men, too lazy or too stupid to follow the appropriate duties of their sex."[5]

The same social changes that pulled men out of the classroom suggested that women take their place, as middle-class mores defined females as nurturing, gentle, maternal, pious, and obedient. Prescriptive gender ideology limited women to the private sphere of domestic life and motherhood, condemning the many poorer and immigrant women who worked in factories and the rare middle-class women who pursued the established professions. As early as 1818, Emma Willard invoked these constructions of gender to argue that teaching children was appropriate for women; nature, she explained, "has given us, in a greater degree than men, the gentle arts of insinuation . . . a greater quickness of invention . . . and more patience." She also mentioned that "women had 'no higher pecuniary object' or ambition than to teach and 'could afford to do it cheaper.' " Three decades later, Beecher proclaimed, "*The educating of children, that* is the true and noble profession of a woman," and went on to argue that filling the teaching ranks with women would solve multiple problems, including the "sufferings" of women who worked in eastern mills and factories, boredom among middle-class women, and the teacher shortage on the western frontier. Mann agreed that teaching was ideal for women, stating that reasons for employing female teachers included "the greater intensity of the parental instinct in the female sex, their natural love of the society of children, and the superior gentleness and forbearance of their dispositions." Indeed, education reformers were attracted to the image of peaceful schoolrooms governed by female teachers' "moral suasion," not to mention the positive budgetary effects of their low salaries.[6]

School leaders still sought to hire male teachers, especially for older pupils, yet they also began to hire greater numbers of women. Massachusetts led the way in the feminization of the teaching force, as it did in school reform in general. As early as 1834, 56.3 percent of Massachusetts' teachers were female. By 1850, 66 percent, and by 1860, 77.8 percent of teachers in the state were women. Beecher's own efforts increased the number of women teachers in the Middle West: between 1846 and 1856, her National Board of Popular Education sent nearly six hundred women from New England to teach on the western frontier. By 1860, women were the majority of teachers nationally; their predominance in the northern states outweighed their relatively low numbers in the South. In each region, the shift to women teachers tended to accompany urbanization and the systematization of schooling in urban and then rural areas.[7]

While employing women was a solution to the problem of finding a larger pool of prospective teachers, it also had indirect effects on the problem of inadequate teacher preparation. Longer terms of employment, education publications, and coordination at the state level, as well as training for teaching, were intended to create a more professional teaching force. The traditional professions, however, resided squarely in the male sphere of middle-class gender ideology. School reformers resolved the potential conflict between professionalization and feminization by emphasizing women's "natural" affinity for teaching, creating gender divisions in the educational structure, and reducing teachers' independence. Superintendents and principals, who were mostly male, would be decision-makers, and teachers, who were increasingly female, would follow their directions. Historian Jurgen Herbst argues that the long-term result was that administrators, and not teachers, became the professionals in education.[8] In the short term, feminization indirectly encouraged the growth of teacher training. Not only would formal training offer another avenue for administrators to direct teachers' actions, but the supposedly innate female characteristics of gentleness, obedience, and patience also must have made women seem especially moldable and thus open to formal training. In addition, women—who had very few options for advanced education—were more willing than men—who had many other career options—to pursue education for a marginal, low-paid occupation.[9] The shift to a feminized teaching force paralleled the intensification of the crusade for common schools and teacher training. Although reformers did not make a connection between feminization and teacher education—in fact, they often spoke as if the teaching force was still composed exclusively of men—perhaps the movement of women into teaching made legislators and the public more willing to heed the reformers' calls.

Education leaders began to make a case for teacher education long before the feminization of teaching. As early as 1750, Benjamin Franklin proposed an academy, one of the purposes of which would be to qualify "a number of the poorer Sort . . . to act as Schoolmasters." In 1789, the *Massachusetts Magazine* published a call to "annihilate all the Latin grammar schools" and replace them with higher-order public schools that would "fit young gentlemen for college and school keeping." The preceptor and board of overseers would "annually examine young gentlemen designed for schoolmasters in reading, writing, arithmetic, and English grammar, and if they are found qualified for the office of school-keeping and able to teach these branches with ease and propriety, to recommend them for this purpose." While Latin grammar schools continued, various academies and colleges in the late eighteenth and early nineteenth centuries made independent, fleeting attempts to add teacher preparation to their offerings. In 1816, Denison Olmstead's commencement address at Yale outlined his plan for "an academy for schoolmasters," inaugurating the movement for separate teacher-training institutions. Seven years later, Samuel Read Hall, a Congregational minister impressed with early institutions for professional training in theology, medicine, and law, opened a seminary for teacher training in Concord, Vermont. This effort continued until 1830, when Hall moved his operation to Andover, Massachusetts, where it became a department in the Phillips Academy. Although he trained only a limited number of young men, Hall also welcomed visiting educators and politicians. "Let the character of teachers be improved, and improvement in the schools will follow of course," he wrote. Hall's efforts, along with earlier advocacy for teacher education, laid the foundation for the veritable crusade in the second half of the 1820s and the 1830s "to awaken the conscience" to the necessity for teacher-training institutions.[10]

The middle to late 1820s saw the publication of a number of pleas for teacher education. Hall himself, in the preface of his 1829 *Lectures on School-Keeping*, argued that, to improve

> the character of teachers . . . institutions should be established for educating teachers, where they should be taught not only the necessary branches of literature, but, be made acquainted with the science of *teaching* and the mode of *governing* a school with success. The general management of a school should be a subject of *much study*, before one engages in the employment of teaching.

He explained that he was publishing this series of 13 lectures as a fallback because there were so few teacher-training institutions; if a future teacher were unable to attend a teacher seminary, at least he or she could read his book. Beginning in the winter of 1824–1825, James G. Carter, a Massachusetts statesman and the Reverend Thomas H. Gallaudet, principal of the school for the deaf in Hartford, Connecticut published a series of articles and essays elaborating on the need for formal teacher training and how to structure it. Carter's "Essays on Popular Education" proposed that future teachers ground themselves in the subjects to be taught, study the as-yet-to-be-developed science of teaching, and have the opportunity both to observe their professors and hone their own skills in a practice school. Gallaudet's "Plan of a Seminary for the Education of the Instructors of Youth" reinforced the need for an "experimental school" and declared that a teacher-training institution should be well furnished with teaching "apparatus" and a library. Gallaudet and Carter, with the backing of Governor DeWitt Clinton of New York, added that these institutions should be state-supported. Gallaudet proposed that a teacher seminary "be established in every state" and "be so well endowed by the liberality of the public that it may have professors of talent who should devote their lives to the theory and practice of the education of youth." Carter opened a teacher seminary in Lancaster, Massachusetts in 1827, and eloquently, yet unsuccessfully, appealed to the state legislature for aid. As these ideas attracted attention throughout the country, newspaper and journal editors wrote supportive reviews and reprinted the writings of Gallaudet and Carter.[11]

Just as the teacher-seminary idea began to receive publicity, education reformers discovered Prussia's successes in establishing institutions to educate teachers for its new centralized school system. Various German cities had established teacher seminaries as early as the mid-eighteenth century, and France and Holland had done so in the 1810s. But it was Prussia's establishment in 1819 of a state-supported system of teacher-training institutions that attracted worldwide attention. Henry E. Dwight's *Travels in the North of Germany in the Years 1825 and 1826*, published in New York in 1829, praised the Prussian government for requiring teacher training and establishing institutions to instruct future teachers in "the best methods of educating and of governing children as well as the subjects they are to teach." Two years later, editor William Channing Woodbridge published commentary and translated reports on Prussian teacher education in the *American Annals of Education and Instruction*. When he noted that the Prussian institutions were for male students and he printed the translation of a female Prussian educator's plea for seminaries for women teachers, Woodbridge was one of the few education leaders to acknowledge unresolved gender issues.[12]

While Woodbridge was spreading the word about Prussian teacher seminaries, French philosophy professor Victor Cousin, commissioned by the French government, spent several weeks visiting the school systems in the German states. He visited many elementary and secondary schools, universities, and the larger and more prominent teacher seminaries. Although he learned of Prussia's smaller, rural teacher seminaries through official documents and conversations with high-level administrators rather than firsthand visits, Cousin lavishly praised their approach to discipline and moral instruction as if he were intimately acquainted. Herbst explains that Cousin's *Report on Public Instruction in Germany*

> was less an objective account of Prussian schools and education than the projection of a French philosopher's ideal version of a school system intended to wed progress and stability. And it was this ideal combination of a nation's progress under responsible conservative guidance that appealed to American education reformers and assured the enthusiastic reception Cousin's report received in the United States.

Widely read in Europe, the report appeared in translation in New York in 1835 and immediately made a splash among school reformers, who began to use the term "normal school," a translation of the French *école normale*. In August, the meeting of the American Institute of Instruction in Boston included a reading of a summary of Cousin's findings. In December, Unitarian minister and school reformer Charles Brooks, who had learned of the Prussian system from a fellow passenger on a trans-Atlantic steamer in 1834 and begun corresponding with Cousin, embarked upon a lecture tour to promote state-supported normal schools in his home state of Massachusetts, as well as New Hampshire, Connecticut, New Jersey, and Pennsylvania. Brooks considered himself the "missionary angel" of normal schools; he appeared in cartoons holding a ferule, leading teachers toward a normal school in the heavens. In 1837, Calvin Stowe's *Elementary Education in Europe*, a report to the Ohio legislature on his own trip to research education in Prussia, strengthened the argument for an American adaptation of the German teacher seminary.[13]

By the middle 1830s, education reformers had certainly gone a long way toward awakening the conscience of the American public and political leaders to the importance of teacher education. A series of small triumphs during the second half of the decade brought them closer to their goal of state-supported normal schools. In 1836, James G. Carter, recently elected to the Massachusetts legislature, was appointed to chair a government committee on education. Not only did he use his new position as a platform for advocating teacher training, but he also helped to create the State Board of Education and appoint Horace Mann its first secretary. Mann and the superintendents in other states, in turn, used their positions to lobby further. Henry Barnard arranged many public meetings in Connecticut beginning in 1839, for the purpose of promoting teacher education. Farther west, throughout the 1840s, 1850s, and 1860s, Missouri's state superintendents of public instruction or secretaries of state made regular appeals to the legislature for the establishment of normal schools. In Plymouth, Massachusetts in 1838, Charles Brooks enjoyed the support of former President John Quincy Adams and United States Senator Daniel Webster, at one of his public meetings on normal schools. Adams declared rhetorically, "We see monarchs expending vast sums, establishing normal schools throughout their realms, and sparing no pains to convey knowledge and efficiency to all the children of their poorest subjects. *Shall we be outdone by kings?*"[14]

Perhaps inspired by Adams and Webster, Edmund Dwight, a philanthropist and leader in industry, donated $10,000 to the state of Massachusetts for "qualifying teachers of the common schools." The legislature matched his donation, and Governor Edward Everett signed the normal-school bill in April 1838. With $20,000 and the expectation that towns willing to host these new institutions would donate buildings and furnishings, the State Board of Education decided to establish three state normal schools. State normal schools would appear in New York within five years, and Connecticut within eleven years. Education crusaders viewed the fruits of their efforts with great anticipation. Mann declared, "Coiled up in this institution, as in a spring, there is a vigor whose uncoiling may wheel the spheres." Governor Everett was more circumspect. In his address to mark the opening of the normal school at Barre, he celebrated its potential to be "an instrument of great good," but also noted, "These institutions are, of course, to some extent experimental." Indeed, the normal schools' future was far from certain in 1839. While reformers had set their sights on state-supported institutions devoted exclusively to teacher education in the mold of the German teacher seminary, several other approaches to teacher training had taken shape, and would overshadow the state normal schools for the next three decades.[15]

TEACHER TRAINING IN ACADEMIES AND COLLEGES, TEACHERS' INSTITUTES, AND NON-STATE NORMAL SCHOOLS

During the 1840s, 1850s, and 1860s, state normal schools educated relatively few teachers; it was in multipurpose institutions of higher schooling, short-term teachers' institutes, and city, county, and private normal schools, that greater numbers of teachers received their training. For centuries, colleges had produced teachers rather unintentionally, in the process of preparing graduates to be ministers, doctors, lawyers, and college professors. By the late eighteenth century, academies were an increasingly popular alternative to colleges for higher-level education. Early teacher education in academies, as one historian explained, was likewise an unintended "by-product. This instruction was incidental, unorganized, unrecognized by the State and even unnoticed for a time by the academy officials themselves." Beginning in the late 1830s, colleges and academies recognized this heretofore-unofficial purpose and made it more explicit. For example, Principal Alvin M. Dixon advertised in 1839 that the curriculum of the fledgling Platteville Academy in Wisconsin embraced "all the branches commonly taught in Academies to prepare youth for college, for teaching, and for filling important stations in life." While Platteville and most other academies were coeducational and antebellum colleges remained for the most part male-only, separate female academies or seminaries proliferated. Following the lead of Troy Seminary in New York State (started in 1821 by Emma Willard), Hartford Seminary in Connecticut (established in 1832 by Catharine Beecher), and Mount Holyoke Seminary in Massachusetts (opened in 1837 by Mary Lyon), many of these institutions articulated teacher education—alongside preparation for motherhood—as a primary mission. Mount Holyoke's healthy endowment enabled Mary Lyon to fulfill her desire to offer tuition assistance specifically for women of limited means who planned to be teachers. Thus, at Mount Holyoke and other multipurpose institutions of advanced learning, teacher education arose as an official function.[16]

Growing numbers of colleges and academies offered teacher training as a specific course of study or curricular track. In the 1830s, Ohio University established a partial, nondegree course for aspiring teachers, and Lafayette College in Pennsylvania established a model elementary school on campus in which students who were future teachers could observe and emulate good teaching practices. By the 1840s, academies commonly offered "teachers classes." In western New York, the 1842 catalog of Brockport Collegiate Institute announced that the formation of a "teachers class" would "receive particular attention," and an 1849 flier for Geneseo Academy declared, "The faculty will make arrangements to give a course of instruction especially adapted to those who are preparing themselves to become teachers." Platteville Academy catalogs listed a "Teachers Class" beginning in the late 1840s, and the 1850 catalog of Castleton Seminary in Vermont included the following description of its "Teacher's Class":

> A class will be formed at the beginning of each term for the benefit of those who may wish to qualify themselves to teach either common schools or those of a higher order. Every facility will be given to qualify them for their important task. They will be required to pursue critically and thoroughly those branches they may design to teach; also to inform themselves as to the best method of imparting instruction and of governing schools.

Meanwhile, those who wished "to qualify themselves to teach" in the Baltimore area could study in "normal classes" established during the 1840s and 1850s at Patapsco Female Institute, Baltimore Female College, or the city's public high schools.[17]

These early teachers or normal classes were usually just add-ons to the regular academic curriculum and rather thin in content. Various institutions took the additional step of establishing separate normal departments. Randolph-Macon College in Virginia added a normal department as early as 1839, and by the 1850s, these departments were becoming common. Platteville Academy's 1855–1856 catalog promised, "Arrangements are being made for a Normal Department to be connected with the School, which will be perfected as soon as the necessary buildings are erected." From 1856 to 1857 into the 1860s, each Platteville catalog listed the subjects to be studied in the Normal Department, as well as the English and Classical Departments. Most of the Normal-Department classes were in academic subjects—and students probably took these classes with the other academy students—but "Theory and Practice of Teaching" was a staple by the late 1850s, and "Science and History of Education" was a third-year course beginning in 1861. Throughout the late 1850s and early 1860s, Geneseo Academy's catalogs listed a "Teachers' Department . . . for instruction in the Science of Common School Teaching." Denominational colleges in Michigan and other states, as well as some private universities, also established normal departments. The first university department, at Brown University in Rhode Island, was established in 1850 and suspended only four years later. Normal departments were especially common, yet also short-lived, at midwestern state universities. Indiana University's normal department existed intermittently between 1852 and 1873, and normal departments appeared at the state universities of Iowa, Missouri, and Kansas in 1855, 1868, and 1876, respectively. The University of Wisconsin had appointed a normal professor as early as the late 1840s, but lack of interest among the male-only student body delayed the establishment of a normal department until 1863, when the Civil War depleted the number of male students and the university sought for the first time to attract women. At Wisconsin and other universities, the normal department offered both academic and teacher-education courses, segregating female students as well as preparing teachers for the common schools.[18]

State support legitimated and privileged multipurpose institutions' role in teacher training. In addition to funding the establishment of normal departments at public universities, several states took explicit steps to encourage and support teacher education in academies and private colleges. New York State's tradition of providing limited financial support for its academies through the "literature find" made them "quasi-public" in the early nineteenth century. While much of this funding was for general operating expenses, the minutes of the New York State Regents noted as early as 1821, "When it is recollected that it is to these seminaries that we must look for a supply of teachers for the common schools . . . the Regents trust they shall be enabled to extend the sphere of their bounty . . . [as far as] the finances and resources of the State may warrant." In 1834, the Regents selected eight academies, one in each senatorial district, to receive $400 each for special teacher-training classes, in addition to their regular share of the literature fund. Seven years later, the number of designated academies doubled and the subsidy decreased to $300. This practice was discontinued in 1844, although academies continued to train teachers. Then, in 1849, the state began to designate one academy per county to receive a yearly grant of $10 or $12.50 per teacher-education student, with the stipulation that no single institution receive more than $250. Geneseo Academy's Teachers' Department received this appropriation, allowing its catalogs to advertise during the 1850s, "The appropriation of the State for the support of this department provides for the instruction of twenty-five [later: twenty] teachers, four months, gratuitously," provided they "give a written pledge to engage in teaching."[19]

While New York's program was probably the most extensive, other states also provided funding for the purpose of teacher education. New York's neighbor to the south began in 1834 to support teacher education through monetary and land grants. Although few would offer actual teacher-education courses, several of Pennsylvania's colleges received land grants that required them to train a certain number of students as teachers, and 29 academies chartered by the state received monetary support and/or land on the same condition. Farther south, the Virginia legislature in 1842 passed an act that started the Virginia Military Institute "on a distinctive mission as a normal school to supply the schools of the Commonwealth with efficient teachers." A state grant covered the board and tuition for teachers in training, or "state cadets," in exchange for their promise to teach for two years following graduation. To the west, Wisconsin in 1857 passed "An act for the encouragement of academies and normal schools," which established a system for dissemination of one-fourth of the proceeds from the sale of state-owned swamplands to academies and colleges, and a board of regents to oversee the process. Instead of establishing normal schools, the board at first preferred to support teacher training in existing institutions. Funds were to be distributed according to students' performance on a state teacher-certification examination; each academy or college was to receive $40 for each student who passed the examination. This funding encouraged the growth of Platteville Academy's Normal Department. Other states also provided some financial support for the education of teachers in existing institutions. For example, Maine passed an act in 1860 that funded the creation of "normal departments" in academies, and Alabama enacted a similar law in 1868. Although these programs were short-lived, they did convey state support for teacher education.[20]

With and without state support, teachers classes and normal departments became prominent at many multipurpose institutions of higher learning during the mid-nineteenth century. At the elaborate commencement ceremonies in which each graduating student presented an oration, essay, or declamation, it became very common for one or more students to address an educational topic. At Platteville Academy, for example, in 1854, N. D. Glidden won a prize for his declamation, "Educational Prospects of Wisconsin"; the following year, A. W. Bell presented a speech entitled "Education" and J. W. Blackstone spoke on "The American Student"; and in 1863, Meta Waters discussed "Practical Education." Meanwhile, at Geneseo Academy in 1863 Jennie E. Buell presented "The True End of Education," and at different end-of-term commencements in 1867, W. S. Peterson's declamation was entitled "The Education of the Masses, the Security of the People," and Mary E. Cole answered the question, "What is Education." As education became a subject of study at these multipurpose institutions, they became an important source of teachers. Randolph-Macon College produced many of Virginia's outstanding male educators of the mid-nineteenth century, while a majority of the teachers in Pennsylvania and other states received their training in academies. Female seminaries such as Hartford and Troy produced huge numbers of teachers. Historian David Allmendinger explained that, between 1837 and 1850, "Mount Holyoke graduates taught in overwhelming proportions." More than 80 percent taught at least briefly, and

26 percent taught for ten or more years. Colleges and academies, as well as growing numbers of public high schools, would continue to play an important role in teacher training into the late nineteenth century; they provided the opportunity to study education alongside academic subjects, for students who could afford the time and cost of sustained higher schooling.[21]

Many teachers in the mid-nineteenth century did not have the benefit of extensive higher schooling. For them, as well as teachers who needed to brush up on or had never studied teaching methods, teachers' institutes provided at least a modicum of professional guidance. As part of his efforts to rally support for teacher education in 1839, school reformer and Connecticut State Superintendent Henry Barnard held an experimental six-week meeting for 26 male teachers in Hartford. The teachers studied pedagogy, reviewed common-school subjects, and observed in the Hartford public schools. Barnard repeated the experiment the following year, inviting women as well as men teachers to attend. In 1843, a county school superintendent applied the term "institute" to the two-week teachers' meeting that he convened in Ithaca, New York, which replicated much of what Barnard had done in Hartford. While the ultimate goal of Barnard's undertakings was permanent, state-supported institutions along the lines of the German teacher seminaries, the more immediate result was the inauguration of teachers' institutes, which historian Carl Kaestle calls "one of the most popular innovations of the reform program."[22]

The idea of establishing teachers' institutes spread quickly throughout Connecticut and New York, and to the rest of the country. By the end of the 1840s, education leaders in Massachusetts, Rhode Island, Maine, Vermont, Pennsylvania, Ohio, Michigan, Illinois, and Wisconsin convened teachers' institutes at least somewhat regularly. The idea reached California by the mid-1850s, New Mexico by the 1860s, and southern states such as West Virginia and Arkansas soon after the Civil War. State laws and funding often supported the spread of teachers' institutes; a law passed in Rhode Island in 1845 called on school commissioners to hold "institutes 'where teachers and such as propose to teach may become acquainted with the most approved and successful methods of arranging the studies and conducting the discipline of instruction of public schools.' " Teachers were strongly encouraged—sometimes required—to attend institutes, which were scheduled in various locations throughout each state in order to ensure accessibility. For example, in 1857 in Vermont, institutes took place in six different towns, ranging from St. Johnsbury in the northeast to Townshend in the south. The first law regarding institutes in Arkansas stipulated that one be scheduled in each county, and the revised law in 1875 required the state superintendent to organize an annual "normal district institute" in each of the state's 14 judicial districts. Most of these Arkansas institutes met in county courthouses, and leaders in other states used whatever facilities they could find. In 1859 and 1860, Wisconsin's Platteville Academy donated its facilities for an institute; Henry Barnard, then serving as chancellor of the University of Wisconsin, directed the 1859 meeting.[23]

Teachers' institutes generally convened once or twice per year, and lasted between two and four weeks. Like other southern states, Arkansas first only provided institutes for white teachers, but later added separate institutes for black teachers. Henry Barnard explained in 1865 that the "exercises" of each institute should include three elements: a review of common-school subjects, with attention to the best methods of teaching them and a discussion of "such difficulties as any member of the Institute may have encountered in teaching the same"; lectures on and discussions of "the organization of the schools, the classification of pupils, and the theory of practice teaching"; and public lectures on and discussions of broader educational topics, to take place in the evenings. Especially in the early years, institutes often fell short of this ideal because the teachers' thin grasp of subject-matter knowledge necessitated thorough reviews. Nevertheless, even the early institutes did give teachers a chance to grapple with some issues. For example, teachers at the 1847 Chautauqua County Institute in New York discussed extensively, "What is the best method of preventing whispering?" and "What is the best method of teaching morals?" Questions for discussion at an institute in Pennsylvania 11 years later included, "Should teachers open their schools in the morning by reading a portion of the scripture?" and "Should the wages of females be equal to those of male teachers?" Guest lectures for the institutes' evening sessions throughout the country included Louis Agassiz and Henry Barnard. By the 1880s, Arkansas teachers also listened to local doctors, ministers, and lawyers address school law, as well as hygiene and morals. Institutes gave practicing teachers a rare and brief chance to reflect just a little on their craft.[24]

What teachers' institutes lacked in serious content, they made up for in abundant enthusiasm for education and teaching. The superintendent of public instruction in Illinois in 1858 likened the importance of the institute for a teacher to the pilgrimage to Mecca for a Muslim—"the source whence he renews the spirit and life of his existence." In only slightly less flowery language, the superintendent in West Virginia in 1870 listed many benefits of institute attendance for the teacher, including, "[It] inspires him with an unwonted enthusiasm, caught by contact with superior and sympathetic hearts; and carries him back to his school room with new impulses and higher aspirations, to work out better results." Henry Barnard was hardly exaggerating when he called the teachers' institute an "educational revival agency." While teachers' institutes did more to engender enthusiasm than to provide in-depth training, they reached a very broad number of teachers. Reports indicated high attendance figures: in 1849, 36 percent of the teachers in Maine, in 1859, about 20 percent of the teachers in Wisconsin, and also in the late 1850s, approximately 15 percent of the teachers in Michigan attended institutes. According to historian Paul Mattingly, in the mid-nineteenth century, "institutes were the most prevalent teacher preparatory agency in America and touched the lives of more teachers than any other educational institution." The institutes' popularity would continue through the end of the nineteenth century. Thus, teachers' institutes did much to "awaken the conscience" of practicing teachers, but fell short of the reformers' goal of permanent institutions for teacher training.[25]

Other approaches to teacher education that began in the early to mid-nineteenth century were less itinerant than teachers' institutes, although some were certainly short-lived. Private, county, and municipal normal schools appeared throughout the country and, like teachers' institutes and programs in colleges and academies, helped to train sizable numbers of teachers. Samuel Read Hall's seminary in Concord, Vermont from 1823 through 1830 was the first private normal school. In the following decades, all sorts of private normals appeared throughout the country. Ohio had more than any other state, at least in part because of fairly strict teacher certification laws dating to the 1820s. In Kirtland, Ohio, Western Reserve Teachers' Seminary trained several hundred teachers during its existence from 1838 to 1853. Established in Lebanon, Ohio in 1855, the National Normal University attracted students from Ohio, Indiana, and Kentucky; it would enroll upwards of 2,000 students per year by the late nineteenth century. The struggles of another of Ohio's private normals are probably representative of the many that left few records. McNeely Normal School, which later became Hopedale Normal College, was established by Cyrus McNeely, a successful businessman with a social conscience. After a couple of false starts, in 1855 McNeely convinced the Ohio State Teachers Association to control the institution, while he served as president of the board of trustees. The normal reopened with normal and academic departments, and would later add an industrial department. Although it suffered financial troubles almost from the beginning, McNeely Normal School survived through the 1870s on low but fairly consistent enrollments of short-term students, and finally closed in the 1890s. A few other private normal schools, including one founded in 1867 by Joseph Baldwin in Kirksville, Missouri, survived by transforming themselves into state normal schools. Most private normals, however, faded away as Hopedale did; although they were more numerous than public normals through the end of the nineteenth century, most private normals struggled for survival and left few records.[26]

With at least a modicum of tax support, city and county normal schools did not struggle for survival as many private normal schools did. Like teachers' institutes, they provided many enrollees with a rather superficial orientation for teaching. Publicly created and supported, municipal normal schools were mainly in large eastern cities. New York's Public School Society created some of the earliest city normals, and turned them over to the city board of education by the early 1840s. The Male Normal School, Female Normal School, and Colored Normal School all met in the board's headquarters; sessions for the men took place in the evenings, ten hours per week, and women's sessions lasted for five hours on Saturdays. Similarly, the normal school in Newark, New Jersey, from its opening in 1855 through the 1870s enrolled only women and met on Saturday mornings in the city high school building. Students reviewed academic subjects and studied basic pedagogical principles, but little else. Other cities established more full-time normals.

The Philadelphia Normal School was the only municipal normal in Pennsylvania with its own building and faculty. Boston established a normal school for females in 1852; two years later it became Boston Girls' High and Normal School, and continued to offer a normal course until the normal and high schools were again separated in the 1870s. By the 1860s, several other eastern cities, as well as San Francisco and St. Louis, had established municipal normal schools, and would continue to train teachers into the twentieth century.[27]

County normal schools grew directly out of teachers' institutes. In Pennsylvania, teachers' institutes grew tremendously in popularity during the early 1850s. Many of the state's county superintendents of education, appointed after the creation of the new position in 1854, sought to institutionalize the itinerant institutes. In Lancaster County, superintendent J. P. Wickersham held three institutes during his first six months in office. Impressed with his efforts, the trustees of Millersville Academy decided to transform the institution into a permanent normal school; under Wickersham's leadership, the Lancaster County Normal School opened in 1855 in the former academy building. While the Lancaster institution would become a state normal school in 1859, several other county normal schools retained much of the temporary nature of the institutes, yet continued to train teachers for decades. Beginning in the 1870s, county normal schools were especially attractive to education leaders in the Midwest who were concerned with preparing teachers for rural schools; these leaders saw county normals as a rural version of municipal normals. By the early twentieth century, Wisconsin would have a system of "county training schools." Thus, county normal schools were yet another popular source of trained teachers.[28]

The growth of teacher education in academies and colleges, teachers' institutes, and non-state normal schools was evidence that the common-school crusade enjoyed a certain degree of success by the mid-nineteenth century. Their "conscience" awakened to the necessity of teacher education—or at least the potential market for teacher-training classes—leaders of academies and colleges, like the founders of private normal schools, added these new programs of study. State and local leaders provided support for some college and academy programs, as well as teachers' institutes and city and county normals. Relatively large numbers of current and future teachers took advantage of these new opportunities for training. However successful, these approaches to teacher education lacked the features that the school reformers so admired in the Prussian system. Academies and colleges were not devoted exclusively to teacher education, and the non-state normal schools were not part of the larger bureaucratic system necessary for extensive oversight and direction of the teaching force. While teachers' institutes were more closely linked to the forming educational bureaucracy, they were too temporary and itinerant to satisfy the reformers. Thus, school crusaders continued to advocate the establishment of state normal schools. While they would enjoy some success in the decades before 1870, it was limited and not due solely to the force of their arguments.

Establishing the First State Normal Schools

During the 1840s, 1850s, and 1860s, the movement for state-supported institutions for teacher education reached far beyond its northeastern beginnings. In Pennsylvania, school superintendents in the late 1830s, 1840s, and 1850s consistently and relentlessly argued for state-supported teacher training, and by the 1850s, virtually every report presented at any sort of educational meeting contained a statement about the necessity of normal schools. Beginning in the 1840s, Missouri's secretaries of state and superintendents of public instruction regularly presented the legislature with formal recommendations for the establishment of state normals. Meanwhile, Wisconsin's superintendent of public instruction, Azel P. Ladd, pleaded in his 1853 report: "That a school of this character is needed, the difficulty of obtaining good teachers for our schools is the best evidence. . . . Until we have an institution of this kind, we cannot reasonably expect the character of our schools to be satisfactory." The superintendents who followed Ladd echoed his pleas, quoting Horace Mann and Henry Barnard in their annual reports. Despite such ardent support, many legislators and citizens did not share the reformers' devotion to the ideology of Mann and Barnard, and states were slow to establish normals. During the decade following Edmund Dwight's donation, New York was the only state to join Massachusetts in passing normal-school legislation. Meanwhile, Massachusetts' first state normal schools were almost as transient as teachers' institutes: the institution established at Lexington moved in 1844 to West Newton, and then in 1853 to Framingham, and the institution established at Barre moved in 1841 to Westfield; only the school in Bridgewater, which opened in 1840, stayed put. Just as Governor Everett had made it clear that Massachusetts' normal schools were experimental, the legislation establishing the New York State Normal School at Albany in 1844 authorized state support for only a five-year trial period.[29]

Albany passed its trial to become a permanent institution, and in 1849 both Connecticut and Michigan authorized the establishment of state normal schools. During the next half-decade, these institutions opened in New Britain, Connecticut and Ypsilanti, Michigan, while Massachusetts established its fourth state normal, in Salem. Before the Civil War, four additional states opened a normal school each: New Jersey (at Trenton, in 1855); Illinois (at Normal, in 1857); Pennsylvania (at Millersville, in 1859); and Minnesota (at Winona, in 1860). During the Civil War years, Pennsylvania added two more normals, and three states passed authorizing legislation and opened their first: California (at San Francisco, in 1862); Maine (at Farmington, in 1864); and Kansas (at Emporia, in 1865). The legislation establishing these first state normal schools was quite similar from state to state. Michigan's act stipulated "that a Normal School be established, the exclusive purpose of which shall be the instruction of persons both male and female in the art of teaching, and in all the various branches that pertain to a good common school education." It also declared that applicants be examined for good moral character and a proclivity for teaching, and that, once accepted, students should sign a declaration "of intention to follow the business of teaching primary schools in this State." Supplemental action by the board of education allowed each member of the state house of representatives to appoint one male and one female student, presumably from his district, to attend for a significantly reduced fee. Such arrangements were very common, and in most states, the students with legislative appointments paid no tuition.[30]

During the years immediately following the Civil War, the number of state normal schools more than doubled. Between 1866 and 1869, New York, Pennsylvania, Maine, and Minnesota all added additional institutions, and Maryland, Wisconsin, Vermont, West Virginia, and Nebraska each opened one or more. Normal-school legislation in the latter states was very much like the legislation passed in other states before and during the Civil War. In the cases of Nebraska and West Virginia, normal schools were included in founding state legislation. The very first session of the Nebraska state legislature, in 1867, established the Nebraska State Normal School at Peru, and the state's first school law, passed in 1869, set forth requirements for and the purposes of the new institution. Discussions of whether to establish normal schools in West Virginia began soon after its statehood in 1863 and continued until the passage of enabling legislation in 1867. The new postwar constitutions of Virginia and Maryland both included provisions for state normal schools; Maryland opened one right away, in Baltimore, but Virginia would wait until the 1880s.[31] At the close of the 1860s, the United States had a total of 35 state normal schools, in 16 different states. The movement had certainly made important strides, yet these institutions continued to have a tenuous existence.

Virtually all state normal schools suffered from some degree of limited funding. In Massachusetts, Governor Everett declared that the normal schools were experimental "of necessity" because the "funds provided for their support, with all the subsidiary aid which can reasonably be expected from the friends of education in the neighborhood of the schools . . . are quite inadequate to the endowment of permanent establishments." While Everett described the situation in most states, a couple were especially extreme in the underfunding of their normals. Pennsylvania's Normal School Act of 1857 simply listed standards for designation as a state normal school, assuming that private enterprise would fund individual schools in pursuit of official recognition. A decade later, the Vermont legislature established normal courses in three county grammar schools, but appropriated no money to fund them. Vermont began in 1870 to provide some funding. Pennsylvania had done the same in 1861, but its normals remained "privately owned stock companies," or semiprivate institutions, into the twentieth century; in this unique arrangement, the state provided some aid, with virtually no oversight. The normal schools in these states, as in many others, remained at the mercy of the legislature from year to year. Some states farther west, such as Michigan, Kansas, and Wisconsin, had a slightly more stable source of funding in the various grants of salt-spring or swampland from the federal government for the support of education. Still, the proceeds from the sale of these lands were not enough to provide a rich endowment, and the normal schools had to compete with other state educational institutions; for example, the University of Michigan was always a higher priority than the state normal at Ypsilanti when it came to appropriations.[32]

Governor Everett assumed that the state normals would receive "subsidiary aid" from the towns in which they were located. Local support and financing were in fact essential to the founding of most state normal schools. Most normal-school legislation, as historian Charles A. Harper explained, "provided for competitive bidding from communities."[33] In towns throughout the United States, prominent community members and business owners rose to the challenge, encouraging their fellow citizens to donate money and land to woo the state normal. Historian David Potts has explained how important town boosterism was in the founding of colleges during the same period: "Particularly important to an analysis of college founding as a folk movement would be a careful delineation of the economic and cultural benefits perceived by many college supporters who never set foot in a classroom."[34] In other words, the boost that the new institution would give to the town was more important than the specific function of the institution itself. When it came to state normal schools, town boosters were less impressed with the vision of Horace Mann than they were interested in drawing attention, interest, and investment to their community. For example, in Platteville, Wisconsin in 1865, the local newspaper declared that the town "should leave no stone unturned to secure" a state normal, because

> the collecting here of from two to three hundred Normal students, with their necessary expenses for board, clothing and books, and the expenditures from the State Normal fund for the various running expenses of the School, would add to the resources of the village. Fifty thousand dollars per annum is a reasonable estimate. Then there would be a certain increase in population and value of property. There is yet another consideration . . . The two or three hundred students educated here every two years would form connecting links between so many respectable, often wealthy, but certainly enterprising families of this State and village of Platteville. The graduates would go forth from us and as teachers in various parts of the State relate with pleasure their school life here, and thus become the most eloquent advertisers of the beauty and healthfulness of the location, and the educational, material and social advantages possessed by our village. This could not fail to be to us a source of prosperity and growth.[35]

Teacher education was clearly less important than economic development; town boosterism probably did more than the arguments of the school reformers to help the early state normals survive.

The passage of normal-school legislation often led to intense competition between towns, many of which went to great lengths to attract a state normal. The Connecticut legislature stipulated in 1849 that normal-school "buildings and fixtures be furnished without expense to the state," and as soon as January of the following year received proposals from Middletown, Farmington, Southington, and New Britain. The "Committee on Location" visited the four communities and chose the latter, which had good railway connections to other parts of the state. In addition, boosters in New Britain had raised $16,250 to build normal-school facilities; donors included the leaders of the town's growing hardware industry. Similar events transpired in many other states, including Michigan, where members of the Ypsilanti community donated four acres of land and $13,500 plus the principal's salary for five years, and Vermont, where the town of Castleton financed the remodeling of a former medical-school building so that it could house the normal school. In Wisconsin, no fewer than 17 communities vied to be the location of one of the first two state normals. While several towns offered a site and $30,000, Platteville and Whitewater, both of which offered smaller sums, were the victors. Politics influenced the decision, as State Senator and Whitewater lumberman N. M. Littlejohn secured a board-of-regents appointment for a judge who was sympathetic to his interest in attracting a normal school. After Whitewater was chosen over frontrunner Racine, a newspaper there stated bitterly, "We rather hope the matter may go into the courts, and be thoroughly ventilated, so that we may know what inducements were offered by the Whitewater politicians." This never happened, and proud Whitewater citizens donated $400 to beautify the grounds of their new normal school.[36]

Just prior to the competition for a state normal school, Whitewater boosters had incorporated Whitewater Seminary, but this institution was never to be because they abandoned their plans in order to pursue the normal; securing *an* institution of higher-order education was more important than whether it would be a seminary or a normal school. Many communities with existing academies shared this sentiment, and offered their academy building and grounds for a state normal school. Pennsylvania's unique normal-school legislation essentially required that existing institutions transform themselves into state normals; following in the footsteps of the Millersville institution, the academies at Edinboro, Mansfield, Kutztown, and Bloomsburg converted themselves into state normal schools in 1861, 1862, 1866, and 1869, respectively. In Maine, four of the five towns that submitted normal-school propositions soon after the passage of the 1863 legislation included an existing academy in their offer; Farmington Academy was chosen for its trustees' promise not only to donate the building, but also to renovate it. The Nebraska legislature established its first state normal school at Peru because Peru Seminary was willing to deed its entire 60-acre campus to the state, and West Virginia's first two state normal schools, in Fairmont and Huntington, had both once been academies. Platteville's successful offer to the state of Wisconsin included the buildings and grounds of Platteville Academy, and the four state normals that opened in New York between 1867 and 1869 were all former academies. Many of these institutions were having financial difficulties by the mid-1860s; Platteville Academy's indebtedness and general decline were of grave concern, and Brockport Collegiate Institute in New York was essentially bankrupt. Academy trustees and town boosters jumped at the opportunity to procure state support, however minimal; they were more devoted to the survival of their local institution of higher-level education than to the cause of teacher education.[37]

In spite of—or perhaps because of the nature of—the support of the communities that housed them, state normal schools of the mid-nineteenth century were subject to constant scrutiny and criticism. Many people simply were not sold on the school reformers' arguments for state support of teacher-education institutions. Beginning as early as 1840, the existence of many state normal schools was under attack. In that year, the Massachusetts legislature's Committee on Education recommended the abolition of the normal schools, and 43 percent of the members of the house of representatives, unconvinced of the need to fund teacher education and fearful of centralized power, voted for the proposal. The normals survived, but the charges against them remained. Acting on similar concerns, the Connecticut General Assembly in 1867 cut off appropriations to the state normal in New Britain, and the school was forced to close for two years. A historian of the state normal in Peru, Nebraska described the "indifference and opposition" that the institution faced: "The press of the state, while not openly hostile, was commonly silent, except when an appropriation was asked for from the legislature. . . . The feeling was prevalent that the state was under no more obligation to found schools for the training of teachers than for the training of physicians and lawyers or other professions." Hostility was more open in Minnesota by the late 1860s. There, Governor William R. Marshall vetoed an 1868 appropriations bill, encouraging opponents of the state normals. Proponents of denominational colleges and academies—who stood to lose students to the normals—joined with opponents of state support for teacher education to back several anti-normal school bills. During the early 1870s, the legislature chipped away at the normals by passing restrictive bills such as one that limited principals' salaries. Later in the same decade, the New York State Assembly appointed a committee to report on the normal schools, especially "whether or not, in the practical operation of such schools, there has been any departure from their original purposes." Although the committee was ultimately satisfied that the institutions were "properly conducted," the scrutiny would continue.[38]

The standing of and outlook for state normal schools during the 1840s, 1850s, and 1860s were ambiguous. The fervent rhetoric of school reformers and enthusiastic efforts of town boosters enabled the opening of 16 normals in 11 states between 1839 and 1865, and a grand total of 35 normals in 16 states by 1869. High expectations and elaborate celebrations marked the establishment of many of these state normal schools. The ceremony for the laying of the cornerstone of Missouri's first state normal at the close of this period, in 1871, was typical. Eight-thousand people attended, and the local newspaper reported, "Yesterday was the grandest day Kirksville ever witnessed. . . . our streets bore evidence that the people of Adair County, were alive to the importance of the occasion." Yet these decades were also marked by intense scrutiny and criticism, as well as varying degrees of state neglect. Many normals struggled in hand-me-down buildings with inadequate support. For example, the normal in Winona, Minnesota occupied small quarters in a city-owned building for most of the 1860s, and even suspended operations between March 1862 and November 1864 because it was unable to replace the principal and one instructor, who had left to fight in the Civil War. The institution laid a cornerstone for a new building in 1866, but construction proceeded slowly, and it did not open until 1869. In the same year that Missourians celebrated in Kirksville, California's first state normal school "was removed to San Jose" because of "the neglect and indifference" it struggled against in San Francisco. State normal schools had gained a toehold, but on shaky ground. The schools' approach to teacher education was as tenuous as their standing as institutions, which contributed to the normals' ambiguous position.[39]

Teacher Education in the Early State Normal Schools

In his address upon the opening of the institution at Barre, Massachusetts Governor Everett outlined the "nature and objects" of a state normal school, including the four "core" elements of instruction. Everett was either insightful or prophetic, for he described the approach of state normal schools throughout the United States for decades to come. First, Everett explained, a normal would offer "careful review of the branches of knowledge required to be taught in our common schools; it being, of course, the first requisite of a teacher that he should himself know well that which he is to aid others in learning." Such academic instruction would indeed be a central component of normal education. In an 1841 letter to Henry Barnard, Principal Cyrus Peirce of Lexington, Massachusetts, stated that the first of his two aims in "instruction in the *art of teaching*" was "To teach *thoroughly* the principles of the several branches studied, so that the pupils may have a *clear* and *full understanding* of them." Academic instruction would soon become very important in the state normals for reasons other than just teacher preparation. Yet teacher education through core elements two, three, and four remained the main concern of Peirce and other principals.[40]

Everett described the second essential aspect of a normal-school education as "the art of teaching," or the principles of instruction and learning and the best methods by which to utilize these principles in particular subjects. The third element, captured under the general heading of "the government of the school," would include the moral influence of the teacher and issues surrounding the teacher's position in the community. Finally, the fourth part of Everett's core was the opportunity to observe and practice good teaching methods. Throughout the mid-nineteenth century and beyond, normal-school principals uniformly adhered to this framework for teacher education, and the informal national network that they created allowed them to share ideas about how to execute it. Although some historians of teacher education have emphasized distinctions between the philosophies of different normal-school principals, students at different normal schools throughout the country had remarkably similar experiences in their teacher-education programs. During the 1840s, 1850s, and 1860s, the substance of instruction in "the art of teaching" and "the government of the school" was uniformly rather thin, and observation and practice took place under less-than-ideal circumstances. The one area in which the normal schools were strong, was the creation of a sense of a calling, which historian Charles A. Harper described as "the spirit of consecration and the tendency to exalt the teacher's work."[41]

Mid-nineteenth century normal-school principals created several channels through which to exchange ideas and encourage one another. These leaders had much in common, and fit the profile of common school reformers in general. Virtually all were middle-class, Protestant men, and many had grown up in modest circumstances. David Perkins Page, principal at Albany, New York from its opening in 1844 until his death in 1848 at age 38, was raised in a New Hampshire farming family. When he attended an academy in his late teens, other students made fun of his plain clothes. According to Horace Mann, "To their gibes and jeers he was subjected, and doubtless his mind here got . . . an unspeakable contempt for the pretensions that are founded on wealth or habiliments, and a profound religious respect for moral worth." Page's experiences were an extreme version of those of many of the early normal-school principals. Cyrus Peirce, Charles Hovey (principal at Normal, Illinois from 1857 to 1862), David N. Camp (New Britain, Connecticut, 1858–1866), Albert Boyden (Bridgewater, Massachusetts, 1860–1906), Edward A. Sheldon (Oswego, New York, 1861–1897), and many others were from farm families. Richard Edwards (Salem, Massachusetts, 1854–1858, and Normal, Illinois, 1862–1876) was born to a poor family in Wales that immigrated to Ohio when he was ten years old; in Ohio, Edwards helped on the family farm while attending the local district school part time, then apprenticed as a carpenter.[42]

For many of these future principals, education was a way to escape the struggles of farming or poverty, and teaching was a way to finance further education. Page alternated between teaching winter sessions in the common schools and attending the academy. Peirce taught in district school during vacations from his undergraduate work at Harvard. Hovey, who had begun teaching in a district school at age 15, entered Dartmouth College at age 21 and paid his expenses there by teaching for three or four months per year. College degrees were common among the early normal-school principals. Peirce and his two successors, Samuel May and Eben Stearns were Harvard graduates, and Edwards earned a bachelor's degree from Rensselaer Polytechnic Institute. At Ypsilanti, Michigan, principal Adonijah Strong Welch (1852–1865) was a graduate of the University of Michigan, and his successor, David Porter Mayhew (1865–1870) was a graduate of Union College. As was often the case with poorer college students, several future normal-school principals became ministers. Peirce was a Unitarian pastor and his counterpart at Westfield, Massachusetts, Emerson Davis (1844–1846) was a Congregational minister. Malcolm MacVicar (Brockport, New York, 1867–1869) had attended Knox College in Toronto, but a lack of funds had kept him from graduating. Ordained as a Baptist clergyman, he soon found that he was more interested in teaching than preaching, and dedicated his evangelistic energy to work in education. Whether or not they became ministers first, other normal-school principals shared MacVicar's devotion to the social institution that had allowed them to improve their lives. An 1868 graduate of Illinois State Normal University reflected, "Dr. Edwards was so enthusiastic himself that every one of his pupils felt, 'Woe is me, if I do not teach.' In his masterly way he so pointed out the great possibilities for bettering humanity that were open to the teacher that the matter of salary seemed a secondary affair." Similar backgrounds and a shared cause linked early normal-school principals to one another. Their movements from one normal to another and communications through publications and national societies strengthened these bonds.[43]

The movement of principals and faculty members among various normal schools linked institutions that were geographically widespread. The most well-known examples of this migration are the "sons of Bridgewater," 26 former students of Principal Nicholas Tillinghast at Bridgewater, Massachusetts between 1840 and 1853, who all became normal-school principals themselves. The group included Edwards—who attended Bridgewater before Rensselaer—as well as Ahira Holmes, the first principal of the California State Normal School in San Francisco, George M. Gage, principal of Farmington, Maine from 1865 to 1868, and others who were scattered throughout the country. On a smaller scale, graduates of other early normal schools also moved on to head their own institutions. For example, William F. Phelps, a member of Albany's first class, served as principal of Trenton, New Jersey from 1855 to 1864 and Winona, Minnesota from 1864 to 1876, and Lyman B. Kellogg, who graduated from Normal, Illinois in 1864, was principal of Emporia, Kansas from 1865 to 1871. Faculty migration also allowed for considerable cross-pollination among normal schools. Gage's faculty at Farmington included graduates of Framingham and Salem, Massachusetts, as well as his own alma mater. The Bridgewater graduates, Julia A. Sears and Helen B. Coffin, both arrived in 1866; Sears left in 1868 to teach at the Minnesota State Normal School at Mankato, and Coffin moved in 1869 to the Eastern State Normal School at Castine, Maine. Graduates of Oswego were especially mobile: Mary R. Pitman made it to Emporia, where she was in charge of the Model Department in 1867–1868, and Anna R. Randall worked as an instructor at both Platteville and Whitewater, Wisconsin in the late 1860s. It was also common for newly appointed principals to visit other normals. For example, the Maryland State Board of Education in 1865 ordered that M. A. Newell investigate operations at state normals in five northeastern states before taking the helm at Baltimore. These principals and faculty members served as conduits for the transfer of ideas and energy between institutions.[44]

Normal-school leaders also communicated through publications and professional organizations. Some principals were rather prolific writers. Sheldon published widely on elementary instruction and object teaching, Edwards wrote a popular series of textbooks for elementary reading instruction, and Newell also authored a series of readers. Page's *Theory and Practice of Teaching*, first published in 1847, was read and reprinted widely. Welch wrote books on English grammar and on object teaching, while Ypsilanti Professor F. H. Pease began publishing on music education. Some faculty members at other normals were authors as well; Hermann Krusi, who served on the Oswego faculty from 1862 into the 1880s, published three books, including *Life and Works of Pestalozzi*.[45]

Involvement and leadership in professional organizations also allowed normal-school principals to share ideas. By the 1850s, many states formed professional organizations for education leaders, and normal principals played an active role. In Michigan, Welch helped to establish the State Teachers' Association in 1852, then served as its first president. In Pennsylvania, James P. Wickersham did the same thing while serving as the first principal of the state normal in Millersville. At the same time, national organizations provided a wider forum for communication. The American Normal School Association, formed in 1855, had as its primary goal the raising of normal-school standards. Phelps served as its president from 1856 to 1863. Wickersham helped to found the National Teachers' Association in 1857; it would become the National Education Association in 1870. Camp was secretary of the association at the time of its 1864 meeting, and Wickersham was president in 1866. In 1870, the Department of Normal Schools became one of the first National Education Association departments. During a time before government bodies exercised much oversight, these less official channels provided informal direction for normal-school principals and faculty members. Professional organizations, publications, and movement of personnel fostered uniformity in different normal schools' approaches to teacher education. Still, the information exchanged through these channels in the mid-nineteenth century tended to strengthen the "spirit of consecration" more than the scholarly or scientific substance of teacher education. Richard Edwards stated at the 1865 meeting of the National Teachers' Association, "In our eagerness we have plucked some immature fruit. The tree is young and has not yet, we are confident, reached its best bearing." The normal schools' approach to instruction in the art of teaching and school governance, as well as practice teaching, was indeed immature in the 1840s, 1850s, and 1860s.[46]

Governor Everett explained that "second part of instruction in a Normal School" was "the art of teaching. . . . those principles of our nature on which education depends; the laws which control the faculties of the youthful mind in the pursuit and attainment of truth; and the moral sentiments on the part of teacher and pupil which must be brought into harmonious action." He added, "Besides the general art of teaching, there are peculiar methods, applicable to each branch of knowledge, which should be unfolded in the instructions of a Normal School." While all normal schools made an effort to cover "the art of teaching" and methods for particular subjects, their approach in the early years was rudimentary. Edwards acknowledged in his 1865 speech, "There are immutable principles here, that ought to be studied and comprehended by every young person entering upon the work of teaching," but added that many theories were still "a trifle unsubstantial." Especially in the 1840s and early 1850s, instruction in this area usually consisted simply of lectures in which the principal reflected on the teaching and learning process. The approach at Westfield, Massachusetts in 1844 was typical—students simply attended a daily lecture by the principal on the "modes of teaching." By 1854–1855, the Westfield catalog listed "Theory and Practice of Teaching" as a course, but it most likely consisted mainly of lectures by the principal. One of the "sons of Bridgewater," Albert Boyden summed up this approach when he reflected, "Mr. Tillinghast was our text book in Theory and Art of Teaching."[47]

At Lexington in 1839 and the early 1840s, principal Cyrus Peirce was the text. He explained that he delivered "to the school a written *Formal Lecture* once a week, in which I speak of the qualifications, motives, and duties of teachers, the discipline, management and instruction of schools, and the *manner* in which the various branches should be taught." These lectures took place on Saturdays, and some consisted of Peirce reading published works on teaching. In his journal, he noted on March 27, 1840, "This day finished reading Abbott's Teacher [*The Teacher*, by Jacob Abbott]. . . . In the Course of Reading we discussed its rules, directions and principles." On Saturday, June 6, 1840, Peirce was "chiefly occupied in reading a piece in the School Journal on the Subject of Managing Schools for Small children; particularly recommending the use of the slate in teaching the pupils to read, spell and write, as well as cipher." The following September, Peirce took his normal students to hear Horace Mann deliver an address at a "School Convention." Peirce noted that Mann "discussed the subject, whether a "Scholar should ever be compelled to study or recite his lesson." Peirce's lecture the following Saturday focused on Mann's address. In addition to his formal Saturday lectures on teaching, Peirce was sure to discuss "more *particularly* and *minutely*, some point or points suggested by the exercises or occurrences . . . relating to the *internal operations* of the schoolroom, or to physical, moral, or intellectual education" during weekday academic lessons. He had students consider hypothetical questions related to teaching, both during these lessons and in debates held on Saturdays.[48]

Mary Swift, a student in Peirce's first class at Lexington, dutifully recorded the content of his lectures and other school activities. Her journal reveals that Peirce touched on many topics related to teaching and learning, but rarely went into much depth. For example, on December 11, 1839, she described a discussion of child development. The question, "at what time does the moral power of a child commence?" had come up in Moral Philosophy class. Among the students, Swift wrote, "it seemed to be the general opinion that it commenced with the intellect. Mr P. said that he agreed," and went on to speak "of original depravity, & said that he thought the principles of Physiology & Philosophy shewed [*sic*] plainly, that the child inherited the predominant traits of character in the parents." Swift wrote that one of the Saturday debates was "upon the question, 'Is there more need of educating the faculties, or of communicating information?' " Regarding the content of the debate, she noted only, "The pupils all seemed to be united on this question, all being in favor of the first." A few of Peirce's Saturday lectures focused on teaching particular subjects. Swift noted that one lecture on methods for teaching reading contained some practical instructions, such as how to conduct "the first lesson in reading: First show them a very simple word with which they are all familiar, and also its representation; associate the name with the picture, and have them to speak its name several times." Peirce illustrated this idea with the word "cat." However, his next lecture on teaching reading was more concerned with the importance of the subject in general, as he reflected, "It is more difficult for a child to learn to read, than for a grown person to learn several languages." His advice for future teachers was to have patience. Similarly, in his lecture on teaching spelling, Peirce dwelled on the importance of the subject—arguing, for example, that correct spelling "preserves the true etymology of the word" and that "ignorance of spelling is a source of great embarrassment"—before quickly discussing when and how to teach it.[49]

David Page at Albany recorded his lectures in *Theory and Practice of Teaching*, published in 1847 and later used as a textbook in many normal schools. Of the book's fifteen chapters, only five covered the teaching and learning process specifically, and they tended to provide many tips and pointers, but little by way of coherent theory on teaching and learning. In chapter VI, "Right Modes of Teaching," Page outlined two incorrect ways of approaching teaching before explaining "the more excellent way" of "waking up [the] mind" by stimulating students' interest in the subject, which he illustrated by outlining a lesson based on an ear of corn. Chapter VII, "Conducting Recitations," explained nine principles for conducting good recitations, including "the teacher should specifically prepare himself for each lesson he assigns," and "Be careful to use language which is intelligible to children, whenever an explanation is given." In chapter VIII, "Exciting Interest in Study," Page discouraged the use of prizes before listing "Proper Incentives," such as the pupils' desire to please their parents and the pleasure of learning. Chapter X, "School Arrangements," discussed lesson planning, stressing the "importance of having a plan" and some tips on planning for everything from the first day of school to recess. Finally, chapter XIV offered "Miscellaneous Suggestions"; "Things To Be Avoided" included attending to "extraneous business" during school hours and losing patience in the face of parental interference, and "Things To Be Performed" included being a friend to one's students and "the use of the decimal or American currency" (rather than the English pound) in lessons.[50]

Henry Barnard's annual report for 1850 suggested that instruction at the New Britain Normal School in how to teach was also immature. Barnard, who served as a figurehead principal at the Connecticut institution from 1849 to 1855 while campaigning for teacher education, wrote that the curriculum was to include lectures on teaching and a list of 58 "topics for discussion and composition," such as "Methods of teaching, with illustrations of each, viz: Monitorial, Individual, Simultaneous, Mixed, Interrogative, Explanative, Elliptical, Synthetical, Analytical" and "Manner of imparting collateral and incidental knowledge." However, the published curriculum for the normal included no official listing of these subjects; Barnard probably expected that the associate principal would include them when possible in the standard academic classes. By the mid-1850s, students did attend formal lectures on teaching; John D. Philbrick, who served as associate principal under Barnard and then took over as principal in 1855, wrote, "The pupils of the Normal School are also instructed each term by lectures in the Theory and Practice of Teaching, and oral and written essays are required on topics relating to school teaching." These lectures were probably very similar to the ones presented by Peirce and Page.[51]

By the late 1850s and early 1860s, the number of courses in the art of teaching and methods began to expand somewhat, and a rudimentary theory of teaching began to take shape. The normal at Millersville, Pennsylvania in 1859 required two terms of "Theory of Teaching" and one term of "Practice of Teaching." In the mid- to late 1860s, "The Theory and Art of Teaching" was a curricular staple from Farmington, Maine to Normal, Illinois. At Bridgewater, this course included "Mental and Moral Philosophy" and "General Principles and Methods of Instruction," while methods, "both elementary and advanced," were also covered "in the different branches of study." At Emporia, Kansas in 1865, among the topics covered in six terms of "Theory and Art of Teaching" were: "The Order, in time, of the development of the Mental Faculties, and the exercises best adapted to encourage their growth. The special purpose of each faculty, and the means to train it." In addition to "Theory and Art of Teaching," students at Westfield in 1866 studied "Mental and Moral Science, including the principles and art of reasoning."[52]

The first wave of education theory, Pestalozzianism or object teaching became entrenched in the normal schools by the mid-1860s. Swiss educator Johann Heinrich Pestalozzi (1746–1827) had a romantic view of childhood, and believed that the teacher's job was to nurture and excite children's curiosity, allowing their natural abilities to unfold. Memorization and recitation of textbook material were therefore inappropriate for young children. Instead, Pestalozzi argued, children should learn from natural objects, pictures, activities, and field trips. Because the senses were at the heart of all learning, teachers should enable their pupils to see, touch, smell, hear, and even taste objects and natural forms. This approach would require teachers to plan lessons carefully and exercise ingenuity in obtaining objects and using them to facilitate learning. Pestalozzian ideas first caught on in Switzerland and Germany, and were a central component of teacher training in the German teacher seminaries by the 1820s. In the United States, Charles Brooks and other normal-school advocates praised the seminaries' emphasis on the Pestalozzian approach. Henry Barnard began in 1839 to distribute pamphlets on Pestalozzian methods to educators in Connecticut, and Horace Mann's seventh report, published in 1843, celebrated the Pestalozzian influence in Prussian schools.[53]

During the 1840s and 1850s, Pestalozzian principles began slowly to make their way into American normal schools. As early as November 1839, Peirce's lecture on methods of teaching reading reflected the influence of Pestalozzi. Swift recorded in her journal that Peirce suggested, "Before you begin to teach them to read, teach them common occurrences, about things around them . . . Let all your instructions be upon visible objects." Peirce instructed his students to use a picture in the lesson focusing on a simple word such as "cat," so that the children would be able to make sense of the word in its proper context. In *Theory and Practice of Teaching*, Page's notion of "waking up" children's minds likewise reflected Pestalozzian influence. Page discredited "book-studies" for young children, then suggested that the teacher "find some object," such as an ear of corn, "which he can make his *text*." He summed up, "Let the teacher but think what department he will dwell upon, and then he can easily select his *text*; and if he has any tact, he can keep the children constantly upon inquiry and observation." At Westfield, Massachusetts, John Dickinson, who joined the faculty in 1852 and became principal in 1856, developed his "analytic, objective method" based on Pestalozzianism. He explained that the method "is objective, as it requires the object observed to be brought into the presence of the mind. It is analytic, as the analytic process is employed in obtaining a knowledge of the object." Dickinson also asserted, "The pupil should acquire the knowledge by his own mental activity."[54]

It was in the 1860s that Pestalozzianism really took hold in American normal schools, primarily due to the influence of Principal Edward A. Sheldon and his staff at Oswego, New York. As superintendent of Oswego schools in the late 1850s and early 1860s, Sheldon became enamored of the Pestalozzian-based approach of the Home and Colonial Institute of London, which he observed in Toronto's schools. He opened the Oswego Training School in 1861, and hired Margaret E. M. Jones of the institute as instructor of object teaching. Inspired by Jones, Sheldon articulated his own theory of object teaching, which came to be known as the Oswego method. Some of its principles were: "Begin with the senses"; "Activity is the law of childhood. Train the child not merely to listen, but to do. Educate the hand"; "Reduce every subject to its elements, and present one difficulty at a time"; and "Let every lesson have a definite point." Object teaching offered a template for teaching virtually any subject, and thus was potentially a theory to impart to future teachers. In 1862, a committee of prominent educators visited Oswego and published a report declaring:

> *Resolved*, That in the opinion of your committee the system of object teaching is admirably adapted to cultivate the perceptive faculties of the child, to furnish

> him with clear conceptions and the power of accurate expression, and thus to prepare him for the prosecution of the sciences or the pursuits of active life; and that the committee do recommend the adoption of the system, in whole or in part, wherever such introduction is practicable.

With the sponsorship of the National Teachers' Association, another committee visited Oswego and reported on object teaching in 1865. The committee cautioned against superficial use of Oswego methods, but heartily endorsed the general principles. Such publicity, along with the migration of Oswego graduates to normal schools throughout the country after the institution became a state normal school in 1866, helped Sheldon's ideas to spread.[55]

At Ypsilanti in 1861, Principal Welch acknowledged the importance of "the objective methods of training the senses of the child." Two years later, the Michigan Board of Education reported that it was "convinced that the time has come when the school can render no greater service to the State, than to so modify its courses of study that all its pupils may receive instruction in the Pestalozzian system of Primary Teaching." The resulting teacher-education curriculum at Ypsilanti included courses such as "Object lessons in Geography" and "Spelling by Object Lessons." When the normal school at Winona, Minnesota reopened in 1864, principal William Phelps stressed object teaching; an early historian of the institution explained that "Theory and Practice of Teaching" reflected "the influence of the 'Oswego movement' or Pestalozzianism in America, of which the writer of the course was an ardent advocate." Emporia's yearly report in 1865 listed object teaching as a subject covered in "Theory and Art of Teaching," and explained that "instruction in Penmanship is based upon the Pestalozzian idea of reducing each subject to its elements." Still under Principal Dickinson, Westfield continued to offer "general instruction" in "object lessons" in the late 1860s.[56]

While object teaching offered essentially the first concrete guidelines for teaching as an art or science, it remained immature as a theory. After all, Edwards reminded the National Teachers' Association that "we have plucked some immature fruit," in the very same year that its committee endorsed object teaching. In one of his lectures at Normal, Illinois, Edwards cautioned, "Object teaching tends to be mechanical, develops verbalism, and makes the work of making a teacher too easy a task." In other words, it was formulaic and atheoretical. A story began to circulate in normal-school circles about a teacher who introduced a lesson on Christopher Columbus by showing the class a covered dish.

> "Now what is this?" inquired the teacher as she lifted the cover from the dish.
> "Cover," came the chorus of replies.
> "What kind of cover? Now remember Columbus," pressed the teacher.
> "Dish-cover," replied the bright boy in the class. "Columbus dishcovered America." That was the answer, of course, that the teacher desired.

Whether hyperbole or true, the story reflects the skepticism that object teaching occasionally inspired. In 1866, the faculty of the new State Normal

School at Baltimore—whose principal had just completed visits to prominent northeastern normal schools—stated, "The science of education is still in its infancy." While the normal schools made progress in instruction in what Governor Everett referred to as "the art of teaching," their work in this area remained tenuous throughout the 1840s, 1850s, and 1860s.[57]

After the art of teaching and methods, Everett explained, "The third branch of instruction to be imparted in an institution" is "the important subject of the government of the school . . . that is, of exercising such a moral influence in it as is most favorable to the improvement of the pupils." Under school government, he also included "the all-important subject of direct instruction in morals and religion, the relations of teachers and parents, of teachers and the higher school authorities, and the duties of teachers to each other and to the community, and of the community to them, as the members of a respectable profession."[58] For most of the mid-nineteenth century, the normal schools' treatment of these subjects could also have been called "immature," because what Harper termed the "tendency to exalt the teacher's work" crowded out more serious content. When Cyrus Peirce, David Page, and other principals and faculty members covered issues related to "the government of the school," they tended to focus on the responsibility and importance of the teacher's position in general and then present suggestions, but without in-depth analysis.

Governing the school generally meant keeping order and providing moral direction. Several of Peirce's Saturday lectures were about various aspects of governance. As Mary Swift recorded in her diary, his lecture on Saturday, August 31, 1839 "was upon the subject of *School Order*. He commenced by calling our attention to the importance of the teacher, with how many & various & invaluable interests it may connect us; if we are faithful how many will be rendered more happy by our instructions." After exalting the position of the teacher, Peirce talked at length about the importance of order, and then the importance of neat personal habits and "gentle movements" on the part of the teacher in keeping order and setting a moral example. Peirce did not present more concrete ideas until the following Saturday, when his topic was again "School Government." In that lecture and others during the following weeks, he presented a romantic vision of maintaining order through kindness and love; on October 26, he advised future teachers, "Avoid scolding and fault finding" and "Make it a point to speak of the *good*, rather than the bad." Direct moral and religious instruction was to stem from the Bible; Peirce mentioned briefly in his journal on February 3, 1841, "Gave my views on the Mode of using the Bible—Speaking of God and attempting to teach children the idea of God at a very early period."[59]

Page included a chapter on "School Government" in *Theory and Practice of Teaching*, but it followed chapters on "The Responsibility of the Teacher" and "Personal Habits of the Teacher." In these earlier chapters, Page listed broad and challenging responsibilities, including religious training in non-sectarian Christianity. He emphasized the power of the teacher's physical presence in the community, exclaiming, "*He teaches, wherever he is.* . . . How desirable then that he should be a *model* in all things!" Like Peirce, Page offered a detailed discussion of keeping order only after thus exalting the teacher's position. Page's treatment of school government, also like Peirce's, was romantic, as he advised teachers to "avoid governing too much" and punish with "kind reproof," although he did support selective use of the rod. In the years following the publication of *Theory and Practice of Teaching*, courses with titles like "General Principles of Government" appeared at normals such as Westfield, Massachusetts. The catalogs usually did not describe the course content, but because normals from Millersville, Pennsylvania to Peru, Nebraska used Page's book as a text by the 1860s, it is likely that such courses followed his approach.[60]

The other elements of Everett's notion of school governance all related to the position of teachers in society. In exalting the role of the teacher, Peirce, Page, and the many principals and faculty members who followed them created an imbalanced and idealized vision of teachers' role in the community. While they heaped a tremendous number of responsibilities on teachers, they also implied that teachers had a great deal of power. Under "Responsibility of the Teacher," Page listed children's physical health, intellectual growth, moral training, and religious education—a heavy load, but one that suggested much potential to have a big impact on many lives. Page, and the many instructors who used his book, also discussed relations with parents. He introduced his specific suggestions for ways to connect with parents by explaining,

> The teacher should consider it a part of his duty, whenever he enters a district, to excite a deeper interest there among the patrons of the school than they have ever before felt. He should not be satisfied till he has reached every mind connected with his charge in such a way, that they will cheerfully co-operate with him and sustain his judicious efforts for good. Being imbued with a deep feeling of the importance of his work, he should let them see that he is alive to the interests of their children.

Again, this was a tall order, but one that implied the teacher would be greeted with nothing but respect.

Page did acknowledge the problematic standing of teachers as professionals; in his chapter entitled "Teacher's Relation To His Community," he noted that teaching had "by no means received the emolument, either of money or honor, which strict justice would award in any other department to the talents and exertions required for this." However, he devoted most of the chapter to what teachers could do to "elevate" their status, implying that it was within their power to do so. Furthermore, the overall tone of exaltation regarding the teacher's role would have interfered with any serious discussion in the normal curriculum of problems with teachers' social standing. In the 1850s and 1860s, courses in school law and history of education appeared at normal schools from Farmington, Maine to Winona, Minnesota. In 1869, Ypsilanti promised, "Lectures on the School Laws of Michigan and upon the History of Education, Oriental, Classic, and Modern, will be given," and the institution also offered coursework in philosophy of education and professional ethics. As normal-school students considered their future vocation in these wider contexts, the normals' "tendency to exalt the teacher's work" likely prohibited them from looking critically at the historical, legal and ethical issues influencing the position of teachers in society. The normals' treatment of the broad "subject of the government of the school" thus tended to be a bit superficial.[61]

Governor Everett also declared, "In last place, it is to be observed, that in aid of all the instruction and exercises within the limits of the Normal School, properly so called, there is to be established a common or district school, as a school of practice, in which, under the direction of the principal of the Normal School, the young teacher may have the benefit of actual exercise in the business of instruction." Richard Edwards echoed Everett's sentiment, declaring before the National Teachers' Association, "Another essential requisite of a Normal School is, that it gives its pupils an opportunity of some kind for practice in teaching, under the supervision and subject to the criticism of experienced and skillful instructors." Many state legislatures concurred, and stipulated in the legislation establishing normal schools that they include model schools. For example, the 1849 act passed in Connecticut called for a "model Primary School—in which the pupils of the Normal School shall have an opportunity to practice." Similarly, the Missouri legislature in 1870 declared it the "duty of the Board of Regents to establish a model school or schools for practice teaching in connection with each state normal school." Whether called "model" or "practice"—model was the more common term in the mid-nineteenth century—these schools within the normal schools were to allow normal-school students both to observe models of good teaching and to practice their own teaching. Although most state normal schools established model schools shortly after opening, the vision of Everett and Edwards for the most part remained an ideal rather than reality during the 1840s, 1850s, and 1860s.[62]

Especially in the 1840s and 1850s, many model schools struggled with limited facilities and staff. While in Lexington, the first Massachusetts normal school occupied the ground floor of the normal building, but after it moved to West Newton in 1844, just one room was available for the model school. Finally, in 1849, the public school across the street from the normal became the model school at West Newton. Although Michigan's 1849 normal-school legislation provided for a model school, one did not begin operation at Ypsilanti until 1853, and was a one-room school that was too small to allow for much practice teaching or observation. This model school was enlarged three years later with the intention that each member of the normal's highest class would teach one model-school class per day, yet even after the model school was divided into nine grades by 1865, student teachers provided little of the instruction. While in the mid-1850s Principal Phelps devoted more energy to planning and constructing the model-school building than the normal-school building at New Jersey's state normal, the situation in nearby Maryland was probably more typical. As late as 1866, the model school at Baltimore struggled with facilities in a rented house, which inspired

Principal Newell to reflect, "The location was by no means a good one"; in the following years, the itinerant model school would move to various locations, including the basement of the normal building.[63]

As many early model schools coped with less-than-adequate facilities, they also made due without a permanent faculty of their own. The most extreme example of this situation was Lexington, where Principal Peirce tested the limits of his stamina by directing both the normal- and model-school programs. His journal entry for October 29, 1839 explains how he juggled the two roles:

> This day I began the Normal School at 8 A.M., remained with the Normalites until 9; then I went down into the Model School, leaving the Normal School to take care of themselves, until 10 o'clock;—at 10 the Normal had a Recess, until 11; at Eleven the Model School was dismissed—and N.S. came together again, & remained in session until 12—P.M. The Model School came together at 1 o'clock—and continued until 3. The Normal School came to order at 2, & continued in Session until 5.

Peirce listed advantages of this arrangement, including "a better opportunity for the Pupils in the Normal School to be in the Model School." By early December, he declared that normalite Mary Swift would be "superintendent" of the model school and her normal classmates would do the bulk of the teaching. A year later, he explained that the model school "was committed to the immediate care of the pupils of the Normal School, one acting as superintendent, and two as assistants, for one month in rotation, for all who are thought *prepared* to take a part in its instruction." Peirce did his best to maintain his supervisory role: "Twice every day the Principal of the Normal School goes into the model school for general observation and direction, spending from one half to one hour each visit," and later telling the student teachers what was "good" or "faulty, either in their doctrine or their practice, their theory or their manner." In addition, he very occasionally took "the whole Normal School" into the model school and conducted a model lesson for the normalites to observe. While this arrangement allowed normal-school students to have what Everett termed "the benefit of actual exercise in the business of instruction," it also exhausted Peirce. Furthermore, Peirce was present in the model school for only short periods of time, meaning that most students' teaching exercises did not actually take place "under the direction of the principal of the Normal School," as Everett had intended.[64]

In the years following Peirce's pioneering efforts, other principals and faculty members also struggled to provide instructive teaching experiences for normal students. At the same time, limited faculty in the model schools forced them to look to the normalites to fill holes in the teaching staffs. At New Britain in the mid-1850s, Principal Philbrick explained,

> At the beginning of each term, a pupil from the Normal School is assigned by the Principal for each recitation in the Model School in which assistance is needed, the Normal pupil having only one class to teach in each branch during

> the term, or for half a term. In the grammar department of the Model School there are four classes, each class having four recitations daily, . . . consequently, about thirty Normal pupils would assist in that department at different hours during the day, not more than four being engaged there at the same time.

The 1866 catalog for Baltimore stated that the model school was "conducted by one of the teachers of the Normal School with the aid of a number of the students who are detailed from week to week to observe and assist." The normalites were thus in the model school both to learn and to help it run smoothly. At New Britain, Philbrick added, students' teaching was "afterward criticized by Professors and Normal pupils" who observed. At Millersville, Pennsylvania, Principal Wickersham was only able to observe student teachers twice per week, but after each observation he held extensive conferences with the students in which they discussed and analyzed their teaching experiences in great detail. Although principals like Wickersham and Philbrick made valiant efforts to make normalites' practice teaching experiences as instructive as possible, limitations in the model schools' facilities and faculties prohibited normal schools from offering an ideal student-teaching experience.[65]

Some early model schools lost the struggle to survive. In Massachusetts, the model school initiated by Peirce closed in 1854 and did not reopen until 1861. Bridgewater was without a model school from 1850 until the late date of 1880. Westfield's model school closed in 1855 when the normal school proper took over its space. In 1868, the town built a district school across the street from the normal, but, because local parents were wary of "experimentation" on their children, normalites observed without being able to practice teaching in the new school.[66]

In situations where there was no model school or the model school was too small to allow normal students to gain much experience, they practiced teaching one another. This practice originated during the first term at the Lexington institution and continued throughout the mid-nineteenth century. At Peru, Nebraska, for example, between the opening of the normal in 1867 and the addition of the model school in 1869, students conducted lessons in which their classmates posed as young pupils. "Mutual instruction" was especially popular at the Massachusetts normals that found themselves without model schools. Principal Marshall Conant at Bridgewater reported, "I have selected individuals to give exercises before the class, after which I have called for suggestions and criticisms from its different members, adding also my own." Also referred to as "play teaching," practice teaching with other normalites received heavy criticism. As early as 1839, Mary Swift noted in her journal, "I was enabled to play the teacher for a short time. I think that he [Peirce] can judge very little about our idea of teaching from the example which we give him in hearing the recitation." In 1865, Edwards addressed this controversy: "It is said that more skill is necessary to teach a class of adults personating children, than to teach an equal number of actual little ones, and that, therefore, this practice is of more value than the other. . . . But this assumption is not true." Practice teaching in an actual class of young pupils was better, he maintained, because "There is no make-believe."[67]

The model school at Normal, Illinois in 1860s, according to one historian of the institution, "was Edwards' particular pride." The experience of teaching actual pupils and having one's work critiqued thoughtfully, he argued, was "an opportunity for improvement in the art of teaching such as offered by no other instrumentality." At Normal, Edwards had a staff and model school of an adequate size to allow him to provide a student-teaching experience that closely approached his ideal. The model school was divided into grades, from a primary year through high school. Student teachers were all responsible for the progress of a particular class, and were observed regularly by both the principal and other normal instructors, as well as, occasionally, other normalites. Observations were followed by extensive suggestions and discussions of methods. By the 1860s, other normal schools were beginning to expand their model-school staff as well. For example, Platteville, Wisconsin opened with Euretta A. Graham serving as principal of the model school, and various normals began to add "critic teachers" in their model schools. Nevertheless, inadequate facilities and small faculties, as at Baltimore, or "mutual instruction," as at the Massachusetts normals, typically shaped students' experiences with practice teaching in the 1840s, 1850s, and 1860s.[68]

The normals' approach to practice teaching, like their instruction in "the art of teaching" and "the government of the school," was worthy of Edwards' admonition, "we have plucked some immature fruit." Immaturity in these three areas contributed to the normals' "spirit of consecration," while "the tendency to exalt the teacher's work" in turn inhibited the maturation of the normals' approach to teacher training. The lack of adequate facilities and staff for model and practice teaching left normal principals to focus on fostering desirable motives and temperaments in their students. David Page's collection of lectures began with a discussion of the "spirit of the teacher," in which he urged future teachers to seek "that highest of rewards, an approving conscience and an approving God" and recognize and revere "the handiwork of God in every child." This spirit, he continued, "is the first thing to be sought by the teacher, and without it the highest talent cannot make him truly excellent in his profession." Principals and other instructors likewise compensated for the paucity of serious content in instruction in the teaching and learning process and subject-specific methods by stressing the importance of the teacher's impulses and personal characteristics. Cyrus Peirce explained, "Upon the pupils of the Normal School I inculcate, much and often, the idea that their success depends much upon themselves. . . . They must be moved by a *pure and lofty desire* of doing good. They must be intelligent and discerning. They must be firm, consistent, uniformly patient, and uniformly kind." The more normal principals and instructors focused on motives for teaching and teachers' personal characteristics, the less room they allowed in the curriculum for developing theory and extensive practice. Exaltation of the teacher so dominated the normals' treatment of school governance and educators' social position, that there was little room for discussion of contentious or problematic issues. Page stated in his preface that he wrote *Theory and Practice of Teaching* "to contribute something toward elevating an important and rising profession"; instead of fulfilling the promise of his title, Page and other principals sought advancement primarily through consecration and exaltation of teaching.[69]

Normal-school students were immersed in the language of consecration. Their instructors consistently placed teaching on a higher plane than other pursuits. Page's list of the rewards of teaching included intellectual and moral growth, the satisfaction of helping others to grow, the honor of joining the ranks of Confucius, Aristotle, and Plato, and *the approval of heaven*. After hearing this list, respect and a good salary would have seemed like mundane concerns. Students during this period left normal school with little knowledge of educational theory and a tenuous grasp of the teaching enterprise, but a very firm sense that they had a mission. Heaven's approval, according to Page, came from employment "in a heavenly mission." Teaching was the

> Heavenly Father's business. . . . Heaven regards with complacency the humble efforts of the faithful teacher to raise his fellow-beings from the darkness of ignorance and the slavery of superstition; and if a more glorious crown is held in reserve for one rather than another, it is for him who, uncheered by worldly applause, and without the prospect of adequate reward from his fellow-men, cheerfully practices the self denial of his master.

Peirce's lecture on the origin and status of normal schools had a revivalist flair; Swift wrote that he discussed "their origin & the expectations of the Board of Education & of the Friends of Education in general," before admonishing, "To us, therefore, all the friends of Education turn, anxious for the success of the first effort to establish such schools." A student at Normal, Illinois, John Calvin Hanna absorbed the lesson that a teacher "must be animated by a higher motive than a mercenary one. He must teach because he loves to teach; because he looks upon teaching as a noble, . . . a holy employment." While normal schools adhered to the template of the four "core" elements of instruction laid out by Governor Everett in Massachusetts in 1839, the notion that teaching was a consecrated mission shaped the normals' approach to teacher training throughout the mid-nineteenth century.[70]

"Grand Beyond Compare": Further Dimensions of Early State Normal Schools

Teacher education was, of course, the official purpose of the American state normal school. This did not exclude instruction in academic subjects, however; as Governor Everett stated, teachers needed to be knowledgeable in the subjects they would teach. Basic liberal-arts subjects were a necessary part of the curriculum at all state normals from the beginning, especially because, much to the principals' consternation, the majority of their students arrived with limited exposure to anything beyond what was offered in the common schools. Normalites soaked up their academic studies and relished opportunities that extended beyond the classroom. For these women and men from unsophisticated backgrounds, daily normal-school life and fledgling student activities in many ways provided entree to a wider world. Lyman Kellogg, who graduated from the state normal in Normal, Illinois in 1864 (and then served as principal at Emporia, Kansas from 1865 to 1871), never forgot the impact of attending normal school. "Having never before attended any school other than the country district and the public school of a small village," he wrote to commemorate the institution's semi-centennial, "Major's Hall and the then Normal School seemed to me very majestic and impressive. . . . I considered them grand beyond compare."[71]

As historians of teacher education have pointed out, several of the early normal-school principals expressed frustration with the need to focus on academic studies. At Lexington, Massachusetts, Cyrus Peirce used his journal to vent his dissatisfaction with his students' prior preparation. On August 27, 1839, he wrote, "I think the scholars have not been much habituated to hard close and methodical studying. There is a great deficiency among them in knowledge of the Common Branches." A couple of weeks later, he granted that "nearly every" student was "of *fair* capacity," but added, "many of them are yet backward," or lacking "language—they want the power of generalization, and of communication." At times, Peirce seemed despondent. In late November, 1839, he confessed, "My heart is heavy, and my hands hang down," and the following September, he lamented, "School drags very heavily. The Head is sick; and the heart is faint. I feel almost discouraged; Cannot get my own feelings interested in the matters and business of the School: The scholars seem dull & lifeless! How long, in all probability, will it be, before some of these Normalites become good Teachers!" A few months later, he added, "What miserable work some of these girls make in Arithmetic, Grammar, Philosophy and Algebra!" In the following decades, Richard Edwards at Normal, Illinois reported that many of his students were "unable to read with any adequate apprehension of nice shades of meaning," and William Phelps at Trenton, New Jersey exclaimed, "How are you to teach them how to teach that of which they know nothing?"[72]

Historian Paul Mattingly has suggested that many of the frustrations of these principals resulted from their inability to "bridge the differences between themselves and their students." Still, historians have generally followed the lead of the principals themselves and have largely overlooked the role that early normal schools played in expanding access to education in the mid-nineteenth century. Education reformers and normal-school principals regarded these institutions only in relation to the crusade to improve the common schools through better teacher preparation. Thus, what Mattingly refers to as "students' educational and class disadvantages" were merely obstacles to the reformers' larger goals. Enrollment incentives such as the stipulation in Maine's 1863 normal-school legislation that "Applicants for admission . . . shall signify their intention to become teachers" and thus "be received without charge for tuition," were intended solely to increase teacher numbers and quality. When students with few other opportunities jumped at the chance to attend normal school, the principals were uncertain of what to do with them. Even though the principals did not acknowledge it and historians rarely have, the provision of post-common school education at little or no cost was an unintended yet profound change in educational opportunity.[73]

Well into the nineteenth century, class, gender, and race advantages translated into educational opportunities. Only in the 1830s and 1840s were young white men with "class disadvantages" just beginning to have some options. Academies were proliferating, and some charged only modest fees. Public high schools were also becoming more numerous, although they were located mainly in cities and larger towns and attracted a predominantly middle-class student base. In addition, education societies began to offer college scholarships for poor young men and structural changes made provincial New England colleges more accessible. The educational opportunities for white girls and women of the middle and, to some extent, lower classes were also expanding just a bit by the 1830s and 1840s. Some academies and public high schools were coeducational, and female seminaries and girls' high schools were growing in number. Colleges and universities, however, would remain essentially closed to women until the late 1860s. In addition, non-white people of both sexes had access to very few educational institutions.[74]

In this educational landscape, the new normal schools began to function as an accessible alternative for both men and women, especially those of limited means. In the 1840s, 1850s, and 1860s, normal schools attracted male and female students alike. While enrollment at Peirce's institution was restricted to women only, most of Pennsylvania's normal schools enrolled close to equal numbers of men and women. Typical of most early normal schools, the student bodies at both New Britain, Connecticut and Westfield, Massachusetts were approximately 30 percent male before the Civil War, and contained fewer men during and after the war. At Westfield, according to institutional historian Robert T. Brown, "students came from the most economically depressed class—small farmers." When the Superintendent of Public Instruction in Minnesota reported in the 1860s, "The students of the [Winona] normal school are composed entirely of persons from [rural] industrial classes," he could have been describing state normals throughout the country. Many normalites worked, often as teachers, before enrolling; the average of entering students at Farmington, Maine from 1864 through 1869 was between 18 and 19 years. And many students left before finishing the course of studies because, as principal George Minns of the San Francisco State Normal School explained, "it was absolutely necessary for them to do something to support themselves." William Stone, who was 20 years old and "had no money" when he enrolled at Mansfield, Pennsylvania in 1866, "swept the halls and attended to the fires in the building" in order to pay for his board and tuition. Struggling Irish immigrants and a few Native Americans enrolled at Albany, New York during this period, and isolated African American students at Massachusetts normals included Chloe Lee, whom Peirce welcomed to West Newton in 1847. No matter what their race, for these students even attending normal school was a stretch; George Martin, who attended Bridgewater, Massachusetts in the mid-1860s, later reflected, "It is hard for us to conceive how provincial these students were."[75]

Although frustrated with his students' prior preparation, Peirce acknowledged in his journal at calmer—or less exhausted—moments, "there is a fair share of Intellect among them," and "They have talent enough, but are deficient in Knowledge, & suffer from the *neglect*, or rather *want* of early discipline." Students were generally eager to overcome their deficiencies, and a bit awed by their new opportunity. An early student at Potsdam, New York remembered, "Most of us were somewhat scared by the dignity of the Normal and academic teachers, by the size of the chapel, and by the tremendous responsibilities which we soon found resting on our shoulders." Stone at Mansfield "gradually began to learn how to learn," realizing "that the studies were to discipline the mind, as a drill disciplines the soldier," and some students at Westfield studied so hard that the 1858 catalog declared, "unseasonable rising and study will be regarded as a violation of the rules of the institution." The expanding academic curriculum and various activities outside the classroom provided normalites with the opportunities they desired to widen their horizons.[76]

During this period, instruction in the humanities, mathematics, and the sciences focused mainly on the fundamentals out of necessity, but in the process broadened students' exposure to these areas. At the heart of the language arts, orthography, grammar, and reading were curricular staples. Westfield catalogs listed the three separately in the mid-1840s, and also included English Literature by the mid-1850s. The 1850 curriculum at New Britain included "Rhetorical Reading, comprising Analysis of the Language, Grammar and Style of the best English Authors, their errors and beauties; Orthography with Phonetic and Etymological Analysis; English Grammar with Analysis of Sentences: Composition and Declamation." Students at Ypsilanti, Michigan in 1863 studied Reading and Spelling as well as Synthetical Grammar and Analytical Grammar, while those at Emporia, Kansas in 1865 took classes in Reading, Rhetoric, Grammar, which included etymology and "critical parsing," and literature. History of the United States was a standard subject from Westfield to Winona; students at the latter institution also studied the constitutions of the United States and of Minnesota. Most courses of study also included Moral Philosophy and/or Mental Philosophy, or Elementary Philosophy in the case of Ypsilanti. The mathematics curriculum usually comprised arithmetic, algebra, and geometry, and several normal schools, including Westfield and Winona, added bookkeeping as well. Natural History and Natural Philosophy covered basic science concepts, and some normal schools offered additional classes in physiology, chemistry, astronomy, or botany; Normal, Illinois offered all four, plus zoology, in 1860. Geography was also a prominent subject; for example, the curriculum at Castleton, Vermont in 1868 included "Geography, with map drawing, and the elements of Physical Geography."[77] Science instruction at Winona also included "elements of agriculture," in keeping with calls for mechanical and agricultural instruction that were part of the founding legislation of many midwestern normal schools. Despite such rhetoric, most normals in the Midwest never offered such courses, and "elements of agriculture" had little impact at Winona.[78]

In addition to the fundamentals, some normal schools did offer more advanced academic studies, especially by the late 1850s or 1860s. Although most institutions had meager libraries and laboratories and few students stayed long enough to complete the entire one- to three-year course, normalites did enjoy occasional opportunities to pursue the liberal arts at a higher level. Alongside more rudimentary work in English, the Westfield curriculum in the mid 1850s included "Analysis of Milton and other Poets." A decade later, Emporia offered Literary Criticism, which included "Critical study of Shakespeare" and "the style of Milton, Addison, Goldsmith, Irving, etc.," and History of English Literature, which encompassed "Rise and development, in England and America, of poetry, history, romance, the essay, oratory and metaphysics," as well as a comparison of English literature with "that of other nations." Ancient History was also available at Emporia. Several normal schools included trigonometry in the mathematics curriculum, and by the late 1850s Pennsylvania's normals offered Differential and Integral Calculus at the end of the third year. By the early 1860s, the Natural History Society of Illinois was amassing sizable collections for the Museum of Natural History at the State Normal University, and beginning in 1865, each graduating student at the institution conducted an original investigation and presented a thesis on the geology, botany, or another aspect of the natural history of his or her home district. In addition, by the mid-1850s, many state normal schools offered optional instruction in Latin and perhaps another language. Westfield's approach, explained in the 1854–1855 catalog, was typical: "Instruction is given to a limited extent in Latin and French, for the accommodation of pupils whose time and attainments enable them to pursue studies not embraced in the required course." Ypsilanti offered Latin and Greek from the time it opened and added French and German by the 1860s.[79]

While the academic curriculum broadened normalites' exposure to the liberal arts, other aspects of attending state normal school further enlarged their worlds. Many students traveled to the schools from remote regions and roomed and boarded with town residents or, in the few cases where dormitory space was available, at the normal. Normal schools had fairly strict rules, yet there was still room for students to plan their own activities. Henry B. Norton, who graduated from Illinois State Normal University in 1861, later explained, "We lived as we pleased, formed our own relationships and associations, made our own calls, and managed our affairs, entirely at our own choice and pleasure." In addition to informal friendships, students formed various formal groups. Initiated by male students, literary societies became popular among students of both sexes at several early normal schools, although men and women had very different roles during this period. At Bridgewater, for example, men formed the Normal Lyceum in the early 1840s and were the only members. Meetings on alternate Friday evenings centered on debates among society members; female students joined the audience and produced a publication called the *Normal Offering*. Ypsilanti's Normal Lyceum and Normal's Philadelphian and Wrightonian Societies were coeducational, but only men participated in debates while women read essays and literary selections, and both sexes performed musical pieces. Such student groups and the structure of daily life enabled students to engage in a considerable amount of geographical, intellectual, and cultural exploration, resulting in an "environment" that Albert Reynolds Taylor, who attended Normal, Illinois in the mid-1860s (and would, like Lyman B. Kellogg, serve as president at the normal in Emporia), later described as "changing and enlarging."[80]

As normalites became acquainted with their new surroundings and companions, their environment expanded in a geographical sense. The towns in which the normals were located were, in most cases, larger and more active than the students' rural homes. The largest town between Springfield and Albany, Westfield in 1844 had a population of 3,700, a canal port and railroad depot, brickyards, mills, and more than a dozen whip factories. Oswego, New York in the 1860s was a busy port with giant grain elevators and factories including the world's largest producer of starch, as well as stately homes and a theater. Taking up residence in such communities was an education in itself, and became a base for even further exploration. As early as 1839, Peirce took his students to Cambridge to attend commencement at Harvard, and to another village to see a fair and visit an observatory. After the latter trip, Mary Swift wrote in her journal, "After viewing the fancy articles we entered the observatory & went into the upper part where Mr P named many of the hills around & showed us the state of New Hampshire, & the commencement of the White Mountains. We also saw the ocean but it was at so great a distance . . ." A. J. Hutton, a member of the first class at Platteville, Wisconsin, called the town "an old-fashioned village," yet learned a great deal about the world from its residents. He later remembered, "I met for the first time Kentuckians and other southern people. . . . There were Germans . . . There were Cornishmen and other Englishmen, Welshmen, Scotchmen, Irishmen, New Englanders, New Yorkers, and Jews."[81]

Normal-school life outside the classroom also enlarged students' intellectual worlds. Literary-society activities and other campus events built on what students learned through the formal curriculum and took them in new intellectual directions. In society debates, students explored everything from American history to educational policy and philosophy. For example, at Normal, the Philadelphian and Wrightonian societies debated, "*Resolved*, That Thomas Jefferson should be ranked higher as a statesman than

William Pitt" in 1864, "*Resolved*, That compulsory attendance is beneficial" in 1858, and "*Resolved*, That the labors of Pestalozzi in the educational field have been of more value to mankind than those of Horace Mann" in 1862. Meanwhile, students in Ypsilanti's Normal Lyceum debated whether "schools should be supported by a direct tax on property and should be free to all," and whether "the reading of the Bible should be made a daily exercise in all our schools." Philadelphian and Wrightonian were among the several societies that began to amass collections of books for circulation; normal-school board member Simeon Wright donated valuable books and over a hundred other volumes to his namesake society. Meanwhile, Westfield's Lyceum collected more than seven hundred books by 1860, enabling students to read literature beyond that covered in their classes. Fledgling student publications were another venue for further exploration of material introduced in the curriculum. Students at San Francisco produced the *Acorn* for a short time beginning in 1867; an 1889 account explained that this "small written sheet . . . consisted of witticisms and poems, and essays on literary, scientific, and educational topics, all in manuscript. It was read monthly before the school society by members of the same."[82]

Extracurricular exploration also acquainted students with contemporary intellectual trends and political issues. In 1840, Amos Bronson Alcott and Messrs. May and Stetson visited the Lexington institution and discussed transcendentalism, peace, and nonresistance with students there; Peirce noted in his journal that it was "quite interesting and instructive." In the early 1850s, Ypsilanti's Lyceum discussed the resolution "That Phrenology is the true science of the mind." Lucretia Mott addressed the Page Literary Society at Millersville, Pennsylvania in 1857 on "Woman's Rights," and Normal students were in the audience when Abraham Lincoln spoke at the Bloomington, Illinois railway station after winning the 1860 presidential election. The Civil War and its aftermath were prominent subjects in all of the literary societies throughout the 1860s. Early in the decade, Ypsilanti's Lyceum debated whether "the North would be better off morally, socially, commercially, and politically without the South." Secretary Mary A. Rice reported, "The discussion was of much interest; gentlemen on the affirmative producing unanswerable statistics, which were nevertheless overborne by patriotic enthusiasm and union sentiment." Normal's Philadelphian debated "Has a State a right to secede?" and "*Resolved*, that congress should declare the slaves free." At Emporia in 1865–1866, the Normal Literary Society debated "negro suffrage" and the Normal Literary Union passed the resolution "That the late report of the Congressional Reconstruction Committee should be adopted."[83]

Attending normal school not only expanded students' understanding of current events, but also "changed and enlarged" their cultural environment. Mary Swift noted during her first month in normal school, "In the morn the Brigade Band from Boston came out & with the Lexington artillery marched up in front of this house and then through the Mall.—The scene was entirely new to me & I liked the order and regularity with which they moved."

Oswego students in the 1860s enjoyed plays at the town theater, including *Rip Van Winkle* and a Shakespearean drama. "Shakespearean readings" were popular with Normal's Philadelphian society, and two instructors and several students staged a reading of *The Merchant of Venice* before Emporia's Normal Literary Society. Normalites took advantage of many opportunities to perform and enjoy instrumental and vocal music. Optional music lessons began at Ypsilanti in 1854, and the institution had a choir by the early 1860s; it performed oratorios and choruses from well-known operas as well as patriotic songs. Philadelphian and Wrightonian each purchased a grand piano in 1864. With or without piano accompaniment, society programs at Normal and elsewhere consistently featured musical performances in between debates and readings. At Millersville by the 1860s, for example, programs included vocal, violin, and piano solos as well as songs by the glee club and string band.[84]

For many students from provincial backgrounds, the opportunities presented by the academic studies and extracurricular activities at normal schools were "grand beyond compare." Geographical, intellectual, and cultural exploration enabled them to enter a new world. After attending normal school, many of these women and men led lives that were, in many ways, also grand. As teachers and, in the case of several of the men, normal-school principals, they continued to expand their environment. Mary Swift, who had been one of Peirce's first students at Lexington, and William Stone, who had arrived at Mansfield with "no money" in 1866, both reached far indeed. Swift's teaching at the Perkins Institution for the Blind in Boston, including work with Laura Bridgman in the 1840s and with Helen Keller in later years, became known around the world. She also married, had four children, and worked for Boston's Young Women's Christian Association for 40 years. After normal school, Stone taught at an academy and studied law in the evenings. As a young lawyer, he married a Mansfield classmate and formed a literary and debating club with his friends. Stone soon became very involved in state politics, was elected to the U.S. Congress in 1890, and was elected governor of Pennsylvania in 1898. He capped off his career by founding a law firm with his son: Stone and Stone. Attending state normal school helped launch Swift, Stone, and many others on trajectories of accomplishment and adventure.[85]

Between the 1820s and the 1860s, common-school reformers and teacher educators did much to awaken the public conscience on the subject of teaching and teacher education. Reformers worked zealously to improve the common schools through a more committed and better-educated teaching force. Academies and colleges, teachers' institutes, and private, municipal, and county normal schools developed new approaches to teacher education and sent many graduates into the teaching ranks. State normal schools, which grew gradually in number with the support of town boosters who helped to sway reluctant legislators, most closely fit the reformers' vision of freestanding, state-supported teacher-education institutions. While reformers believed they were replicating the much-admired Prussian teacher seminaries, the

Americans overlooked problematic social-class and gender issues. The Prussian seminaries admitted only men from lower social classes and, as Jurgen Herbst explains, were characterized by "a benevolent paternalism that never questioned the basic rightness and justice of social and economic arrangements."[86] While Prussian normal schools sought to maintain the social order, normalites on the other side of the Atlantic, most of whom were from lower social classes and female, were not interested in conforming to benevolent paternalism. They seized opportunities to learn about and venture into the larger world, lightly brushing class and gender boundaries. As the following chapters explain, teacher education, academic studies, and campus life at the state normal schools matured and thrived in the decades following the 1860s. In the process, class and gender boundary-blurring intensified, took on greater significance, and, along with teacher education, became one of the important legacies of the American state normal school.

Part II

The Heyday of the State Normal School, 1870s–1900s

2

"The Masses and Not the Classes": A Tradition of Welcoming Nontraditional Students

During their first three decades of existence, state normal schools were on shaky ground. They struggled against public skepticism and scrutiny, limited state funding, and the popularity of teacher education in other types of institutions. Making only small advances toward developing and teaching educational theory, normals were successful in fulfilling their intention to instill their students with a sense of teaching as a calling. They also began—for the most part unintentionally—to expose their students, many of whom would not otherwise have had access to advanced education, to a wider intellectual world. Beginning in the 1870s, these intentional and unintentional successes of the early years became defining characteristics of the state normal schools. Between the 1870s and the 1900s, state normals found themselves on much firmer ground as they gained in number and size and their opposition faded somewhat. In many ways, this 40-year period would be the heyday of the state normal schools, when they offered a unique educational environment for a distinct student body. The normals not only attracted students who a century later would have earned the label "nontraditional," but they also served these students quite effectively. S. Y. Gillan, who graduated from Illinois State Normal University, summed up the "democratic spirit" of the normal: "it was a school of the people existing for and representing the masses and not the classes."[1]

Firm Footing for the State Normal Schools

By the 1870s, many of the common school reforms initiated during the antebellum period were entrenched in the Northeast and Midwest; growing state and local administrative bureaucracies ran public, largely tax-supported school systems, in which curricula were more standardized and sessions were longer than earlier in the nineteenth century. During the decades that followed, education leaders in the North sought to further systematize and bureaucratize the schools by instituting reforms such as uniform standards for teacher hiring and pupil promotion, compulsory attendance, and additional administrative positions. Meanwhile, in growing newer communities in the West, school leaders quickly formed bureaucratic systems according to northern standards. In the South, such developments occurred at a slower pace and a more fundamental level; the region made strides toward state-supported education in the late 1860s and 1870s, and segregated common schools became nearly universal in the early twentieth century, mainly due to the efforts of former slaves. Throughout the country, the spread and intensification of the common-school agenda both reflected and supported the belief in progress and modernization that took hold after the Civil War and flowered in the late nineteenth and early twentieth centuries. At the heart of the Progressive movement was the new middle class' "search for [bureaucratic] order," and "culture of professionalism" that was incubated in American universities.[2] State normal schools, which had been slower to take hold than other antebellum reforms, were very well suited to this new ideology, as they prepared students to assume positions in the school bureaucracy through occupational training. These wider social beliefs were the bedrock for the expansion and growth that assured firm footing for state normals between the 1870s and 1900s.

Other types of institutions continued during this period to train teachers as they had in previous decades, although they were eclipsed by state normal schools for the first time. Multipurpose institutions continued to educate teachers. As academies gave way to public high schools, the high schools assumed a greater role in teacher training. For example, New York increasingly redirected the flow of state funding for teacher education from academies to high schools, until the program was discontinued in the early 1930s. Colleges and universities prepared teachers in normal or education departments, while the larger universities established schools or colleges of education, which trained some elementary teachers and many high-school teachers but prioritized administrative training and educational research. By the early twentieth century, the schools of education at Columbia, Stanford, Iowa, Michigan, and several other universities earned distinguished reputations for their graduate education and research.[3]

Single-purpose teacher-education institutions that were not state normal schools also forged onward. Private normal schools continued to appear—and often to quickly disappear—on the educational landscape. Indiana, Ohio, and Kentucky housed many of these institutions; Kentucky alone chartered 19 private institutions whose names included "normal" during the 1870s and 1880s, and 5 more between 1890 and 1905. Throughout the South, religious missionaries established normals to prepare black teachers. The American Missionary Association supported 43 such institutions by 1900; in Georgia, these included Ballard Normal School in Macon and Allen Normal and Industrial School in Thomasville. Meanwhile, municipal normal schools spread far beyond their northeastern roots; by the 1910s, the vast majority of school systems in cities with populations over 100,000 would have their own normal school or teacher-training classes. The rural equivalent of the municipal normal was the county normal school; especially popular in Wisconsin, county normals offered bare-bones preparation for rural teaching. School leaders also continued to utilize shorter-term teachers' institutes, especially in thinly settled, frontier-type states and territories. Arizona organized institutes at the county level in the 1880s, and Oklahoma did the same in the 1890s. In the Indian Territory of Oklahoma, the Cherokees, Creeks, and other nations worked with federal supervisors to conduct summer normal institutes. In Nebraska, summer county teachers' institutes grew longer until, beginning in the early 1900s, they lasted ten weeks and were called junior normal schools. Junior normals ended in 1914, when teacher education was widely available in Nebraska's high schools, not to mention normal schools. In fact, teachers' institutes throughout the country were in decline by the 1910s, and most municipal and private normal schools disappeared by the 1930s.[4]

Despite the persistence throughout the late nineteenth and early twentieth centuries of teacher education in other types of institutions, state normal schools were a dominant force in teacher education during this period. State normals were established and attended at an unprecedented rate. In the 1870s, the number of institutions nearly doubled, from 35 at the end of 1869 to 69 at the end of 1879, as 10 states established their first normal school, bringing the total number of states with normal schools to 26. During the following decade, 28 new normal schools opened their doors, and the number of states or territories with normal schools rose to 33. More normal schools were established in the 1890s than in any other decade: 42 opened, including normals in 10 new states; at the close of the century, 139 state normal schools operated in 43 states or territories. By 1910, 170 state normal schools had opened in 44 states or territories. The establishment of the last state normal schools in 1927 brought the total number founded since 1839 to 210, of which close to 200 remained in operation, in 46 states.[5]

With increases in the number of normal schools came increases in enrollment, both overall and at individual institutions. Table 2.1 summarizes the enrollment statistics reported by the U.S. Commissioner of Education. The statistics are somewhat unreliable as it is difficult to determine whether totals include model-school pupils or how they were tabulated. Public normal schools include municipal and county institutions as well as state normal schools, although the latter enrolled the majority of students who attended public normals. Even allowing for some imprecision, the table indicates the public normal schools' domination of all normal-school enrollments by 1880. Thus, state normal schools enrolled well over half of all normal-school students throughout this period, and their portion of the total enrollment increased over time. Meanwhile, individual normal schools' enrollments grew considerably. One of the biggest, Illinois State Normal University in Normal enrolled over 450 students in its normal department by the late 1860s. Its enrollment remained in the 400s throughout the 1870s, and topped 500 by the early 1880s and 600 by the late 1880s.[6] Probably more typical are the four geographically diverse state normal schools whose normal-department enrollments appear in table 2.2. The enrollments at Geneseo, New York, Florence, Alabama, Oshkosh, Wisconsin, and San Jose, California climbed upward between the 1870s and the 1910s, from numbers in the 100s to 300s in the 1870s, to the 300s to 600s in the 1900s. Large and climbing enrollments, combined with the rapid growth in the number of state normal schools, placed these institutions in a prominent position in the teacher-education landscape of the late nineteenth and early twentieth centuries.

While enrollments increased and the pace of the establishment of new state normal schools quickened, state legislatures continued to debate

Table 2.1 Enrollments in normal schools in the United States, 1875–1915

Year	Normal-school enrollments			
	All normal schools	Public normals	State normals only	Private normals
1874–75	29,100			
1879–80	43,100	25,700		17,400
1884–85	55,100	32,100		23,000
1889–90	51,700	40,400		11,300
1894–95	64,000	36,800		27,200
1899–00	116,600	67,900	60,300	48,700
1904–05	131,300	95,600	76,600	35,700
1909–10	132,500	111,100	94,100	21,400
1914–15	119,000	110,800	102,700	8,200

Source: *Report of the Commissioner of Education* (Washington: Government Printing Office, 1875–1915).

Table 2.2 Enrollments in four state normal schools, 1875–1915

Year	Enrollments in four state normal schools			
	Geneseo, New York	Florence, Alabama	Oshkosh, Wisconsin	San Jose, California
1874–75	347	126	212	328
1879–80	349	201	332	410
1884–85	495	224	334	528
1889–90	880*	336	360	592
1894–95	1085*	346	521	675
1899–00	746*	279	701*	768*
1904–05	488	358	551	608
1909–10	370	403	650	619
1914–15	472	368	571	946*

* Very high figures may include model-school enrollments.

Sources: School catalogs and bulletins, and *Report of the Commissioner of Education* (Washington: Government Printing Office, 1875–1915).

whether to support their first or additional normals. After 1870, normal-school proponents had more success with lawmakers; arguments such as the Iowa state superintendent's in 1875 that there was "no good reason why Iowa should continue to be an exception to the general rule" of states supporting teacher education in normal schools, reflected the increasing momentum of the movement. Swayed by the superintendent's argument—as well as the availability of the state-owned land and buildings that had been the home of a soldiers' orphans' home in Cedar Falls—the Iowa legislature passed a bill to establish its first normal school in 1876.[7] As other states passed numerous bills establishing normal schools in the late nineteenth and early twentieth centuries, they adhered to the pattern established earlier, stipulating that the normal school would be awarded to the city or town that submitted the best bid. Thus, town boosterism continued to play a large role in the establishment of state normal schools. The movement's momentum insured local support as large numbers of business and community leaders enthusiastically pursued the establishment of a normal in their town. By the 1870s, this process was unmistakably a nationwide phenomenon; with enthusiastic local support, state normal schools were for the first time on firm footing in all regions of the country.

The earliest strongholds of state normal schools, states in the East and Midwest added new normals throughout this period. From Gorham and Presque Isle, Maine, to New Paltz, New York, eastern towns donated existing academy buildings to the cause. Milwaukee presented the building and site of its city teacher-training school to the state; after renovations, the new state normal school opened there in 1885. A local citizen's gift of a three-acre plot of land in the early 1900s enabled Danbury, Connecticut to secure a state normal school. In Willimantic, Connecticut, a group of local citizens entertained visiting members of the legislature's education committee in 1889; impressed by their warm welcome as well as the general surroundings, the legislators selected the borough over rival Norwich as the location of the state's second normal school. Competition between towns was especially intense in the Midwest. When Illinois decided to establish a third normal school in the mid-1890s, 12 communities actively pursued the legislature's favor. Charleston's offer of 40 acres of land, approximately $40,000, city water, sidewalks, and more, secured its victory. In nearby Mattoon, which had lobbied equally hard for the normal, the newspaper reported that Charleston's victory caused "real, genuine, heartfelt profanity." Mattoon citizens initiated a legal appeal of the decision, but were unsuccessful. Similar stories unfolded as state legislatures established normals in River Falls, Wisconsin in the early 1870s and Kalamazoo, Michigan in the early 1900s. By the late 1910s, every state north of Pennsylvania's southern border and east of the Dakotas' western border had at least one normal school, and the four states that housed the most normals were located in this northern region: Massachusetts had ten, New York had eleven, Pennsylvania had thirteen, and Wisconsin had nine.[8]

Unlike the North, the South had barely begun to establish state normal schools by the 1870s. During the next four decades, however, it would more than make up for its slow start as state legislatures and both white and African American communities worked together to bring state normals to towns throughout the region. When Missouri's 1870 act stipulated that normal schools for white students would go to the counties that made the best offers, residents began to assess and pool their resources; within three years, Kirksville, Warrensburg, and Cape Girardeau would celebrate their good fortune. In Florence, Alabama, the Methodists donated the buildings and grounds of Florence Wesleyan University, which had been occupied by Union troops during the Civil War and had been struggling since that time, to the state in 1872, just as the legislature decided to establish normal schools. The first one opened in the former university's castle-like building the following year. Meanwhile, Texas established its first normal school for white students, while West Virginia—which, along with Maryland, had in the 1860s established the first southern state normals—added several additional institutions. During the 1880s and 1890s, Florida, Louisiana, Virginia, North Carolina, Georgia, and Oklahoma opened their first normals for white students, while Alabama opened two more—in Jacksonville and Troy. The passage of the first authorizing legislation for a normal school in Georgia was due largely to the efforts of white women residents of the state who showered their legislature with petitions and pleas for the female-only institution that opened in 1890 in Milledgeville. Donations from local residents enabled Greensboro, North Carolina to offer the state $30,000 and thus rise above other contenders to secure a normal school, which opened in 1891. Ten years later, the rivalry between two of the several towns that sought to host the normal for the southwestern region of the Territory of Oklahoma became extremely contentious. After intense lobbying, one legislative committee recommended Weatherford while another recommended Granite. When the territorial supreme court finally selected Weatherford, whose citizens had raised $10,000 to pay for legal action in favor of the normal, the town celebrated with whistles, shotgun blasts, and a bonfire. Residents of many other southern towns also had cause to celebrate in the 1900s and 1910s, as the South opened more normal schools for white students than any other region.[9]

The first in a wave of state-supported normal schools for black students was the State Colored Normal School in Holly Springs, Mississippi, which opened in 1871. Upon the urging of the governor, North Carolina in 1877 authorized state support of a normal school "in connection with some one of the colored schools of high grade in the State." Competition ensued among various African American communities, including Fayetteville's active and relatively literate black community. Its Howard School was selected because of its excellent record in preparing teachers, and the State Colored Normal School would occupy the Howard School's upper story for two and a half decades. By the end of the 1870s, there were seven African American normals in six southern states, including Arkansas' Branch Normal College, the enabling legislation for which stated euphemistically that it was to be for "the poorer classes." During the 1880s and 1890s, southern communities established almost as many normal schools for black as for white students. North Carolina alone opened several additional normals, although by the mid-1900s it would maintain only those at Fayetteville, Elizabeth City, and Winston-Salem. The efforts of local black citizens were instrumental in establishing a state normal school in Langston, Oklahoma. They raised $500 in small donations to purchase a 40-acre tract of land to donate to the state university regents, who in turn established the Colored Agricultural and Normal University there in 1897. The institution's name reflected the added incentive that the second Morrill Act of 1890 provided for southern states to establish educational institutions for African Americans by stipulating that federal funds could be distributed to segregated institutions on a "just and equitable basis." As a result, many states added some element of "agricultural and mechanical" instruction to the functions, or at least the names, of their existing or new black normals. By the late 1910s, 13 different states had established upward of 20 African American normal schools, most of which were still operating.[10]

While the South rapidly made up for lost time in opening normal schools between the 1870s and the 1910s, the still sparsely populated Far West and Southwest were a bit slower to join the movement. California's state normal, which had opened in San Francisco in 1862 and moved to San Jose nine years later, remained the only such institution in the West until the early 1880s. From then on, though, the western states followed the familiar pattern, as local boosters worked with willing legislatures to bring state normal schools to their towns. In Oregon, residents of Ashland, Monmouth, Drain, and Weston saw the state-normal designation as a way to save their local college or academy. In Ashland, for example, supporters of the struggling academy known after 1879 as Ashland College and Normal School hoped that state recognition would attract students. Between 1882 and 1885, the four towns' institutions received the state normal-school designation, but no financial support. The 1880s also saw the establishment of Arizona's first state normal, in Tempe, and an additional California institution, in Los Angeles. When California decided in the late 1880s to establish a normal in the northern section of the state, the residents of Chico banded together to make the visit of the state superintendent and normal trustees a memorable one, complete with a ride through the countryside in buggies decorated with flowers. Their efforts were successful, and Chico's normal opened in 1889. The 1890s were the most active decade of normal-school founding in the West; 13 state normals opened in 8 different states or territories. In Washington, a long-standing rivalry between Cheney and Spokane Falls came to a head over the normal-school issue. Cheney's citizens ultimately won, perhaps because they petitioned the state legislature to accept the gift of the buildings and grounds of their local academy for the normal. By the late 1910s, a few dozen normal schools were scattered throughout the West, from Montana to New Mexico and from Colorado to Hawaii.[11]

The number and geographic distribution of state normal schools made their existence less tenuous in the late nineteenth and early twentieth centuries than it had been before the 1870s. In addition, they faced less outright opposition during this later period. Attacks on New York's and Minnesota's normal schools that had begun in the 1860s continued into the 1870s, but to no avail. In Minnesota, the final bill to close the normal schools was soundly defeated in 1875, and in New York, the special legislative committee appointed to investigate allegations against the normals reported in 1879 that support for normals was "deeply rooted" in state policy. State normals in West Virginia and Missouri also faced organized opposition in the 1870s: the West Virginia legislature dragged its feet on appropriations and considered eliminating the normals, and opponents in Missouri introduced bills—unsuccessfully—to repeal the Normal School Act in nearly every legislative session until 1883. In the decades that followed, attempts to eliminate state normal schools were rare and generally limited to the newest normals in the Far West. In the 1890s, Arizona's normal in Tempe dodged several attempts to close it. During the following decade, Oregon's normals were not so fortunate when the governor and legislature leveled charges against them, including: "The schools were merely local. The schools were usurping functions belonging to the public schools and to the commercial colleges. The admission requirements were too low. . . . The graduates and students did not become teachers, or else remained in the profession only a short time. There were more schools than needed." And, "They were not well located." After much political wrangling, all four of Oregon's state normals closed in 1909, Drain and Weston permanently. While these Oregon institutions were knocked off their feet, their situation was highly unusual; throughout the rest of the country, the dissipation of outright hostility helped to secure the footing of state normal schools.[12]

Aside from isolated incidents, from the 1870s onward criticism of state normal schools was just a murmur. Critics did not speak out against teacher education in general or even the principle of state support; instead, their concerns revolved around normals serving only local students or creating competition for other institutions. Private academies in New York and colleges in Missouri persisted in voicing such opposition to the state normals, while Iowa's private normal schools and colleges banded together to ensure that Cedar Falls would remain the state's only normal school. Outside of Iowa, the only significant result of such criticism was limited state financial support, which afflicted normal schools throughout the late nineteenth and early twentieth centuries. Funding was especially low in the wake of the defeat of proposals to eliminate normal schools, as in Minnesota and West Virginia in the 1870s, Missouri in the 1880s, and Arizona in the 1890s. But virtually all state normals complained to one degree or another about insufficient resources. Cheney was hit especially hard by low appropriations after a fire destroyed its main building. Throughout the mid and late 1890s, the normal barely scraped by, and even shut down for the 1897–1898 school year. Secure in their right to exist, the normal schools still faced the challenge of limited resources.[13]

As they had in founding normal schools in the first place, the efforts of boosters and local citizens often shored up their local normal schools in the absence of state funding. In Cheney, for example, residents voted in favor of bonds to fund the normal school and to build a new public-school building to be loaned to the normal for temporary use. The town rallied after the normal closed in 1897; local contributions of money and services helped it to reopen the following year. When construction of the normal school in Las Vegas, New Mexico halted in 1896 because the board of regents found itself without the necessary funds, local residents donated money to bridge the gap. The normal opened in 1898, and the building was completed the following year. The siege on Oregon's normal schools lasted for much of the 1900s, during which time gifts from supporters and local residents enabled the normals to get by. When Monmouth found itself without funding in 1905, voluntary contributions allowed it to remain open until voters approved appropriations in 1906. After the governor and legislature finally managed to shut down the normals in 1909, a statewide citizen effort to pass an initiative ballot in 1910 saved the Monmouth institution. In the towns where the other normals remained closed, citizens continued throughout the 1910s and early 1920s to try to revive them; only Ashland residents were successful.[14]

Limited resources most certainly constrained construction and maintenance on normal-school campuses. When Thomas Duckett Boyd arrived in 1888 to assume the principalship of Louisiana's state normal in Natchitoches, he was frustrated that he did not have enough money to repair the school's dilapidated buildings. Although many principals shared his plight, the state normal schools' physical plants were, on the whole, quite stately and impressive by the late nineteenth century, reflecting local pride and appreciation of their educational mission. A central building, often known as Old Main, was the only or the primary campus building. Constructed in the early 1870s, Peru's Normal Hall was three and one-half stories high, with a 90-foot bell tower that afforded a view of the surrounding Nebraska countryside, as well as Iowa and Missouri. The original building at Kirksville, Missouri and the 1882 main building at New Britain, Connecticut were both constructed of ever-popular red brick, with imposing stone ornamentation. In 1875, the normal in Indiana, Pennsylvania opened in a red-brick building that was the biggest in the entire state; an observer called it "a town complete within itself." A few years later, Gothic structures at Worcester, Massachusetts and Los Angeles impressed passers-by with arched windows and decorative pinnacles. Beginning in the late 1880s, commanding Romanesque structures housed many normal schools, including Oneonta, New York, Chico, and Las Vegas; Oneonta's Old Main had plenty of gables, turrets, and arches. Completed in 1898, Tempe's Old Main, which was three stories high, surrounded by a ten-foot-deep porch, and covered with a fancy roof, was the largest building in the surrounding valley. San Diego, California constructed its first normal-school building in 1899; modeled on the Fine Arts Palace at the 1893 Chicago World's Fair, it had a Greek facade and an Oriental dome.[15]

The campuses surrounding these impressive main buildings grew in size and scope during this period. State normals began to add all sorts of new

Figure 2.1 Students posing in front of Sutton Hall at Indiana State Normal School in Pennsylvania, ca. 1905. When Sutton Hall opened in 1875, it was the largest building in the state of Pennsylvania.

Credit: Special Collections, Indiana University of Pennsylvania.

buildings to their physical plant. For example, Iowa's normal moved into a new building in Cedar Falls in 1882, and then added an administration building in 1895, an auditorium building in 1900, and five more structures in the 1900s. In the late 1890s, turrets and battlements in the style of a medieval castle decorated the new science building at Carbondale, Illinois and the new gymnasium at Normal, Illinois. Greeley, Colorado added wings to the main building during the 1890s, then built a separate president's residence in 1904 that also served as a meeting place for school and community organizations. When a new library opened on the Greeley campus in 1907, the student newspaper reported, "If it be true that as one's surroundings are so will his thoughts be, then surely the standard of Colorado teachers has been raised by this addition." Some state normals also lavished attention on their grounds. A contemporary description of the new institution in Plattsburgh, New York soon after it opened in 1890 raved, "The grounds are laid out in walks and drives, and a miniature lake fed from a fountain that throws water forty feet into the air forms a pleasing contrast to the green lawns." A visitor to San Jose, California in the late 1890s mentioned the normal's "commodious buildings, which are surrounded by large and well-cared-for grounds." Growth and ornamentation of the normals' physical plants and grounds, along with increased enrollments and their spread throughout the country, reflected and confirmed the ascendancy of the state normal school in the late nineteenth and early twentieth centuries.[16]

"Nontraditional" Students

During their heyday from the 1870s through the 1900s, state normal schools served a distinct student body, which S. Y. Gillan at Illinois Normal referred to as "the masses and not the classes." Indeed, normalites during this period were far from society's elite or favored classes: the majority were women, and among both the female and male students, many were from minority ethnic and racial groups and families that were struggling financially. Furthermore, many normalites also were mature in age, had work experience, and lacked sophistication or worldliness. In other words, normal students as a group fit the profile of students who would still be considered "nontraditional" a century later.[17] The state normal schools opened a form of higher education to those types of students who would struggle for full access to mainstream higher-education institutions for decades to come.

At a time when women were an unwelcome minority on college and university campuses, they were a visible majority at state normal schools. By the 1870s, women had finally made some important inroads in American higher education beyond academies and normal schools. The Civil War, which had caused declines in male enrollment and thus encouraged many colleges to enroll women for the first time, the growing popularity of public education, and the overall growth of universities, all worked to open college and university doors to women. They gained access to an increasing number of women's colleges, especially in the East and South, and coeducational private colleges and state universities, especially in the Midwest and West. Still, the pioneer women students in the 1870s and 1880s studied in a climate of hostility. Opponents such as physician Edward Clarke argued that college education, whether in a single-sex or coeducational institution, would destroy women's reproductive systems and keep them from performing their proper role in society. On coeducational campuses, women were virtually invisible in the classroom and unwelcome outsiders in extracurricular life. By the 1890s, the pioneers had paved the way for greater toleration of female students; as increasing numbers of women attended colleges and universities, they created their own activities, separate from male-dominated campus culture. In the 1900s, women were becoming such a presence in higher education that some institutions attempted to bar them once again, or resegregate them in separate academic programs. Throughout the late nineteenth and early twentieth centuries, women's admission to institutions of higher education did not include their acceptance as full-fledged students or their access to the many dimensions of college life.[18]

At state normal schools, women were such a clear majority of the students that they could not be overlooked or dismissed. The normals in Framingham, Massachusetts, Milledgeville, Georgia, Greensboro, North Carolina, Rock Hill, South Carolina, and Farmville, Virginia restricted their enrollment to women only throughout the late nineteenth and early twentieth centuries, while the institutions in Salem, Massachusetts and Livingston, Alabama did so until the turn of the century. All of these normal schools were located in the Northeast and the South, where women's colleges were most popular.

Other southern and northeastern and all midwestern and western normal schools were coeducational, and enrolled greater numbers of women with each passing decade. Table 2.3 summarizes the percentages of students who were female at five geographically diverse state normal schools: Geneseo, New York; Florence, Alabama; Pine Bluff, Arkansas; Oshkosh, Wisconsin; and San Jose, California. Before the turn of the century, enrollments at these five and other coeducational normal schools were between 25 and 90 percent female. The numbers of women enrolled were on the low side at coeducational southern normals, including Florence and Pine Bluff, as well as Kirksville, Missouri; only one-quarter to one-half of their students were women. Geneseo and Oshkosh, as well as Cedar Falls, Iowa and Greeley, Colorado, were typical of normals outside the South, with enrollments that were approximately 60 to 70 percent women, while a few institutions, including San Jose as well as Winona, Minnesota, and Castleton, Vermont enrolled more than 80 percent women. After the turn of the century, throughout the country the fraction of students who were women was consistently well over one-half. A few coeducational schools, such as San Jose, Castleton, and most of the normals in Connecticut, Massachusetts, and New York, enrolled more than 90 percent women.[19] It is hardly surprising that normals, which prepared students for the female-dominated profession of teaching, primarily enrolled women. In addition, however, normal schools brought higher

Table 2.3 Female enrollments in five state normal schools, by percent, 1875–1915

Year	Enrollments in five state normal schools: Percent women				
	Geneseo, New York	Florence, Alabama	Pine Bluff, Arkansas	Oshkosh, Wisconsin	San Jose, California
1874–75	59.1	24.6	52.5	59.4	82.6
1879–80	not reported	34.8	38.9	58.4	85.9
1884–85	not reported	50.9	32.0	64.1	85.2
1889–90	65.9	39.6	43.2	61.4	93.6
1894–95	70.5	59.0	27.5	64.5	92.6
1899–1900	71.8	61.6	42.9	68.6	90.4
1904–05	74.2	57.3	52.5	72.8	93.4
1909–10	80.8	56.1	52.1	79.8	96.3
1914–15	83.3	58.2	57.9	66.5	94.3

Sources: School catalogs and bulletins, and *Report of the Commissioner of Education* (Washington: Government Printing Office, 1875–1915).

education to women, as well as men, from backgrounds that were unusual in higher education at the time.

State normal schools made a form of higher education available to a significant number of racial- and ethnic-minority students. Like women, minority students by the 1870s had some access to colleges and universities, but with severe limitations. In the South, segregation restricted African Americans to black-only institutions. Philanthropists founded hundreds of black schools between the end of the Civil War and the turn of the twentieth century. Northern white missionaries, black religious societies, and industrial and corporate foundations all funded educational endeavors, but those initiated by the former two groups struggled financially while those under the control of the latter emphasized industrial rather than academic training. The most prominent industrial-education institutions, the Hampton Normal and Agricultural Institute and the Tuskegee Normal and Industrial Institute focused primarily on making their students work hard so that they would appreciate the value of labor, and thus did not offer higher education of any sort. While many state normal schools, especially after the passage of the second Morrill Act in 1890, also offered—or at least paid lip service to—industrial education, this was not their primary focus. Thus, the twenty or so black state normal schools played an important role in expanding opportunities in advanced academic education for African Americans in the South.[20]

Northern blacks and members of other minority groups in the South and the North had only limited access to institutions of higher education in the late nineteenth and early twentieth centuries. Black colleges had operated in the North and African American students had attended majority-white Oberlin College in Ohio since the antebellum period, but the latter part of the nineteenth century brought few additional opportunities. In 1899–1900, only 88 African American students earned degrees from white colleges, mainly from Oberlin. For other minority groups, majority-white institutions were the only option. From Texas to California, some Latino students enrolled in Catholic colleges, while a very few were able to attend state universities. On majority-white campuses, pioneering black and Latino students, as well as Jews, Catholics, and members of other ethnic minority and immigrant groups, remained outside mainstream campus life.[21]

While the numbers of minority students were not overwhelming, evidence suggests that several northern, majority-white normals served students from minority ethnic and racial groups. At the Rhode Island Normal School in Providence in the 1880s, over 17 percent of the students were Catholic, and many of them were probably recent immigrants. At Mankato, Minnesota in the 1870s and 1880s, almost 8 percent of the students were born in European countries; most were from Norway and probably Protestant, but a few were from Ireland and Russia and likely Catholic and Jewish. At Worcester, Massachusetts, the appearance by the early twentieth century of names such as Murphey, Sullivan, Orfanello, Cohen, and Lewandowski on student rosters indicated the arrival of Catholics and Jews. In New York State, Albany matriculated 26 Native American students in the late nineteenth century and many South European, Polish, and Jewish immigrants in the early twentieth century, while Oswego graduates of the 1890s remembered "an amiable American Indian girl," "popular young men from the Hawaiin Islands," "a shy, quiet Negro girl," and a "much respected" African American man; and Plattsburgh graduated its first black student, Augustus Wilson, in 1902. African American students, many of whom were from the South, had begun appearing at state normals from Massachusetts to Illinois soon after the Civil War. As early as 1871, the governing board at Illinois Normal University spoke out against any effort to "recognize distinctions of race or color in determining who shall or who shall not be admitted," and affirmed "the equal rights of all youth of the state to participate in the benefits of our system." The normal schools in Pennsylvania welcomed several students from Puerto Rico and South America beginning in the 1890s. Even southern normals occasionally admitted non-black minority students; Cecil E. Evans, who began his presidency at Southwest Texas State Normal School in San Marcos in 1911, once noted, "Very few Mexican students ever get high enough in the grades to reach us," implying that at least a few Mexican students did "reach" the institution.[22]

Regardless of their race or gender, most normalites shared rather low socioeconomic status; they were, for the most part, the daughters and sons of working people, many of whom were struggling financially. While students like them were not completely absent from mainstream higher-education institutions in the late nineteenth century, they were rare and not particularly welcome. As class differences intensified in American society, social boundaries became sharper on college campuses. Some less-advantaged families looked to higher education as a means of advancement, and their children worked their way through college. At the same time, however, the middle and upper-middle classes began to use colleges and universities to secure their status. Many of the small colleges that had accommodated poorer students earlier in the century grew increasingly exclusive. The newer state universities and land-grant colleges educated many children of farmers and lower middle-class workers, but middle-class and growing numbers of upper-class students outnumbered them. Wealthy students joined fraternities and sororities and treated poorer students as outsiders. By the turn of the twentieth century at both public and private institutions, tuition and living costs as well as elitism in campus life were on the rise; such developments further assured the marginalization of students from the lower classes.[23]

Students from the lower classes attended state normal schools in such large numbers that they could not have stayed on the margins. Such a full range of nonprofessional and nonlucrative occupations was represented among normalites' parents, that Principal Blair at Fairmont, West Virginia in the 1870s said the students "represent the industry and spirit of the people." The class of 1888 at Oshkosh described its "home influences" as, "such as the home of the farmer, the blacksmith, the wheel-wright, the mill-wright and the minister can exert upon their sons and daughters." Between 1872 and 1898, the most common occupation among the parents of normalites at

Providence was farming: 36 percent listed their parents' work as farmer; carpenter (12 percent) was second, jeweler (7 percent) was third, and machinist (almost 7 percent) was fourth. During the same period, skilled, semiskilled, unskilled, and agricultural workers headed the homes of two-thirds of normalites in Massachusetts. The normal in Worcester reflected its location in a growing industrial city, drawing primarily the children of skilled workers and laborers. During the first decade of the twentieth century, Southwest Texas kept very detailed records of the occupations of its students' parents. They were more likely to work in agriculture than any other trade: 47 percent were farmers, and ranchers, fruit growers, stockmen and dairymen were another 10 percent. The next largest proportion of parents, 16 percent, was commercial businessmen: merchants, real-estate agents, bankers, salesmen, jewelers, grocers, insurance agents, cotton buyers, lumber dealers, hotel keepers, butchers, drummers, "officers," and "agents." Only 6 percent of these Southwest-Texas parents engaged in the high-status professions of medicine and law.[24]

The 1890s football cheer at Geneseo, New York, "We came to the gridiron fresh from verdant farms," summed up the background of the majority of normalites, but put a positive spin on difficult circumstances. Farming crises in the late nineteenth and early twentieth centuries shaped the lives of normal-school students from Providence to San Marcos and beyond. In the late 1870s, the 1880s, and the 1890s, nearly two-thirds of the normalites at Cedar Falls, Iowa, were from farming families. Likewise, in 1889, the parents of 428 of the 639 students at Emporia, Kansas, were farmers. Farming was especially hard-hit by economic downturns and depressions in the 1870s, 1880s, 1890s, and 1900s, while the rising urban, middle-class, professional consciousness had no place for agricultural workers. Farming families were left to struggle with debt and poverty. One such family was the Pegrams in northwestern North Carolina. Born in 1872, Phoebe Pegram became used to hard work in the fields at age 5. After her father died and her mother's health began to fail when Phoebe was in her early teens, she had to run the struggling farm and look after siblings. For her, attending the normal in Greensboro was the only available means to escape the hardship of farming; she enrolled soon after the institution opened in 1892, and found that one-quarter of her classmates were also the daughters of widows, or orphans. Also struggling against racism, the farming families of many Pine Bluff students were so hard-pressed that it was necessary for their sons and daughters to arrive at the normal late in the fall and leave early in the spring, so that they could help with harvests and plantings.[25]

The families of many normalites lost the battle against poverty. In his biography of Southwest-Texas graduate Lyndon B. Johnson, Robert Caro described San Marcos as " 'a poor boy's school' . . . It had been a desperate struggle for many of them to raise even the tiny San Marcos tuition; they had made the struggle because they felt the education they could obtain at San Marcos was their only chance of escaping a life of physical toil." Kate Berry Morey, who graduated from Winona, Minnesota in 1872, later remarked,

"I could tell of work done in poorly heated rooms, in stone cold rooms, with tea and crackers for breakfast, crackers and tea for lunch, and supper the only hearty meal of the day . . . expedients to make a scanty allowance reach till the end of the term." J. S. Nasmith left the normal at Platteville, Wisconsin in the early 1870s and later wrote, "It was an awful blow to have to leave. Mother understood the feeling. She wept when lack of money forced me to drop out." Some normals gained a negative reputation for serving the poor, especially in the Northeast. A Vermont newspaper described the normal in Castleton as catering to the "calico-attired country girl of limited means," and in Westfield, Massachusetts, where the normal school enrolled an especially high number of the children of small farmers, "normal" was a disparaging name for a poor person. Similarly, some residents of Oswego, New York, referred to normalites there as "state paupers." Pine Bluff students made light of their reputation in the following lines from a late-1920s school cheer:

> State School, State School, yes we are the state school
> Nothing new or formal, no Sir!
> Our hair is shaggy and our clothes are baggy
> But they'll soon be raggy, Yea!

Throughout the country, many normal-school students could have yelled along.[26]

Financial struggles and the limitations of race and gender often caused delays in the education process. Thus, many normalites—as well as some students at colleges and universities—were mature in age. Throughout the late nineteenth and early twentieth centuries, normal-school students were usually older than the minimum state-stipulated 15 or 16 years. Upon arrival at Providence during the 1870s, 1880s, and 1890s, students were, on average, between 17 and 19 years old; during most years, the oldest students were in their late 20s or 30s. Michael Dignam, who graduated in 1882 from Westfield, Massachusetts, remembered, "The pupils were all mature, no one under 20; the ages ranged as high as 26 or 28." In 1886, the first class at Tempe, Arizona averaged 19.4 years in age; at 26 and 30 years old, respectively, Julia A. McDonald and James M. Patterson were among the oldest students. Oshkosh's class of 1888 lyrically explained: "In age the class varies all the way from the blushing maiden in her teens, filled with anticipation of the future, and the aspiring youth yearning for independence, to the retrospective and reflective minds of maturer years." From Greeley, Colorado, to Oneonta, New York, normal schools in the late 1880s and early 1890s reported that, on average, their entering students were more than 20 years old. Throughout the 1890s and 1900s, women students at Oneonta continued to enter at an average age of over 20 years, while their male classmates were closer to 22. In 1903, the first class at San Marcos signed in at 20.1 years; at 19.7 years, its women tended to be very slightly younger than the men. Very unusual for the time and probably indicative of a certain amount of maturity, some female normal students were married, or perhaps widowed. Roughly between one and eight students per year at Oshkosh used the title "Mrs." during the 1890s and 1900s, and at San Marcos, as few as 3 and as many as 18 students per year were addressed as "Mrs." The normal school in Plattsburgh, New York had at least a couple of mature students during the 1910s, as there were two mother–daughter pairs on campus.[27]

It is hardly surprising that many of these older students arrived at normal school with significant work experience, and that students of all ages found it necessary to work during their studies. Many students had worked as teachers prior to matriculation. Of the 2,500 students who enrolled at Oswego between its opening in the mid-1860s and 1880, more than 1,000 had teaching experience, with an average of three years in the classroom. During the late 1880s, more than half of the students at Farmington, Maine arrived with teaching experience, and throughout the late nineteenth century, approximately 30 percent of Westfield's entering students had teaching experience. Meanwhile, at Mankato during the 1870s and 1880s, close to 70 percent of the students entered as experienced teachers and/or took a leave from their studies to earn money by teaching. Before graduation, Calista Andrews of the class of 1871 taught for five years in four different states, and Mary Faddis '80 taught for eleven years in Minnesota schools; both were 26 years old when they graduated. The typical Mankato woman taught for one and one-third years and was 21 when she graduated, while the typical Mankato man taught for one year and was 22.5 years old when he graduated. At Oshkosh in 1879 as well as in 1884, half of the students had taught, and among them, the average time spent in front of a class was 2.6 years. Between 1892 and 1908, the collective student body at Oshkosh averaged one year of experience. In their class histories, students conveyed the same idea a bit more creatively: the class of 1895 taught "more than a thousand months," and the brains of members of the class of 1896 bore "unsightly marks . . . caused by patient efforts to impress the American youth in our rural districts with the precept, knowledge is power."[28]

Outside the Northeast and Midwest, it was also very common for normalites to have work experience as teachers. Tempe's Julia McDonald and James Patterson were former teachers, and 36 percent of Phoebe Pegram's classmates at Greensboro reported the same. More than 34 percent of the students at Florence, Alabama around the turn of the century had teaching experience. Similarly, at San Marcos in 1903, 29 percent of the students had served in the classroom; two years later, 24 percent of matriculants had taught. Among the experienced teachers in each of these years, the average time spent in front of a class was 2.6 years. When Gordon Wilson enrolled at Bowling Green, Kentucky, he had already taught in two small Kentucky towns, where parents were happy with his performance. Wilson, however, viewed himself as a "tragic, perhaps pathetic failure" as a teacher, and hoped that attending normal school would improve his performance.[29]

To meet their financial challenges, many normalites worked while enrolled, and self-supporting students were not unusual at the state normals.

In 1882, President Edwin Hewett of Illinois State Normal University, in Normal, reported, "Many of our students . . . are dependent upon their own exertions for means." A decade or so later Principal John Mahelm Berry Sill of the State Normal School in Ypsilanti, Michigan said, "Our students are working young men and women who earn their little money by the hardest toil." Just after the turn of the twentieth century, one-third of the students at Florence, Alabama "earned their own money to pay expenses," while Oregon's State Superintendent of Public Instruction expressed concern about the large numbers of "self-supporting" normalites, urging, "Greater precaution must be exercised to prevent the ambitious from overworking than to rouse the sluggards." Gender often determined the term-time jobs normalites were able to find; in New York, female students at Oneonta worked as babysitters or maids, while male students at Oswego shoveled snow, sold various products, or worked as janitors. J. S. Nasmith remembered classmate Aaron Newcomb's sacrifices: "With determination in his eye he brought a sister and brother to Platteville that all three might get an education. He sawed and split wood, did any other kind of work that offered. Had to leave for awhile but came back again." Throughout the country, it was common for normalites to teach in rural schools during breaks or extended leaves of absence in order to earn the necessary money to continue their education. Kate Berry Morey reported that at Winona, "The opening of the summer term of the country schools in April or May made many vacancies in our ranks, for in order to earn money for another year's work, many taught through the long vacation." These students just scraped by financially.[30]

Although age and work experience fostered a certain level of maturity among normal-school students, they were hardly worldly-wise. They tended to be limited in perspective simply due to a lack of exposure. A chronicler of Massachusetts' Bridgewater State Normal School observed, "It is hard for us to conceive how provincial these students were. Most of them had never been far from their own towns." Sarah A. Dixon, who graduated from Bridgewater in 1885, called herself "a sip of a girl from an isolated shore home." Throughout the country, normalites were predominantly from very small, often very rural, towns and villages. A sizable number of the students at Florence, Alabama were from Florence itself, but many of the others were from areas so remote that they reported a home county rather than a town. At Greensboro, North Carolina in 1900, more than 60 percent of the students came from towns with populations under 2,500, and more than half of these students came from remote areas with fewer than 500 inhabitants. Between the 1880s and the 1900s, 23–30 percent of all Oshkosh, Wisconsin students hailed from Oshkosh, which was a booming lumber town. Another 8 percent or fewer of the students came from other lumber towns in the area, and only a few hailed from the bigger cities of Milwaukee, Madison, and, occasionally, Chicago; more than half of the students were from much smaller, remote towns. Meanwhile, many of the students at California's state normal in Chico were from very remote mountain settlements in the northern part of the state; when they arrived at the normal, Chico was the largest town they had ever seen.[31]

Students from such remote areas were unfamiliar with the larger world. An extreme case was the student whom instructor Henry Johnson remembered at Charleston, Illinois:

> One of the most interesting students who ever sat under my instruction was a young man who on his way to Charleston had seen for the first time a railroad train. He could read quite well and he spoke fairly correct English. But he had never read an entire book. He was unacquainted with newspapers. He did not know the name of the man who was then President of the United States. His study of American history had never been carried beyond George Washington. He had attended one of those rural schools in which classes always began at the beginning of a subject and stopped when the term stopped.[32]

Similarly, biographer Caro explained that at San Marcos:

> . . . to many of these young men and women from the land of the dog-run, Old Main was the largest building they had ever seen, larger even than their county courthouses; its spires loomed over them, taller by far than the spires of their churches. The crowd of students in which they stood in front of Old Main on Registration Day was the largest crowd they had ever been in—the most people they had ever seen gathered together in one place.[33]

Figure 2.2 Old Main at Southwest Texas State Normal School, ca. 1909.
Credit: University Archives, Texas State University.

Limited experiences left normal-school students without polish. For many years after the 1868 establishment of the State Normal School in Peru, Nebraska, the many students there who had lived "isolated lives" tended to be "ignorant of the social ways incident to more thickly settled portions of the country; hence, they sometimes appeared reserved and awkward." Early in the twentieth century, the principal at Willimantic, Connecticut complained, "Many of our students are crude. Their manner of talking, their table manners, their actions often show a decided lack of culture." In same vein on the other side of the country, the president at Tempe, Arizona reflected, "So many girls come to us from ranches, mining camps and other places with little or no knowledge of what is the proper thing to do on various occasions." Normalites were a veritable object lesson in rusticity. A student at Peru in the late nineteenth century remembered, "Girls with brown faces and plain clothing" and "boys with calloused hands." Similarly, Irene M. Mead, who graduated from Winona in 1884 and then served on the faculty, described men students who were "raw and awkward" when they arrived at the normal and women students who "had not a knack of dress, and the riot of color which the Assembly room displayed was not an unmixed joy."[34]

Mead's description of "the riot of color" was somewhat pejorative and the students' lack of polish frustrated Willimantic's principal, but most people who staffed the state normal schools were more charitable. They understood that attendance at state normal schools was a significant departure in the theretofore-unsophisticated lives of many students. Students arrived at state normal schools hungry for inspiration, and found it. An Oshkosh student declared:

> When one for the first time beholds an imposing structure, whether reared by man's stalwart arm or nature's majestic art, impressions are made upon the delicate parchment of the mind which age can not dim nor time obliterate. Such is the character of my first impressions on beholding the Oshkosh Normal. Its architectural symmetry symbolizes the noble educational system in which it forms an important factor . . . its spacious assembly room and its commodious recitation rooms silently insinuate to the pupil the possibilities of mental expansion, while each high ceiling proclaims the aphorism, "There is always room at the top."[35]

The state normals not only accepted students from nontraditional backgrounds, but they also engaged their sense of awe to create an atmosphere that embraced them.

Welcoming Nontraditional Students

Unpolished, rustic students did not stand out at state normal schools as they would have at colleges and universities in the late nineteenth and early twentieth centuries. Not only were unsophisticated students—many of whom had faced the challenges of poverty and gender and race discrimination—in the majority, but they also studied under faculty and principals with whom they had much in common. Normal-school faculties included many normal-school graduates, and thus many of their members fit the basic profile of the normalites. As in the mid-nineteenth century, principals in this later period were usually men, many of whom had grown up in modest circumstances. By the late nineteenth century, approximately one-third to one-half of the principals had attended normal school. When the future principal of Kirksville, Missouri (1899–1925), John R. Kirk entered Kirksville as a student in the mid-1870s, it was after many years of plowing on the family farm and a stint as a rural-school teacher; financial problems plagued him during his normal course. Future principal of St. Cloud, Minnesota (1875–1881), David L. Kiehle grew up in an impoverished family in western New York and worked his way through the Albany State Normal School. Kirk and Kiehle both went on to attend other higher-education institutions. Likewise, William E. Wilson, who would head the state normals in Providence, Rhode Island from 1892 to 1898 and Ellensburg, Washington from 1898 to 1916, began as a student at Edinboro Normal School in western Pennsylvania, and went on after teaching for six years, to enter the sophomore class at Monmouth College in Illinois and graduate in 1873. Before becoming a normal principal, Wilson taught at the normal and college levels, and studied and traveled in Europe. Experiences as students at state normal schools must have shaped the sensibilities of Wilson, Kiehle, and Kirk, throughout their long careers in education.[36]

Whether or not they had attended normal school, many principals were sensitive to financial hardship because of personal experience. While a majority of principals during this period had attended college and some had earned advanced degrees, they had generally not been privileged students. For example, George Colby Purington (Farmington, Maine, 1883–1908) alternated between teaching in district schools and working on a farm beginning at age 17, and then prepared for college while serving as an associate principal at an academy. While attending Bowdoin College in his 20s, he also worked as a high-school principal. Albert Salisbury (Whitewater, Wisconsin, 1885–1912) was born in 1843 to a pioneer family in Lima, Wisconsin, and worked hard on the family farm as a child. In the early 1860s, he attended Milton Academy in the winter and worked on the farm in the summer. After fighting in the Civil War, Salisbury returned to the Wisconsin farm and worked in the nursery business, and then returned to Milton and graduated in its first college class in 1870, at age 27. Several future principals followed the well-established pattern of poorer students becoming ministers. Kiehle worked his way through Union Theological Seminary in New York and served as a Presbyterian Minister in Preston, Minnesota before being appointed county superintendent and then president of the St. Cloud normal. The first principal at Cedar Falls, Iowa (1876–1886), James C. Gilchrist financed his academy and college education with farm work and teaching, and was ordained in the Methodist Episcopal Church. By the late nineteenth century, quite a few normal principals had law degrees. After attending the Universities of Kansas and Missouri, Kirk was admitted to the bar in 1885 and practiced for one year. Charles D. McLean (Brockport, New York, 1869–1898) was born in Ireland and immigrated to western New York with his widowed mother. With degrees from the University of Rochester and the Albany Law School, he was admitted to the bar only to realize that teaching was his true calling. McLean's successor, David Eugene Smith, a graduate of the Michigan State Normal School at Ypsilanti and holder of Doctorate of Philosophy degree from Syracuse University, had also practiced law but found it unfulfilling. Although their college degrees placed MacLean, Salisbury, Purington, and other principals in a small elite, they were not strangers to hardship and financial struggle.[37]

Personal experience also acquainted some normal-school principals with the gender and racial struggles of many normalites. A very small number of principals, including Annie E. Johnson (Framingham, Massachusetts, 1866–1875), Ellen Hyde (Framingham, 1875–1898), Julia Sears (Mankato, Minnesota, 1873–1874), and Sarah E. Richmond (Baltimore, Maryland, 1909–1917) were female. Johnson, whose father was a minister in Maine, received her early education from him and professors at nearby Bowdoin College, where women were unwelcome as students. She taught at the grade- and high-school levels before attending Framingham in the early 1860s. Hyde, Sears, and Richmond were also normal-school graduates. Some of the principals of black normals were African American. For example, E. E. Smith was principal of the State Colored Normal School in Fayetteville, North Carolina from 1883 to 1933, aside from leaves of absence to serve in the U.S. State Department and the Army. Smith was born in North Carolina during slavery to a free black woman. During Reconstruction, he received some instruction from sympathetic whites, and walked three miles to a Wilmington night school run by missionaries. From 1873 to 1878, he taught while attending Shaw Collegiate Institute, then continued to teach and served as a principal until his appointment at the normal.[38]

Although they had not personally confronted the barriers of gender or race, some other principals clearly expressed or acted upon their sympathy for the plight of those who had. At Normal, Illinois, Richard Edwards (1862–1876) asserted the intellectual equality of women to men, asking a group of male educators, "is it not the acme of absurdity for you and me, because we happen to grow beards, to step forward, with our little measuring strings, and attempt to fix beforehand, the scope of women's investigation of truth?" Similarly, Thomas Duckett Boyd (Natchitoches, Louisiana, 1888–1896) implored the State of Louisiana to provide extensive educational opportunities not only for its sons, but also for its daughters, "whose education," he declared, was "equally important," and necessary to insure "them independence for life." And Charles Duncan McIver (Greensboro, North Carolina, 1891–1906), whose wife had studied medicine and was a champion of women's rights, spoke in favor of women's need to be "independent and self-supporting." William Burns Paterson, who would serve as principal of Alabama's state normal school for black students in Marion before it moved to Montgomery in 1887, grew up in a rural village in Scotland, far from the racial apartheid of the American South. Perhaps the poverty of his family of farm and brewery laborers and his experience of leaving school at the age of 12 to take care of his siblings helped Boyd to sympathize with the freed slaves when he immigrated to Alabama after the Civil War. Self-educated, he was happy to share his learning with the black laborers he met. Teaching African American men, women, and children soon became a cause, and he started a school for black students. Strong convictions, often shaped by life experiences, guided many normal-school principals in their work. These leaders worked with—or occasionally worked around—governing boards and state leaders to maintain policies that welcomed students from nontraditional backgrounds to state normal schools.[39]

The first steps were to ensure that the normals were accessible and affordable. Throughout the 1870s, 1880s, 1890s, and 1900s, minimal admission standards allowed the educational level of each school's incoming students to parallel growth of public education in the area in which it was located. In Massachusetts, by 1880 two-thirds of the students admitted to Bridgewater, located in the well-developed eastern part of the state, were high-school graduates. At the same time, however, few students at Westfield, in the more rural western part of the state, had attended a high school or academy; the principal remarked that his students came "from country schools many of whose teachers possess limited qualifications." By the early 1890s, approximately 40 percent of students entering Westfield were high-school graduates. The state of Massachusetts began in 1894 to require high-school graduation, or the equivalent, for normal-school admission. As high schools became more widespread in Minnesota, the percentage of normalites at Winona who had diplomas jumped from 8 in 1885 to 83 in 1900. Eleven years later, only one-quarter of the students at Hays, in remote western Kansas, had completed high school. In Washington State, the 1903 ruling by the Board of Education that the completion of ninth grade was required for admission proved to be unenforceable, and the normals soon returned to requiring completion of eighth grade. California's San Jose State Normal School began requiring high-school graduation for admission in 1901, after legislative acts for state funding had resulted in dramatic high-school enrollment increases during the 1890s. In the South, where public education grew at an especially slow rate, there were few non-private high schools for either white or black students throughout this period. Thus, education leaders in Kentucky agreed in the mid 1900s that it would be "futile" to require high-school graduation. The black normal at Fayetteville, North Carolina required only completion of the eighth grade into the 1920s. Florence, Alabama waited until 1925 to require a high-school diploma, and in 1930 San Marcos, Texas still made provisions for non-high school graduates. Throughout the country, admission policies did not stipulate that matriculants had to be high-school graduates until education at that level was attainable by most residents of the state.[40]

Both before and after they officially required high-school graduation, normal schools' admission standards remained fairly loose and flexible. Prospective students had multiple options for demonstrating academic readiness, including diplomas, teaching credentials, or scores on admission examinations. In 1886, Adelaide Pender spent the summer before she entered New Britain State Normal School in Connecticut preparing; she explained, "I reviewed arithmetic, language, history and geography, doing stunts that would cram my mind with facts, facts, facts. But I found my High School diploma would be accepted in lieu of an examination." After Massachusetts began to officially require high-school graduation in 1894, the normals still administered admission exams. In addition, as the 1896–1897 Westfield catalog explained, the Board of Education ruled "successful experience in teaching is allowed to be taken into account in the determination of equivalents in the entrance examinations." In Washington, students who did not have an eighth-grade diploma could present a teaching certificate or pass an entrance exam. Previous years' admission exams were often published in normal-school catalogues, making them available to students preparing for future admission. In the late 1870s, San Jose published questions in arithmetic, grammar, geography, and spelling, for admission to the junior class, and additional questions for admission to the "middle" and senior classes. Applicants to Oshkosh, Wisconsin during the 1880s had to score 70 percent or better in reading and spelling, arithmetic, grammar, geography, and U.S. history. In Vermont, Castleton's admission exam in the early 1890s covered spelling, arithmetic, physiology, grammar, geography, Vermont history, U.S. history, and civics.[41]

In addition to academic qualifications, other admission requirements were fairly easy to meet. Generally, applicants had to be 15 or 16 years in age; most, of course, were quite a bit older. Some normal schools required a statement of the applicants' intention to teach, although paying a fee could exempt a student from this requirement. Under "Terms of Admission," the Florence catalog for 1876–1877 stated, "Students entering the Normal School will be required to give their obligation to teach for two years in the Public Schools of the State." However, students who decided they did not desire to teach could "discharge all obligations to the school by paying for their tuition while connected to the school." When the Branch Normal College for African American students opened in Pine Bluff in 1876, it required incoming students to meet age and academic requirements, and to be of strong moral character. Statements about morals and health became increasingly common, presumably as prerequisites for being good teachers in the future. Florence required "good moral character" by 1883, and added "good health" by 1886. Similarly, the 1887–1888 catalog for San Jose listed "A good moral character" and "Good bodily health" as the first two requirements for admission. The 1894–1895 Westfield catalog went into a little more detail, stating, "Candidates for admission to any one of the [Massachusetts] normal schools must . . . be free from any diseases or infirmity which would unfit them for the office of teacher. They must present certificates of good moral standing." Four years later, the Oshkosh catalog called for "approved moral character and good bodily health." Because normal-school policies were rather vague regarding what constituted "good" character or health, it seems unlikely that these requirements were hurdles for most people seeking admission.[42]

Principals and other leaders rarely enforced the few other stipulations as to who could or could not gain admission to state normal schools. Several states passed legislation limiting how many students from each county or legislative district in the state could attend a particular institution, in order to assure equal access or representation on campus. Principals exercised discretion in this area, however, and few students were ever excluded as a result of such policies. In Iowa, the normal-school board abided by a very liberal interpretation of the law stating that each county was to be represented at the Cedar Falls institution in proportion to its population; the board declared, "admission shall not be refused to any applicant who can be admitted without prejudice to the rights of others." The normal's first catalog stated, "Practically, the School is open to all." Once students gained admission, school policies on the terms of their attendance were quite flexible. While it undoubtedly was easier to conduct instruction when all students were present at the beginning and throughout the semester, normals were very willing to make exceptions. For example, the board governing California's normals in 1883 passed a resolution that students would "be admitted only at the beginning of the terms of the schools." The 1885–1886 San Jose catalog added, "The Faculty have, however, the power to suspend this rule in cases which, for good and sufficient reasons, they may consider exceptional." The Millersville Normal School in Pennsylvania took this flexibility even further, stating in 1903, "students may enter this normal school at any time. To make the most advancement in their studies, they should enter as early as possible in the season." Flexible rules for attendance and minimal admission requirements made it easy for students from less-advantaged backgrounds to enter the normals.[43]

State normal schools also assured accessibility by providing detailed directions to campus and individual assistance with settling in. For several years, Geneseo State Normal School in New York did everything short of printing train schedules in the catalog. The following instructions in 1902 were typical:

> Students living on the line of the New York Central Railway will take the cars to Avon, and thence to Geneseo. Those who come by the Delaware, Lakawanna & Western Railway, should come to Mt. Morris, and thence by the Erie Railway to Geneseo. Those who come by the Western New York and Pennsylvania Railway should come to Mt. Morris, and thence to Geneseo by the Erie Railway, or to Piffard or Cuylerville, and thence to Geneseo by carriage.

Oshkosh similarly described the "numerous lines of railroad and river steamers entering the city, as well as its favorable location"; and Florence explained that the "main line of the Southern Railway from Chattanooga passes through Sheffield, and all passenger trains are met there by electric cars which deliver passengers in Florence in about twenty minutes."[44]

After careful instructions helped many students reach campus, the staff, usually in the form of the principal, literally reached out to individual new students. During the 1870s, 1880s, and 1890s, Oshkosh's Principal George Albee helped each new student get situated and plan a course schedule. At Emporia, Kansas in the 1880s and 1890s, President Albert Reynolds Taylor took very seriously his responsibility to help students settle into residences. He later wrote, "I set about finding homes in a very methodical way, visiting and inspecting many of them myself that I might have first hand knowledge of their condition and facilities for meeting the demands of the different classes of students." The 1907–1908 Geneseo catalog declared, "students both old and new are urged to consult the Principal freely regarding their work and their plans for the future." Clark Davis, who attended Ypsilanti, Michigan during the 1900s, remembered that upon arrival in Ypsilanti, "I made my first call upon the man whom I had been writing—namely, President Jones. He treated me cordially and courteously, received me at his home, took me to his office, and walked over some of the city streets to show me rooming houses and boarding houses." Like many other principals, A. J. Matthews at Tempe Arizona (1900–1930) was known for his open-door policy, both at his office and at his home.[45]

While attainable admission requirements and approachable principals eased many students' adjustment to normal school, probably the most important factor in accessibility was affordability. Most state normal schools charged a modest tuition, which they waived for state residents who signed a pledge to teach in the state after graduation, usually for no more than a few years. While Florence made the pledge mandatory for admission but then exempted students willing to pay tuition, most normals waived tuition for those students who signed the pledge after admission. The 1879 catalog for Bridgewater explained: "Tuition is free to all who comply with the condition of teaching in the schools of Massachusetts, wherever they may have previously resided. Pupils who fail to comply with this condition are charged a reasonable sum for tuition." The Normal School Act in Rhode Island specified that students "admitted to free tuition" would be bonded to teach in the state for "at least one year after graduation," while the tuition waiver at the State Colored Normal School in Fayetteville required graduates to teach for three years. The terms were a little more vague at Tempe, where the following pledge exempted students from paying tuition: "First: That our purpose in coming to this school is to prepare ourselves to teach. Second: That it is our present intention to teach in the public schools of Arizona for a reasonable length of time." Such contracts were in the states' interest because they increased the teacher supply, but tuition waivers also made a normal-school education affordable for many students.[46]

Students who signed pledges to teach had only to buy or rent books, and pay for supplies, "incidental" fees, transportation, room, board, and perhaps activities fees or music lessons. And at many institutions, financial help from the state was also available for these expenses. The Bridgewater catalog also explained:

> For the assistance of those students who are unable to meet the expenses of the course of instruction in the school, the State makes an annual appropriation of eight hundred dollars, one-half of which is distributed at the close of each term among pupils from Massachusetts who merit and need the aid, in sums varying according to the distance of their residences from Bridgewater . . .

In other words, financial aid was tied to travel expenses. Similarly, the Act to Equalize the Benefits of the State Normal School, Etc., passed in Rhode Island in 1871, provided for travel allowances for normalites, and until 1897, Geneseo students who signed the declaration to teach received reimbursement for travel costs. Upon the recommendation of President Taylor, the state of Kansas began in the 1880s to reimburse Emporia students three cents per mile for travel beyond one hundred miles. The Nashville-based Peabody Fund paid travel expenses for Fayetteville students. Finally, New Mexico's territorial assembly in 1909 passed an act that provided for reimbursement for railway fare in excess of six dollars to or from the normal schools in Las Vegas and Silver City, provided students were territorial residents and promised to teach in New Mexico.[47]

Other forms of financial assistance were not tied to travel expenses. Ypsilanti's Principal Sill persuaded the Michigan legislature to provide free textbooks, arguing that "the cost of books is often 'the last straw that breaks the camel's back.' " San Jose provided free writing paper to students planning to teach, and many institutions provided campus jobs for financially hungry students. Fayetteville reduced costs by requiring all students to work for one hour each school day, but it was more common for normals to hire and pay individual students for particular jobs. At Greensboro, North Carolina, helping in the dining hall earned free room and board, and many young women applied for the 16 available positions. From the time it opened, Tempe employed many students, all around campus; one graduate remembered, "Me and my sister both had to have jobs. We worked in the dormitory, but a lot of the girls worked in the dining room or in one of the offices." In several states, especially in the South, government officials were directly involved in granting subsidies for boarding and other costs. Beginning in 1897, each member of the Alabama legislature was permitted to nominate a student for a two-year normal-school scholarship, which covered tuition and incidental fees; those students appointed by senators also received a grant for boarding costs. Between 1903 and 1909, San-Marcos students could also earn "scholarship appointments," which covered boarding costs, through appointments by their senators, congressmen, or even the governor.[48]

Increasing numbers of campus scholarships and loan funds helped normal students pay expenses that were not covered by the state. Beginning in 1878, Pine Bluff had an "honorary scholars" program, which provided students who passed an exam with scholarships that were not limited to study in the normal department, and carried no obligation to teach after graduation. For many years beginning in 1882, the Peabody Fund provided for 16 annual scholarships at Florence, and additional scholarships at other southern normals. By the early twentieth century, San Jose and other normals had scholarships named for former faculty members. Fund-raising efforts within normal-school communities also established various loan funds. At Charleston, Illinois, revenues from the sale of tickets to the senior recital and the senior class play, beginning in 1900 and 1904 respectively, established a student loan fund. A faculty committee approved loans to students who had attended the normal for at least one year. The classes of 1904 and 1910 at Troy, Alabama each raised the money for a loan fund to assist the students who followed them. Similarly, at Greeley, Colorado in 1903, senior students and faculty members donated money to assist a student who had run short of funds; after that student repaid it, that pool of money became a long-standing loan fund.[49]

If all else failed, the personal efforts of principals and faculty members occasionally enabled individual students to overcome remaining financial obstacles. For example, a student at Peru State Normal School who found himself short of money in 1873, later recalled:

> My graduation looked far off to me and my limited means made it necessary to quit school for a while and return to the farm. But Dr. Curry, then principal learning my predicament found me a country school in Otoe county that enabled me to return again after I had finished my school. This I continued to do, alternating between teaching and going to school until I finally graduated in the spring of 1879.

At Brockport, Principal McLean was known for his generosity in finding financial aid for students in need. Many principals, including Edward Sheldon at Oswego, New York (1861–1897), Percy Bugbee at Oneonta, New York (1898–1933), Cecil Evans at San Marcos, Texas (1911–1942), and Smith at Fayetteville, lent their own money to students in distress. Smith and Pine Bluff's Joseph Corbin (1875–1902) were purposely lax in collecting tuition and fees from students they knew to be struggling; Smith also accepted farm products in lieu of currency. These individual efforts, which filled some of the cracks between state-sponsored tuition waivers and subsidies, as well as campus scholarships and loan funds, helped to make normal schools affordable—so affordable that, during the early 1910s, Geneseo catalogs included the assurance that: "No worthy student ever leaves Geneseo because of lack of funds with which to complete the course."[50]

Between the 1870s and the 1900s, state normal schools spread throughout the United States as local boosters and state legislatures worked together to establish new institutions and augment existing ones with additions and enhancements to their buildings and grounds. On firm footing, the normals welcomed unsophisticated, rustic students, many of whom were female, members of minority groups, and poor. Policy makers and principals, many of whom intimately understood gender, race, and socioeconomic struggles, used flexible admissions standards, personal guidance, and financial support to help these students enjoy easy access to a form of higher education. Principal of the normal in Whitewater, Wisconsin, Albert Salisbury described normalites as "people who have missed early advantages but have finally rallied by their own force, to make a final effort at personal development."[51] State normal schools would prove to be an ideal setting for this "effort at personal development," as the normals not only welcomed nontraditional students, but also enabled them to thrive. Normalites in the late nineteenth and early twentieth centuries would partake in an all-embracing intellectual life, professional socialization, and numerous opportunities for leadership and involvement in public life.

3

"Substantial Branches of Learning" and "A Higher Degree of Culture": Academic Studies and Intellectual Life

Attending any form of "higher" education was a stretch for the types of students who populated nineteenth-century state normal schools throughout the United States; normalites' gender, race, or families' financial struggles in many cases prohibited them from traveling to prestigious high schools, academies, or colleges. Affordable and accessible, normal-school training for teaching was within their reach. Because many students in the latter part of the nineteenth century—like those in the previous decades—arrived with little more than a common-school education, normal-school leaders found it necessary to continue to offer instruction in academic subjects. While this situation frustrated some normal-school principals, others looked upon academic instruction as another means of producing well-prepared teachers, even as increasing numbers of normalites were high-school graduates by the turn of the twentieth century. In 1904, Principal William E. Wilson at the state normal in Ellensburg, Washington advocated educating teachers through "general scholarship and broad culture." He explained,

> throughout the Normal school course the student needs to be pursuing energetically substantial subjects for the strengthening and sharpening of the intellect, for the enlarging and liberalization of the mind, for the enrichment and invigoration of the whole life. The education of the teacher must not be narrowed down to mere training in the work of school teaching. The Normal School must cultivate a lively interest in study, it must promote the spirit of investigation, it must beget enthusiasm for learning. To accomplish this it must provide for the vigorous pursuit under able instructors of substantial branches of learning.[1]

Although not very advanced in content, the required studies in academic subjects fostered growth in the areas Wilson considered prerequisites for teaching. Furthermore, many students relished the opportunity to broaden themselves intellectually, and used the required curriculum as a launching point for further exploration. Normalites took advantage of optional advanced courses of study and created a vibrant extracurricular life in which they journeyed far from their unsophisticated backgrounds. Officials at the Kansas State Normal School in Emporia captured this enthusiasm for wider horizons when they observed in 1889, "There is a marked increase in the demand for a higher degree of culture."[2] Indeed, the required curriculum, advanced courses, and student activities at the state normal schools between the 1870s and 1900s offered both "substantial branches of learning" and "a higher degree of culture."

The Required "Elementary" Curriculum: "Substantial Branches of Learning"

The structure of the curriculum at most normal schools throughout this period provided the foundation for a vibrant intellectual life on campus. Work in academic subjects fostered reasoning powers, analytical skills, and interest in intellectual matters, while academic requirements ensured that students gained exposure to a range of subjects and that all students studied the same core of "substantial branches of learning." Whether state normal schools' official offerings included one or multiple courses of study between the 1870s and the 1900s, the structure of the curriculum usually required that all students take the same classes for a couple of years. Especially in the 1870s, some normal schools listed only one course of study, which was usually two or three years in length. When the Rhode Island State Normal School opened in Providence in 1871, its catalog listed two years of required academic subjects: arithmetic, algebra, geometry, geography, history, chemistry, physiology, hygiene, short lessons in botany and zoology, and "Grammar and Analysis of the English Language" in the first year, and more algebra, bookkeeping, more geography and history, natural philosophy, short lessons in mineralogy and geology, and "Rhetoric and English Literature" in the second year. Providence's program mirrored that of nearby New Britain, Connecticut. The curricula at the state normals in Winona, Minnesota, Florence, Alabama, and San Jose, California were very similar to those of the eastern normals in the mid-1870s; however, Florence's and San Jose's curricula were spread out over three years. Winona conformed to the State of Minnesota's "uniform course of study" for its normal schools. Called the "Elementary course" not because of elementary-level content but because it prepared teachers for elementary schools, it included two years of subjects along the lines of those in Providence's curriculum, plus "advanced studies to be pursued when practicable."[3]

As such single courses of study grew longer in the 1880s, 1890s, and 1900s, the normal schools admitted students to different years, depending upon their prior preparation. Providence's course of study was for three years by the mid-1880s, and high-school graduates took only the third year. Normal, Illinois added a fourth year to its course of study in 1896, while maintaining unusually high entrance standards for the first year and providing a preparatory course of study. The state normal that opened in 1903 in Weatherford, Oklahoma also offered preparatory, or "sub-normal," studies; the "regular" course lasted four years and high-school graduates entered as juniors. Not only did students enter normal programs at different points in the course of study, but it was also common for them to leave before finishing the entire curriculum. Some normal schools recognized this phenomenon by awarding different diplomas or certificates at different points in the course. For example, the African American state normal in Fayetteville, North Carolina in the late 1870s granted different teaching certificates for completion of different years in the course of study: a "third grade," or temporary certificate after the "junior" (first) year; a "second grade" certificate after the "middle" (second) year; and a "first grade," or more permanent certificate after the "senior" (final) year.[4]

Most normal schools advertised more than one course of study. In 1874, the board of normal-school regents in Wisconsin mandated that "in the several normal schools in the state there shall be two courses of study, known respectively as the 'elementary course' and 'advanced course.' " The regents outlined the subject matter for the two-year elementary course and the first two years of the four-year advanced course as follows:

> In the elementary course: Arithmetic, 30 to 40 weeks; elementary algebra, 12 to 20 weeks; geometry, 16 to 23 weeks; book-keeping, 6 to 10 weeks; reading and ortheopy, orthography and word analysis, 30 to 37 weeks; English grammar, 28 to 39 weeks; composition, criticism and rhetoric, 20 to 24 weeks; geography, physical geography, 26 to 40 weeks; physiology, 10 to 15 weeks; botany, 10 to 13 weeks; natural philosophy, 12 to 17 weeks; United States history, civil government, 30 to 40 weeks; penmanship (time undetermined); drawing, 20 to 26 weeks; vocal music (time undetermined); theory and practice of teaching.
>
> In the advanced course, the studies of the first two years shall be the same as those of the elementary course, with the addition of Latin for 20 weeks, which shall take the place of rhetoric.

Students who finished the elementary course earned a certificate that entitled them to teach in the state for five years, while graduates of the advanced course earned a diploma that functioned as a permanent teaching certificate. These courses underwent slight modifications in later years, but the overall structure remained the same for a couple of decades. Beginning in the late 1880s, Oshkosh's catalogs explained that high-school graduates would "be admitted to the Junior Class of the Normal School, and will receive the Diploma of the Advanced Course upon completion of the 'professional work' of the Elementary Course and that of the Junior and Senior years."[5]

Especially by the 1880s, Wisconsin's approach represented the curricular structure of many normal schools throughout the country; it was most common for normal catalogs to list an "elementary" as well as an "advanced" course. Although officially divided into two different courses of study, these curricula were actually very much like Weatherford's and Fayetteville's; the elementary course was simply the first one to three years of studies, and the advanced course was the elementary course with minor or no modifications, plus an additional year or two of studies. When Winona began offering both a three-year elementary and a four-year advanced course in 1877, the elementary course included the subjects in the old curriculum with a few additions and the advanced course was the elementary course plus a final year of studies including those previously listed as "advanced studies to be pursued when practicable." Modifications in 1895 clarified that high-school graduates would begin in the third year of the elementary course and added an additional year to the advanced course, but Winona continued to advertise two courses while essentially offering one. Emporia, Kansas, which earlier had imitated the single course at Normal, Illinois, by the early 1880s offered the standard elementary and advanced courses. Similarly, Florence's catalogs began in the mid-1880s to list an elementary and an advanced course; the elementary was the old three-year course, and the advanced was the same three years plus a fourth year of studies. Eastern normals in the late nineteenth century including Farmington, Maine, Castleton, Vermont, and Westfield, Framingham, Salem, and Bridgewater, Massachusetts all offered a two-year elementary or "regular" course and a three- or four-year advanced or "higher" course that began with the elementary course. Washington's board of normal-school trustees in 1893 spelled out a course of studies for its new institutions in Cheney and Ellensburg; the first two years were to be the elementary course and the entire four years the advanced course.[6]

Instead of just "elementary" and "advanced" courses, a small number of state normal schools offered multiple parallel courses of study during this period. Like the two official courses at other institutions, these multiple parallel courses of study in fact shared most subject requirements. Very unusual in the 1870s, the Michigan State Normal School in Ypsilanti had four courses: the English Common School Course, the Full English Course, the Classical Course, and the Course in Modern Languages. All four courses were exactly the same in the "preparatory" year. The English Common School Course lasted one additional year, in which students studied algebra, arithmetic, natural philosophy, botany, and "professional instruction." These requirements were identical to those in the "first year" of the Full English Course, which continued for two additional years. The Course in Modern Languages and the Classical Course were one year longer than the Full English, or four years beyond the preparatory year. In the "first" year, they shared with the English courses all subjects except "professional instruction," which Modern Languages replaced with German and Classical replaced with Latin. These courses saved "professional studies" for the final year; in fact, during the second through fourth years, the Modern Languages and Classical courses included all of the same subjects as the Full English Course, although classes in one or two subjects per semester were shifted to the following year in order to make room for continuing instruction in German and the addition of French, or continuing instruction in Latin and the addition of Greek.[7]

In the 1870s and 1880s, the few other normal schools with parallel curricular tracks were mainly in states like New York and West Virginia, where most of the state normals were converted private academies. Differing course titles were a vestige of communities' expectations for their academies, yet subject requirements in the parallel courses showed little variation. For example, Brockport, Geneseo, and other normal schools in New York listed both English and Classical courses in the 1870s and 1880s and added a Scientific course by the 1890s. Because students could take these courses through either the "academic" or the "normal" department, the normals appeared to offer six different courses of study. However, the courses had much more in common than not. The only difference between courses in the "academic" and "normal" departments was the lack of teacher-education classes in the academic-department courses of study. In both departments, Elementary English lasted two years (and was discontinued in the mid-1890s), Advanced English and Scientific were three years long, and Classical lasted four years. In all of these courses of study, students had the same academic schedule in the first year and many of the same classes thereafter; the Classical and Scientific courses included languages while the English courses did not. Similarly, the normals in Fairmont, West Liberty, and Huntington, West Virginia offered separate yet overlapping normal and "academic" or "classical" courses; Huntington also annexed a business college in 1895.[8]

By the 1890s, increasing but still small numbers of other state normals began to differentiate courses of study. The Wisconsin regents in 1892 reorganized the normal curriculum to include a two-year elementary course, a four-year English—soon to be called English-Scientific—course, and a four-year Latin course. The elementary course was the first two years of the English course, and the four-year courses had the same requirements in the first year, after which the Latin course slightly reduced the requirements in English and science in order to make room for instruction in Latin. The regents also stipulated that German could "be elected instead of Latin"; the result at Oshkosh was a four-year German course, which basically substituted German for Latin in the Latin course. High-school graduates in the mid-1890s were admitted to a two-year Graduate Course that was very similar to the junior and senior years of the English Course, and beginning in the late 1890s, graduates completed two years of work in the English-Scientific, German, or Latin courses. A very small number of normal schools, including Indiana, Pennsylvania, and the all-black institution in Prairie View, Texas, added parallel courses in manual or industrial arts. An even smaller number, including the all-female normal in Greensboro, North Carolina, added domestic science courses. Even in such specialized courses, students took their basic academic classes with the rest of the school. More typical of the parallel-course approach, Emporia had begun in the late 1880s to offer a four-year English Course, as well as three slight variations of it: The Elementary Course consisted of the first, second, and fourth years of studies, the Latin Course substituted Latin for "designated subjects," and the Academic Course consisted of the first three years with only a couple of modifications. Graduates of the latter course earned a certificate of graduation, while students who completed the other courses earned a school diploma that was a lifetime teaching certificate in Kansas.[9]

Whether they offered single, elementary and advanced, or multiple parallel courses of study, state normal schools between the 1870s and 1900s channeled all students through the same core of academic subjects. These institutions essentially had one curricular stream, which students were able to enter and exit at different points. Thus, no matter what their gender, race, or social-class background, all students in the first two years of the curriculum studied the same subjects and tackled the same assignments. This curricular structure conveyed the implicit message that all students were legitimate, capable participants in academic pursuits. Other aspects of the academic climate reinforced this idea, especially as it related to gender. Although separate study rooms for male and female students were fairly common at normal schools throughout the country, there is no indication that students were separated according to gender—or any other attribute—for any academic classes.[10] Surviving grade books from Oshkosh indicate that men and women earned similar marks.[11] Scattered anecdotal evidence also suggests that instructors were equally demanding of their male and female students. For example, a Geneseo alumnus remembered that when students made mistakes in the class of instructor John Milne (who would become principal in 1889), "he used to browbeat and bellow at the girls quite as whole-heartedly as he did at the boys."[12]

The presence of female as well as male instructors, even occasionally in nontraditional fields for women, probably contributed to the relatively equal treatment of men and women in normal classrooms. In 1874, the governing board at Normal, Illinois adopted the following resolution:

> Whereas a majority of the students attending the Normal University are female students; and whereas women have demonstrated their ability to compete with men in the schoolroom; therefore, be it Resolved; That in the judgment of the Board, the interests of the University would be promoted by filling at least three of the nine regular professorships with female teachers, at as early a date as practicable.

Within two years, Martha D. L. Haynie was appointed Professor of Modern Languages at Normal. Nationwide in 1900, 57.5 percent of professors and instructors at normal schools—and only 17.1 percent of college and university faculty members—were women. The representation of women on individual normal-school faculties varied, roughly according to the gender composition of the student body. Thus, the fraction of the faculty that was female was lowest where women were a small percentage of the student body, such as at Pine Bluff, Arkansas, and highest where women were a very high percentage of the student body, such as at San Jose, California. At the former it was rare for more than half of the faculty to be women, while at the latter the percentage of faculty members who were women often approached or exceeded 70.[13]

The subjects taught by different faculty members tended to follow typical gender divisions, with several exceptions that defied tradition. It was common at state normal schools for women to teach courses in education, English and language arts, French, history, civics, geography, music, and art. But it was also not unusual for women to teach ancient languages, mathematics, and science, subjects considered by many to be male domain. Two especially long-serving examples were Emily Webster and Mary Apthorp at Oshkosh. Webster was appointed to teach Latin in 1875. Eight years later, Apthorp took over as Latin instructor and would remain for nearly 30 years, while Webster embarked upon 45 years of teaching math. In the sciences, women often taught physiology, botany, and zoology. Juliet Porter taught physiology as well as math at Worcester, Massachusetts from 1874 until 1894, while Mary V. Lee taught science at Oswego, New York for 18 years after earning her medical degree in 1874. A succession of female instructors at Castleton, Vermont covered mineralogy and geology during the 1880s and 1890s, and at the turn of the century women taught chemistry at San Jose and physics at Oshkosh. Meanwhile, Mary Macy Petty began teaching chemistry at Greensboro, North Carolina in 1893 and continued for more than thirty years.[14]

The composition of normal-school faculties, like the structure of the courses of study, hardly differentiated among who could pursue various academic subjects, which helped to establish a foundation for a lively intellectual life at these institutions. The faculty and curricular requirements not only suggested that all types of students were capable academically, but also ensured that all students shared an understanding of core subject matter. Normalites thus learned a common academic language, which would allow them to engage together in intellectual exploration. While the content of this common academic experience was not very advanced, the foundation it formed was quite solid. With no choice but to meet students where they were academically, the state normals' "elementary" curriculum fell somewhere between common-school and college level. Still, the normals' approach to this subject matter answered Principal Wilson's call for "substantial subjects," as it stressed foundational knowledge, reasoning, and inquiry skills. Colleges and universities by the latter part of the nineteenth century were beginning to replace recitations with seminar discussions and laboratory and library research, offering more elective and fewer required subjects, and reducing the amount of instruction in ancient languages while adding more in modern languages, science, history, and English.[15] Although the elective system would arrive later at the state normals, their basic academic curriculum was a sort of scaled down version of that taking shape at many colleges. The elementary course, or first couple of years of instruction, in mathematics, the sciences, history and civics, and English and language arts at normal schools answered Principal Wilson's call for "the strengthening and sharpening of the intellect" and "the enlarging and liberalization of the mind," as well as the cultivation of "the spirit of investigation" and "enthusiasm for learning."

Required instruction in mathematics at state normal schools focused on fundamental rather than advanced subjects. Algebra, plane geometry, and solid geometry were standard fare from the 1870s through the 1900s, while bookkeeping also appeared in many "elementary" curricula during all or some of the same decades. Normals such as Farmington, Maine required surveying. Florence, Alabama, Normal, Illinois, San Marcos, Texas, and several others included trigonometry in the required curriculum. The normals in Washington, Wisconsin, and many other states required arithmetic, while Ypsilanti offered standard arithmetic only as a review course. All students there in the 1870s studied Elementary Algebra, Analysis of Arithmetic, geometry, and Higher Algebra. More creative titles of required courses included "science of arithmetic" at Winona, Minnesota and "inventional geometry" at Willimantic, Connecticut. Some catalogs listed "mental arithmetic," and Ashland, Oregon included both "mental" and "written" arithmetic in first-year studies. The former probably focused on mental calculations and drills; Adelaide Pender remembered that the "custom" of a math instructor named Miss Atwook at New Britain in the late 1880s was "to enter the assembly hall just prior to the noon dismissal . . . she paused before a poor freshman's desk and began a combination like, 'Take forty-five, divide by nine, add twenty, multiply by four. Result.' "[16]

The mental drill at New Britain notwithstanding, thorough understanding and sophisticated reasoning were intended outcomes of these foundational math classes. As stated in the first Oshkosh catalog in 1872, the goals were "habits of correct and definite thinking . . . rather than to impart quantities of information labelled, but not understood." Even Pender reflected that Miss Atwook "knew all that there was to know about mathematics, and I learned more in her course of a few months than I ever had in all the previous years of my education." Perhaps Atwook shared the intentions of mathematics instructor Arthur Curtis at Albany, New York, who aimed "to train the student to think more logically, and, consequently, more accurately, both as to the fact, and the expression of fact." When Westfield revised its course of studies in 1894, official documentation reflected a desire to move far beyond rote learning. The description of classes stated that arithmetic students "are encouraged to seek information at the post office, at lawyers' offices, banks, stores and the teacher's desk, and thus to become familiar with the practical applications of arithmetic in the affairs of every-day life," and geometry students "are required to work out and teach most of the definitions, theorems and constructions of the course." Turn-of-the-century students at Oshkosh learned "the science of the equation," while those at Florence "increased" their "power of abstract analysis and reasoning." In the same vein, math instructors at San Marcos during the 1900s aimed "to develop in the pupil the power to reason clearly and logically." A catalog explained, "Students are led to derive the rules and formulae before using them; to study principles, not simply to learn facts." Indeed, normal instructors expected all students to begin to think like mathematicians.[17]

Normal-school science curricula covered a remarkable number of subjects—so many that it would have been impossible for students to have developed real expertise in any particular area. Instead, in "elementary" science classes, they learned how to discover and interpret knowledge while gaining practical exposure to many fields. Between the 1870s and the 1900s, virtually all normalites studied physiology, botany, geography, and natural history or natural philosophy, which became nature study by the early twentieth century. And the list hardly stopped there; most required curricula also included one or more of the following subjects: chemistry, geology, mineralogy, biology, zoology, physics, and astronomy. In the early 1870s at Ypsilanti, as professor Jessie Phelps explained later, science courses "were all taught in one laboratory or suite of rooms on the first floor of the only college building. Physiology, Natural Philosophy (a potpourri of explanations of common phenomena such as gravitation, the formation of dew, etc.), Geology, and the physical sciences were presented by some one or two men faculty, while Botany was assigned to the preceptress." Later in the same decade, Ypsilanti added classes in zoology, physiology, chemistry, and astronomy. Westfield's revised two-year course in 1894 included physiology, botany, geography, chemistry, mineralogy, zoology, and physics.[18]

As in math instruction, the aim of this plethora of science courses was "the strengthening and sharpening of the intellect." In junior-year chemistry at Florence in the mid-1880s, "the fundamental principles of chemistry" received "especial attention." Each student was "led to turn his attention to these instead of committing to memory the properties of bodies." Later in the decade, natural-science courses there intended to "develop a spirit of investigation, train the powers of observation, and produce habits of independent thought." In the 1894 course revision at Westfield, botany students were to "observe, draw, describe, experiment, teach," and physics was to include "special attention . . . to elementary phenomena and to practical applications. In this subject everything is taught experimentally, pupils being required, as far as possible, to perform all important experiments for themselves." Meanwhile, Oshkosh's biology department aimed "to familiarize the student not so much with the facts of the subjects as with the ideas of science," and its chemistry department fostered "skill in arriving at the right conclusion from observed facts." Edwin J. Saunders, who taught physical science and geography at Ellensburg from 1898 through 1909 described his goals as, "to cultivate the power of observation, independence of thought, and the spirit of scientific inquiry into the phenomena of nature, and not to have the student memorize a portion of text every day." In science classes at state normal schools, students honed their reasoning skills. While they did not become experts in particular subjects, they did become familiar with "the spirit of investigation" and gain a great deal of "enthusiasm for learning" in many branches of science.[19]

Because scientific analysis was based on observed or collected data, first-hand investigation was integral to teaching normalites to think like scientists. By the 1870s, most normal schools had laboratories, which grew as the institutions were able to add new classroom space. At Bridgewater, a building addition in 1872 made room for a chemistry laboratory, and a separate laboratory annex completed in 1881 housed several labs. In chemistry and physics, equipment for lab work was somewhat rudimentary, especially in the early years; instructors often taught students to construct simple apparatus as well as conduct experiments. Arthur Boyden, whose father was principal at Bridgewater (and who would later take the reins), explained that in the annex, "Each student had a place at the tables, and was taught how to make and use simple, inexpensive apparatus such as could be used in schools. . . . One of the physical laboratories was arranged for sixty students to work in classes at the tables, with a dark room for measuring the candle power of lights, one for photography, and one for spectroscopic work." Emporia's 1888–1889 catalog described physics instruction as follows:

> This subject is pursued, as far as the size of the class permits, exclusively in the laboratory. The laboratory work during the past year has embraced the manufacture of both physical and chemical apparatus, including electric batteries, electro-magnets, permanent magnets, helices, electroscopes, electric bells, electrophori, and other instruments.

The same catalog explained that chemistry included "Laboratory work during the whole of the first term. The class determines the effect of the ordinary reagents on twenty-five common elements, and from their experiments devise and put in practice a scheme of qualitative analysis." By 1895, the African American normal in Pine Bluff had "an air pump, electrical machines, barometers, thermometers, induction coils, Geissler's tubes, reagents, etc."[20]

With such equipment, normalites were able to construct chemical and physical knowledge, and enthusiastically seized the opportunity. Martha M. Knapp of San Jose's class of 1880 remembered that a classmate, "desirous of performing more experiments than she had time to attempt at school, took home some chemicals. During dinner the family were startled by an explosion, and discovered that a fire was the result. Part of the hall carpet and her brother's new overcoat were sacrificed to her zeal." After joining the Westfield faculty in 1903, Professor Louis B. Allen developed a class called "Kitchen Chemistry," in which students analyzed the chemical composition of common items. When several local children complained of sore mouths in 1906, his students' investigation revealed that coal tar was present as a coloring agent in lollipops. The children and their parents must have been grateful for the dedication of these students to their work in chemistry.[21]

Normalites also investigated the astronomical world firsthand. Student Lottie Matthis recalled "star gazings" at San Jose in the early 1880s; they "generally took place on the normal grounds, and on one occasion at the uncanny hour of three A.M." Along with other scientific apparatus, many normal schools acquired telescopes. The normals at Ypsilanti, Michigan, and Peru, Nebraska added observatories in the mid-1880s. Although San Jose did not have an observatory, from 1891 through the 1910s each senior was able to use a large telescope during a yearly trip to Lick Observatory on nearby Mount Hamilton. Some studies in astronomy did not even require a telescope; at Peru in the mid-1880s, instructor J. M. McKenzie led his students in a study of sunspots. He explained that two times each week, he took them to a darkened room and "projected a picture of the sun's disk, on a white card, size eight by ten inches. A student stood by the card, pencil in hand, to mark the exact position of any sun-spot projected, and to draw as nearly as possible the shape and character of the spot." After much sunspot activity in October and November of 1884, Principal Farnham displayed the cards at an exposition in New Orleans.[22]

Laboratory work in other scientific fields involved all sorts of "specimens," and normal schools collected them in abundance. Like colleges at the time, normals amassed sizable collections or even established museums for the study of natural history. Boyden explained that at Bridgewater in the early 1880s,

> The removal of the chemical and physical laboratories out of the main building made it possible to reorganize these old rooms into natural-history laboratories with abundant apparatus and specimens for individual work. One laboratory was fitted up for mineralogy and geology, with cabinets and typical specimens arranged for study; another laboratory was fitted up for zoology, physiology, and botany, with cabinets of specimens.

Collections of various rocks, minerals, and fossils enhanced instruction in mineralogy, geology, and natural history. As early as the mid-1860s, Principal William Phelps at Winona began collecting minerals and fossils from nearby quarries and railway cuts, and in 1875 local citizens purchased a collection of minerals and fossils, as well as corals and shells for the school. Instructor Lucy Osband at Ypsilanti improved the normal's collection of geological specimens with minerals from upper Michigan and Colorado. At Westfield, Joseph Scott, who taught natural sciences and then served as principal from 1877 to 1887, added Syrian fossils and geological artifacts from Pennsylvania coal mines to the collections. Geology and geography also took students outside the classroom. At Oshkosh, San Marcos, and Vermont's three normal schools, they studied nearby landforms on field expeditions. In the 1900s, normalites in Vermont conducted "a first-hand study of soils and their formation, the work of streams, and the weathering of rocks." Aspiring geologists at San Jose conducted some very localized studies; in 1886, the campus magazine explained with tongue in cheek: "If a thoughtful looking student be seen busily scratching among the gravel which now abounds on our sidewalks, do not imagine him or her to be searching for a lost diamond. It is only a geology student."[23]

Students also gained practical scientific understanding through field exploration in the biological sciences. The botany curriculum included excursions to study local flora. Nature walks were integral to Arabella Tucker's botany classes at Worcester between 1888 and the 1910s. Such excursions often involved adding to one's personal or school's collections; San Jose's Martha Knapp remembered returning from a daylong expedition "laden with specimens that more than compensated us for the weariness we felt." As other San Jose students sought specimens on campus, the student magazine reported that the gardener was "completely overwhelmed by the young ladies of the Botany class. These fair Botanists number about one hundred and twenty, and apply to him daily for Botanical specimens, vieing [*sic*] with each other in their captivating speeches." After collecting specimens on botany excursions, Florence and San Marcos students dried and arranged them for their own herbaria. At Peru, Principal Farnham encouraged instructors to have their students gather specimens of all sorts of animals. When one class collected and mounted local birds, Farnham procured large display cases for them. "In collecting specimens," remembered Helen Kelley, Oswego '06, "not a stone was left unturned—literally. We even skinned a snake." Normal schools also purchased mounted animals for their zoological collections. The Illinois normals in Carbondale and Normal were especially proud of their museum collections; as early as 1870, Normal listed 100,000 botanical, 900 ornithological, and 300 mammal specimens (not to mention fossils and minerals). To replace items lost in an 1894 fire, Oneonta, New York purchased 1,200 stuffed birds and 300 reptiles and mammals. A collection at Plattsburgh, New York included 1,600 species of shells and 800 insects.[24]

After the institution purchased or students and faculty collected biological specimens, normalites studied them extensively in the laboratory. The 1888–1889 Emporia catalog explained that in botany: "Laboratory work includes the complete description and analysis of at least fifty species of phanerogams, the making of a herbarium, drawings of cells and tissues, the study of the life history of yeast, moulds, amoebae, etc., window cultures, and experiments with artificial soils." Students examined all sorts of plant and animal samples under microscopes. Beginning in the 1870s, botany and zoology classes at Ypsilanti utilized a compound microscope, which Professor Phelps described as "a very up-to-date method." The 1884–1885 Florence catalog reported "an excellent Crouch's histological microscope which is in daily use," and the 1888 San Jose catalog announced that students used microscopic analysis to deduce "physiological functions." The microscopic study of insects was prominent enough in the science curriculum at San Marcos in the 1900s to inspire the seniors to compose a poem:

> Bugs, bugs, bugs under the glass;
> Weeds and flowers found in the grass;
> These are the joys of the Senior Class.[25]

Dissection was another popular method of investigation in biology, zoology, and physiology. By the late 1870s at Oshkosh, according to Principal George S. Albee, "the true method of biological study was begun against great prejudice on the part of some officers and many good citizens, who were sure that the nature of students must be coarsened and hardened by contact with flesh and blood in scientific study." The 1888–1889 Emporia catalog stated that dissection reinforced what students read in their physiology and zoology textbooks, and explained that in physiology, "one or more afternoons in each week are spent in the laboratory, studying the art of dissecting and preserving for laboratory study or class illustration the parts and organs of mammals," and "dissections performed in the zoological laboratory extend to all the branches of the animal kingdom except protozoans." In Human Physiology at Normal, Illinois, students dissected "a cat or a rabbit, paying special attention to the organs of digestion, circulation and respiration, muscles, and nervous system." Students dissected "the lower animals" at Florence, and starfish, holothurian, crayfish, frogs, and other "larger forms" at San Marcos. Lottie Matthis remembered that while she was a student at San Jose, "many representatives of the species feline were sacrificed to science," and Oshkosh students reported in 1895, "Prof. Browne's Physiology class have been dissecting the cat for the past two weeks. The circulation has been studied from a specimen injected with a preparation of plaster of paris."[26]

The students about whom the officers and citizens in Oshkosh had been concerned were most likely the women, whom gender norms forbade from gory endeavors. Such notions were probably behind the report of Oshkosh's junior class in 1897 that, "Miss Scardia is very proud of her progress in the art of surgery. She is able to dissect a frog without a scream." At Oswego, Helen Kelley remembered "some squeamishness" among female students regarding tasks like "cat-carving." Overall, however, female students seem to have taken to this method of scientific investigation with enthusiasm. Florence Doolittle Bean, who graduated from Oswego in 1896, remembered dissections in Dr. Mary Lee's zoology class: "Under her expert guidance, we dissected many creatures, from flies and angleworms to a lone dog. She personally took charge of the dog. We learned the wonders of animal anatomy, the function of its parts, and the animal's habitat." Elizabeth Allen, Oswego '94, also remembered Lee's dissection of the dog, and reported, "while at home in Westford, I stewed a cat which my brother shot for me, and removed the flesh from its bones, so I could take the bones back to Oswego and mount them. In mounting them, I had the help of a cat's skeleton loaned me by another girl." Classwork in the biological sciences allowed Allen, Bean, and other normalites to discover knowledge in the laboratory and in the field. Although probably somewhat superficial, the many areas of scientific inquiry at normal schools acknowledged students as researchers and thinkers. It is hardly surprising that Lewis H. Clark, who graduated from Whitewater, Wisconsin in 1879 and later served on the faculty there, recalled the institution's "decided bent toward natural science. . . . Because of this prevalence of the scientific spirit, many a student received such an impulse toward the natural sciences that it developed into a passion . . ."[27]

History and civics classes at state normal schools contributed to "the enlarging and liberalization of the mind" as well as furthered "the spirit of investigation." At virtually every institution, United States History was a required subject and many also required European or English history. Ashland included ancient history in second-year studies when it opened in 1882. At Farmington in the late 1880s, General History in one term was followed by U.S. History in the next. Most normal schools also required a class in civics or the U.S. Constitution; at Ypsilanti in the 1870s it was

Science of Government, and at Farmington in the late 1880s it was Civil Government. Ashland in the early 1880s required American Ideas, which presumably dealt with government and other societal institutions. Classes in general or European history began with Greece and Rome and continued on to cover the highlights of Western civilization. Bridgewater imported from London "casts, models, and flat copies" of great art works, which must have enhanced instruction in European history as well as art. The plan for Westfield's revised two-year course in 1894 stated that United States History would cover "Periods of discoveries; explorations; settlements and colonies, with the included wars; revolution, constitution; civil war and events following; collateral reading; civil polity." In such classes, students became familiar with the broad outline of American and Western history, culture, and legal structures.[28]

Beyond conveying historical facts and the duties of citizenship, history and civics classes were designed to "arouse a spirit of inquiry, to instill a love of original research, and to train the judgment to independence of thought, instead of requiring a mere memorizing of the text," as stated in a San Jose catalog. At Ypsilanti from 1881 into the 1910s, Julia Ann King taught and expanded offerings in history and government, in the interest of making students more socially thoughtful and aware. She explained, "History is Society becoming conscious of itself . . . One could hope that social consciousness, understood in history, might become in time an idea in the individual sufficiently clear and strong to determine his thinking and conduct." While independent thought was most important to King, instructors at Florence in the early 1890s stressed research skills; history was "taught by outlines, each student contributing in recitations such facts bearing on the topic under consideration as may have rewarded his labor." The catalog explained that this approach, in "addition to encouraging original research on the part of individual members," gave "to the entire class the benefit of every author consulted." A required class in "library readings" at Oshkosh during the 1890s and 1900s included a section on "biography and history," in which each student read a biography and reported on it to the group.[29]

In the history, civics, and other curricula, the library often played a role similar to that of the laboratory in the sciences. In 1890, Greeley, Colorado's first catalog described the library as "a fountain of knowledge, a source of discipline, and a means of culture. The room is fitted up to serve the purpose of a 'literary laboratory.' " Normal-school faculty routinely expected students to spend time in the library, doing research or reading beyond their assignments. From the early 1870s into the twentieth century, Oshkosh's catalogs described the normal's "excellent Reference Library," where the student would "readily carry his investigation beyond the text book." Library collections varied considerably in size and scope. In 1888, Pine Bluff reported "an excellent beginning made toward securing a good Library by the collection of about one thousand volumes," while Geneseo advertised "a complete text-book library, containing, besides the works used in the School, others for reference" and a "public reading room." At the same time,

Emporia subscribed to seven daily papers, over sixty weeklies and monthlies, and over thirty other periodicals ranging from *Art Amateur* to the *Weekly Times* of London; the catalog boasted:

> the library is located in a handsome suite of rooms in the new wing. It contains about 5,000 volumes of choice books, most of them selected with special reference to the needs of the School. The list embraces a fine line of cyclopedias, lexicons, gazetteers, and educational reports; works on the theory, the art, and the history of education; and the standard works on history, literature, science, philosophy, etc. . . . The annual appropriation of $500 for the purchase of new books, is rapidly making the library one of the most valuable features of the School.

It was 1899 before Troy, Alabama's library had 5,000 volumes. A decade later, Winona, Minnesota had over 9,000 books and a "Federal library of over 4,000 public documents," collected since it had become a depository for the First Congressional District in 1884. Isabel Kingsbury Hart, Oswego '07 reported that the library was less than ideal, as it

> was open for only a few hours a day. There was no librarian; faculty members rotated this function among themselves. . . . My own memories are restricted to the reference shelves which were always accessible, and to the periodical room which was just plain funny. One big piece of equipment stood in the center of the room—a long narrow desk. . . . A superstructure ran above the middle of the desk similar to a modern clothes rack, from which hung strong but fine steel chains. Each chain, in turn, was attached to a current issue of a magazine. A student could read the magazine only as long as he could endure standing.

Whether they housed sparse or extensive collections, normal-school libraries were an important resource for intellectual exploration.[30]

Normal schools encouraged students to explore Western literature as well as history. In 1890, the library at Oneonta gladly received a donation that included works by Milton, Shakespeare, and other classic authors, which must have helped instructor Charles Schumacher work toward his goal in teaching literature: for students to "understand it, love it, live it." The 1888–1889 Emporia catalog listed books "suggested for general reading during attendance at the school," such as Charles Dickens' *A Tale of Two Cities*, George Eliot's *Romaola*, and the poems of Goethe, Tennyson, and Longfellow. Exposure to literature usually came through reading assigned texts for classes in English and, occasionally, American literature. The "elementary" curriculum at many normal schools included at least one literature class, most commonly called English Literature. Pennsylvania's normal schools required English Classics, while English and Fiction was a sophomore class at Greeley. Westfield's plan for the revised two-year course in 1894 included the following description of English Literature: "History of the language; study of the eighteenth and nineteenth century authors;

reading (in addition) 'The Merchant of Venice' and 'Julius Caesar.' " In addition to English literature, students at Bridgewater studied American poets and writers such as Longfellow, Whittier, and Hawthorne. Willimantic, according to the 1890 catalog, scheduled American Literature during the first term, after which English Literature continued for three more terms and covered "Dickens to Burns" during the second term, "Burns to Bacon" during the third term, and "Bacon to Chaucer" during the fourth term. At nearby New Britain, Adelaide Pender's literary studies introduced her to Robert Browning, which, she later reflected, "opened a lovely new field in my life. When I left school I continued to study this poet, memorizing many of the selections and reciting them in season and out to my mother."[31]

English curricula sought to sharpen students' reasoning skills through analysis of literature. The English department at San Jose in 1900 centered on "a process of investigation carried on by the students in connection with a study of carefully graded English and American classics." Works "for critical study" at San Marcos included Longfellow's *Evangeline*, Gray's "Elegy in a Country Churchyard," Shakespeare's *Julius Caesar*, and Emerson's "Essay on Behavior." During the 1870s, every second-year student at Castleton wrote a "critical exposition of Milton's *Paradise Lost* or Bacon's essays," while students at Normal, Illinois dissected Shakespeare's *Hamlet* and wrote orations or essays on the subject. In the late 1880s, Pender's literary studies resulted in essays entitled "Introduction to Evangeline," "Wordsworth's Characters," and "Irving's Literary Work." Later in the century, students at Florence discussed their literary analysis in "essays upon the life, times, style and works of each author" studied.[32]

In addition to honing their skills in analysis, these assignments to compose essays on literature refined students' writing skills. Indeed, the main focus in the English curricula at state normal schools was the expression of ideas in both writing and speech. Language-arts were more numerous than literature classes, as virtually all normals required more than one class in rhetoric, grammar, composition, and/or orthography, as well as elocution. For example, Emporia in 1888–1889 required Grammar and Composition, Elements of Rhetoric, and Declamation and Elocution in the first year, Essay in the second year, and Oration in the third and fourth years. Catalogs for institutions throughout the country explained that such courses covered fundamental language-arts concepts in order to give "practical training in the correct and effective use of our mother tongue," to "cultivate the individual student's powers of expression in both oral and written language," or "to teach pupils to think, and give both oral and written expression to their thoughts." "Thrilled" when her elocution teacher "rendered a choice bit of literature," Pender was inspired to do the same. She later wrote, "after graduation, I learned wholes and parts of many fine classics and spent hours declaiming them to my long-suffering mother. Shakespeare was a favorite."[33]

Most normal schools required each student to present some sort of school-wide public declamation at least once, but often on a weekly or monthly basis. At some institutions, such as Normal, Illinois, men generally presented "orations" and women "essays," which meant that men orated while women simply read; but all students appeared on the stage and spoke publicly. Other institutions made fewer gender distinctions. At Cortland, New York by the late 1890s, for example, both men and women presented orations, such as "Blessed be the drudgery" by Marie A. Carberry and "The Decline of England" by Clyde Griswold, at Friday afternoon "rhetoricals." Similar Friday Afternoon Exercises began at Geneseo in 1877; from that time until the 1910s, classes took turns presenting programs of essays, readings, and recitations. At Florence, class-wide Weekly Rhetorical Exercises prepared students for monthly all-school exercises in the chapel, "at which," the catalog stipulated, "every student is expected to appear at least once during the year with an essay, declamation or oration." Oshkosh also required public readings during the early years, and from 1895 until 1912 had a program similar to Florence's. Meanwhile, at Pine Bluff "public rhetoricals" continued into the 1920s. Rhetorical exercises and classwork usually culminated in commencement addresses by normalites. Most members of each graduating class presented an academic speech, either as part of the graduation ceremony or in a separate ceremony during commencement week. Students and instructors took commencement speeches, as well as rhetorical exercises, quite seriously. Mary Hendrix, San Jose '73, remembered "the drill on our commencement exercises. We were taken, one at a time into the large, unfinished audience hall to practice." Oshkosh's student magazine regularly published the names of students who earned honorable mention in all-school rhetoricals.[34]

As normalites prepared and presented their graduation speeches, they drew upon their solid grounding in the "substantial subjects" of mathematics, the sciences, history and civics, and English and language arts. In addition to the academic content of these subjects, students had learned important analysis, research, and presentation skills. Most "elementary" curricula also included courses in drawing and vocal music, which honed other skills useful for elementary-level teaching. Students thus shared a degree of exposure to artistic and musical endeavors along with academic subjects. As manual training arose as a new area of instruction in the public schools in the late nineteenth century, a small number of state normals added it to their required curricula, creating another set of shared experiences. In the late 1890s, Manual Training appeared in the sophomore and junior years at Greeley and in the "general" two-year course for all Massachusetts normal schools. In 1888 and 1900 respectively, the legislature in New York and the State Normal Board in Minnesota declared that manual training was to be part of the curriculum at all normal schools. Many institutions did not have adequate resources to comply immediately, but by the early to mid-1900s most normalites in the two states studied manual training. Providence included Domestic Science in its regular course beginning at the turn of the century. Many, but not all African American normal schools required manual training and/or domestic science, often depending upon the educational philosophy of the institution's leader. For example, at Tallahassee, Florida, in the 1900s President Young instituted industrial education although his predecessor, President Tucker, had opposed such studies and concentrated on the liberal arts. Where manual training was required, normalites of all races and both sexes learned a new set of skills. Don McGuire, who attended Oswego from 1904 to 1907, later recalled "spending a semester with the needle. . . . Meanwhile, the girls learned to use rip and crosscut saws in a definitely masculine manner."[35]

While manual training and domestic science probably would not have fit Principal Wilson's definition of a "substantial subject," ancient and modern languages would have. Most normal schools did not require languages, but a minority did. The Elementary Course agreed upon by the Board of Normal School Principals in Pennsylvania in 1871 incorporated Elements of Latin, "including the First Book of Caesar." When the course was lengthened in 1899, more readings by Caesar, Virgil, and Cicero appeared in the official Latin curriculum. Latin was required at Florence in the 1870s and 1880s, and at Castleton at the turn of the century. African American normal schools occasionally required Latin as well: Pine Bluff included Latin in the regular course of studies in the 1870s and 1880s, and Fayetteville, North Carolina added Latin to the basic course in 1882, but discontinued it in 1902 after receiving criticism for offering subjects "for which most of the colored teachers will never have any practical use." Modern languages were extremely rare as required subjects, but not unheard of; at the turn of the century, the "elementary" course at Las Vegas, New Mexico included Spanish.[36] At the few normals where they were required, ancient and modern languages added to the firm foundation students shared in the core academic subjects. The goals of instruction in mathematics, the sciences, history, and English, as well as the enthusiasm with which students pursued their studies testify to the normal schools' efforts to realize Principal Wilson's vision of "substantial branches of learning." At Normal, Illinois, president Richard Edwards agreed wholeheartedly with Wilson's sentiments, and added that vigorous academic training for future teachers would benefit the communities in which they taught: "Every particle of culture imparted to them will be so much clear gain for the schools."[37] The intellectual polish normalites gained through the "elementary" curriculum was just the beginning; in advanced studies and extracurricular activities, students would seek an even "higher degree of culture."

ADVANCED STUDIES AND STUDENT ACTIVITIES: "A HIGHER DEGREE OF CULTURE"

By the late nineteenth century, "a higher degree of culture" had clear social class connotations. The class divisions that were becoming more pronounced as the nation became more industrialized, had created a new distinction between elite (or high) and mass (or low) culture. In 1871, Walt Whitman denounced "this word Culture, or what it has come to represent."[38] Indeed, during the final decades of the nineteenth century, social class came to define one's lifestyle and tastes as well as standard of living. Popular entertainment, including vaudeville and Barnum's curiosities, no longer attracted the middle and upper-middle classes, who instead enjoyed Shakespeare, symphonic music, the fine arts, the natural sciences, and travel. Beginning in the 1870s, those wealthy enough to do so sent their children to college so that they could adopt a status-appropriate "scientific" outlook and gain cultural knowledge in classic literature and ancient languages. A bachelor's degree came to symbolize not only academic achievement, but also possession of cultural capital and membership in elite society.[39] Normal schools could not provide such certification, both because the vast majority did not offer bachelor's degrees and because the name "normal" conferred nonelite status.[40] Still, students at the normals sought through their studies and activities to gain prestigious knowledge and cultural refinement, or elements of cultural capital. Despite their lowly status, state normal schools enabled students to cross cultural boundaries associated with social class.

Advanced Studies

While all normalites gained a solid foundation in academic subjects through required classes, some sought more esoteric knowledge through elective subjects. However they structured their course(s) of study, virtually all state normal schools by the 1870s offered opportunities for students to move beyond the "elementary" curriculum. When institutions had only one course of study, they invariably offered a smattering of additional non-required classes. The Rhode Island State Normal School's first catalog stated that, in addition to the required course, "Latin, Greek, French, German and other advanced studies may be pursued." Normal, Illinois in 1874 listed seven subjects, including Analytical Geometry and Greek, as "optional studies," and by 1905, there were more than fifty "electives," including everything from English Poetry to Advanced Construction. Multiple courses of study offered various options for the pursuit of advanced subject matter. The board of normal-school regents in Wisconsin in 1874 outlined the "advanced course" as follows:

> In the advanced course, the studies of the first two years shall be the same as those of the elementary course, with the addition of Latin for 20 weeks, which shall take the place of rhetoric. In the advanced course, the studies of the last two years shall be: Higher algebra, 20 to 28 weeks; geometry and trigonometry, 17 to 23 weeks; Latin, 80 weeks; rhetoric and English literature, 10 to 28 weeks, zoology, 6 to 12 weeks; geology, 12 to 17 weeks; universal history, 12 to 23 weeks, political economy, 15 to 17 weeks, mental and moral science, 20 to 30 weeks; theory and practice of teaching.

The Advanced Course at Farmington, Maine in 1889 was similar; it included further studies in the core academic subjects, and "Latin, French, or German." When normal schools offered various parallel courses, the titles of the courses indicated loosely the areas of advanced studies offered. At Ypsilanti, Michigan and Brockport and other normals in New York, the Full or Advanced English Course included some additional studies in English and much extra work in math and the sciences; the Classical Course focused on Latin and Greek; and the Course in Modern Languages or Scientific Course focused on German and French, or one of these modern languages and an ancient language.[41]

Few normalites actually completed these more advanced courses of study. Of those who graduated from Millersville, Pennsylvania between its opening and 1905, for example, 95 percent had finished only the Elementary Course. At Normal, Illinois, fewer than 10 percent of students graduated from the full course. However, many more students pursued advanced studies than graduation statistics would indicate. In New York, a legislative committee appointed to investigate the normal schools in the late 1870s found that 35 percent of students took advantage of the opportunity to study an ancient or modern language. At Oneonta by the 1890s, Classical was the most popular course. After Cedar Falls, Iowa added Latin as an optional subject, for an added fee, in the early 1880s, many students elected to study it despite the expense. Officials at Emporia, Kansas described normal schools throughout the country when they observed in 1889, "There is a marked increase in the demand for a higher degree of culture, as indicated by the greater size of the classes in optional studies, and the larger number of pupils who take Latin."[42]

Even when they would be unable to complete all of the requirements for graduation, students took advantage of opportunities to further refine themselves through advanced elective studies in the subjects in which they had acquired a firm foundation. In mathematics, advanced courses typically included higher-level work in algebra, geometry, and trigonometry. In the 1870s, the Scientific Course at Pennsylvania's normal schools included Geometry, Higher Algebra, Plane and Spherical Trigonometry, Surveying, Analytical Geometry, and Calculus. The latter subject was also an optional class at institutions such as Normal, Illinois and Pine Bluff, Arkansas. In the sciences, if a normal school did not require chemistry, geology, mineralogy, biology, zoology, physics, or astronomy in the "elementary" curriculum, then the subject was likely part of an advanced course or elective offerings. At Winona, Minnesota in the late 1870s, the advanced course allowed students to add astronomy and geology to their "elementary" science studies. At Cheney, Washington in the mid-1890s, students in the advanced course studied geology, mineralogy, astronomy, and chemistry in addition to earlier work in physics and the biological sciences. The optional science class at Geneseo, New York in the 1900s was simply "advanced science." At many normals, advanced science students were able to conduct independent in-depth studies or accompany their instructors on extended expeditions. For example, under Professor Cockerell at Las Vegas, New Mexico at the turn of the century, Ada Springer studied animal and plant life in nearby Pleistocene beds, Emerson Atkins focused on bees, and Mary Cooper studied ants. Springer also accompanied Cockerell and other faculty members on a trip to the southern California coast, where they collected insects and marine fauna. On such expeditions and in advanced science classes, normalites immersed themselves in the middle-class culture of science. As they collected specimens and worked in the laboratory, they demonstrated a scientific understanding of and outlook on the natural world.[43]

While all normalites gained exposure to Western culture in their required history and literature classes, some sought more in-depth cultural knowledge through optional subjects. When advanced courses included further historical study, it was usually in the form of a general history course, or "universal history" in the case of Wisconsin. Westfield, Massachusetts' 1894 course revision stated that, after studying U.S. history in the two-year course, students who continued in the four-year course would take General History: a survey of "Oriental Nations" and "Greece and Rome . . . including development of characteristic political, social and religious institutions," as well as "Europe from the beginning of the Middle Ages to the present time: Study of Europe as a whole by periods, giving a connected account of the leading events of each period, but devoting special attention to the main forces at work, the formation of new States, the growth of nationality and constitutional government, and the relation of Europe to America." Similarly, further literary study in advanced courses was usually in an additional course called English Literature. At Westfield in 1894, a second English Literature class appeared in the four-year course; in this advanced class, students studied Shakespeare's *Hamlet* or *Macbeth* "and one or two comedies," Milton's *Paradise Lost*, Tennyson's *Idylls of the King*, Thackeray's *Henry Esmond*, works by Coleridge, Shelley, Keats, Byron, and more. A few normals also offered elective classes that looked in depth at a particular author; at San Jose, California in the early twentieth century, such classes included Shakespeare, Chaucer, and Poetry of the 18th Century. Thus, in advanced history and literature classes, normalites added to their stock of prestigious cultural knowledge.[44]

While ancient and modern languages were rarely required, the majority of state normals did offer optional language instruction. Most advanced courses of study included Latin, the traditional marker of an educated person. Winona's advanced course incorporated Latin as early as the 1870s, and Cheney's included it by the late 1890s. In parallel courses of study, Latin was generally part of "scientific" and, of course, "classical" courses, but not advanced English courses. Latin was an elective subject at normal schools from Geneseo, New York, to San Marcos, Texas, to Tempe, Arizona. The African American normal at Prairie View, Texas added a Latin department in 1896; it offered an optional two-year sequence. In the 1890s, students who elected Latin at both Florence, Alabama and Westfield read and translated Caesar, Cicero, and Virgil. A few normal schools offered the most erudite of subjects, Greek. Classical courses, such as at Ypsilanti and at Plattsburgh, New York included both Latin and Greek. The academic department at Fairmont, West Virginia also offered instruction in Greek as well as Latin. Oregon's normal schools offered one year of Greek in addition to two years of Latin by the late 1880s, and Prairie View added a Greek department in the early 1900s. Some advanced and parallel courses of study allowed students to choose between ancient and modern languages. In New York, for example,

Oneonta's Classical Course required Latin, and Greek, French, or German, while Brockport's Scientific Course required any two of the four languages. German and French were the most common modern languages offered as elective subjects, and they were at the heart of Ypsilanti's Course in Modern Languages. French and German, as well as Latin, appeared in Westfield's 1894 course revision, but the state of Massachusetts would eliminate such advanced studies in late 1890s. Throughout the rest of the country, instruction in ancient and modern languages continued to serve as an important channel for normalites' acquisition of the trappings of the well-educated. Through advanced studies in languages and other subjects, students built on their solid academic foundation.[45]

Student Activities

While the formal curriculum enabled them to cross many of the cultural boundaries associated with social class, students' largely self-initiated activities outside the classroom were even more representative of their "demand for a higher degree of culture," as well as intellectual enthusiasm in general. The normal schools' intellectually encouraging atmosphere extended far beyond the classroom, as students expanded upon their shared understanding of "substantial subjects" to further educate themselves and each other through a variety of activities. By the 1870s and increasingly each year thereafter, normal students founded and participated in countless lecture series, clubs, performing groups, literary societies, and publications, many of which had significant cultural dimensions. These organizations helped create a vibrant intellectual life on campus while enabling students to reach a still "higher degree of culture."

Normalites had numerous opportunities to hear renowned speakers and performers, who visited their campuses regularly. Visiting lecturers and musicians increasingly appeared in well-publicized series. During the 1876–1877 academic year, the state normal at Castleton, Vermont offered "Twelve Nights Entertainment," a series of lectures and concerts. Also in the 1870s, student groups at Geneseo began to sponsor lecture series, and by 1888, the catalog stated, "A course of LECTURES and CONCERTS by the best talent available gives the students an opportunity of becoming acquainted with the representative scholars and musicians of the country." The 1890–1891 catalog at Slippery Rock, Pennsylvania described "lectures" as "opportunities for culture," which enabled students "to become stronger intellectually through mental contact with the strong minds of the country." Lecture courses featured university professors, writers, political figures, and social activists. Noted historian Frederick Jackson Turner delivered a series of lectures on "American Development from 1789 to 1829" at Oshkosh in 1895. At Framingham, Massachusetts poet Sidney Lanier presented his work, and his wife read his poetry after his death in 1881. Hamlin Garland's lecture at Weatherford, Oklahoma in 1905 apparently did not go over well; one student later reflected that the speech "left his audience of pioneers stone cold," while a talk by the Indian agent who had accompanied Garland "entranced us by the hour with true stories of Indian life and his own pioneer endeavor in his own Plains country." Among the speakers at the all-female normal in Greensboro, North Carolina were Theodore Roosevelt, William Jennings Bryan, and Helen Morris Lewis, president of the North Carolina Equal Rights Association. Jane Addams of Chicago's Hull House appeared on the rostrum at a few normals, including San Jose in the 1890s. Reformer and photojournalist Jacob Riis spoke at several institutions, including Whitewater, Wisconsin, where he gave an illustrated presentation on "The Battle with the Slum" in 1902.[46]

Academic clubs, which focused most commonly on science, foreign languages, or the arts, also contributed to the vibrancy of intellectual life. Among the many short-lived science clubs was the Agassizean Society founded in 1874 by a science professor at Whitewater. Despite the popularity of activities such as searching for new species in creeks and marshes, the club disappeared soon after the professor's death in 1876. Meanwhile, members of the Agassiz Club at Emporia, Kansas from 1874 to 1877 discussed student papers on various topics in the natural sciences. In the late 1870s, meetings of the Natural History Society at Brockport included reports on "Meterories," "Ruffed Grouse," and "Motions of Plants," as well as musical performances with titles like "Laughing Leaves." A decade later, a similar club at Bridgewater, Massachusetts organized excursions to collect plants, minerals, and animals from the surrounding area, while members of Oshkosh's Science Club reviewed scientific journals and attended talks on evolution and astronomy. San Jose's Astronomical Club in the early 1890s studied the moon and Jupiter.[47]

While natural-history and astronomy clubs fueled students' passion for scientific investigation, foreign-language clubs deepened their cultural sensibilities while strengthening their agility in other tongues. German clubs thrived in the 1890s and 1900s at several state normals, including Ypsilanti and Oneonta, where members of Hermania, founded in 1893, practiced speaking and published a monthly paper in German. At San Marcos, according to the 1908 yearbook, the Germanistiche Gesellschaft stood "for the cultivation of a deeper feeling of sympathy among the students of German, and a truer appreciation and broader knowledge of German life, history, literature, and music." Because many Wisconsinites, like Texans, were of German ancestry, it's not surprising that several normals in Wisconsin had German clubs. In 1895, students at Oshkosh revived an earlier club; called Der Deutsche Literarische Greis [*sic*], it conducted its business in German and occasionally produced German plays. Deutsche Gesellschaft formed in 1897 at Milwaukee, and a few years later a German reading group at Whitewater focused on German newspapers and quotations. Students' extracurricular interests also extended to an ancient language; organized in 1906, the Latin Club at the state normal in Peru, Nebraska was "a Roman Republic with a full corps of consuls, quaestors, aediles and triumvir," and was "especially helpful in the study of Roman history," the cornerstone of Western culture.[48]

Science and language clubs fostered deeper study of topics introduced in the classroom; other student organizations, as well as visiting speakers and performers, exposed students to areas of high culture that were generally outside the formal curriculum, especially classical music and art history. The Pease Musical Art Club at Ypsilanti covered both areas, and other clubs focused on one or the other. In 1899–1900, the History of Art Club at Geneseo "studied German, Dutch and Flemish art," and exhibited a "large loan collection of examples of the art of those schools." A few years later, San Jose's Art Club studied "American painters, sculptors and illustrators." Meanwhile at Southwest Texas, members of the Mendelssohn Club studied "the best available works of the great composers." State normal schools commonly offered private instrument lessons, and by the 1890s, most normals had various vocal and instrumental performance groups, which further enriched the cultural backgrounds of their members as well as other students who heard them perform. Visiting popular and classical musicians allowed students to experience higher-quality performances: the Mendelssohn Quartette Club performed at Geneseo, Brooke's Marine Band Orchestra appeared in the "lecture course" at Whitewater, and the New York Philharmonic presented a concert at Normal, Illinois. Lecture series also included discourses by well-traveled professors and visiting speakers on art history. At Oshkosh in 1899, instructor Harriet Magee presented a series of lectures, "fully illustrated with steroptican views," on "the world's most famous masterpieces of painting, sculpture and architecture," many of which she had seen on her travels. Several years later at the same institution, a visiting lecturer presented a talk, also "illustrated by steroptican views, on 'The Characteristics of Nineteenth-Century Painting.' "[49]

Campus visitors and academic clubs helped some normal-school students reach "a higher degree of culture." But it was in the literary societies, by far the most long-lived, popular, and far-reaching student organizations, where students worked hardest to refine themselves. Literary societies had flourished on college campuses throughout the country before the Civil War, when they provided a forum for intellectual engagement beyond the classics-focused curriculum. By the 1870s, college societies were giving way to Greek letter fraternities (and later, sororities) as growth in the formal curriculum satisfied student interests and wealthier students looked for a means of distinguishing themselves socially.[50] At the state normal schools between the 1870s and the 1900s, however, the literary societies could not have been more integral to student life. Writing about Illinois State Normal University, historian Charles A. Harper remarked upon "the extreme loyalty and deadly seriousness" of literary-society members. "The intellectual and social poverty of their daily [home] environment, their new-found importance in attending such a 'cultured' institution, [and] their desire to try their wings" at intellectual discourse, he explained,

> all combined to produce in the students the feeling that participation in the literary societies was the final emancipation from the crudeness, the uncouthness

> and the humiliating boorishness of backwards frontier life. There was no extreme in effort or sacrifice that these farmer boys and girls would not attempt in their devotion to the . . . societies. They strove to make them as far removed as possible from the narrow, sordid, pinch-penny monotonous grind of their daily existence. They wanted them noble, inspiring, refined, brilliant, glamorous, luxurious, dignified, and powerful.[51]

Throughout the late nineteenth and early twentieth centuries, virtually all state normal schools had literary societies, and majority of students belonged. Some societies were well-established by the 1870s and continued to grow in popularity; these included the Normal Lyceum at Bridgewater, Lyceum at Ypsilanti, and Philadelphian and Wrightonian at Normal, Illinois. Many more societies were founded throughout this period, at both long-standing and new institutions; as late as 1903, students established societies almost immediately at brand-new institutions: Athenian and Aurora at Weatherford, and the Polymathean Literary Society at Presque Isle, Maine. Societies met weekly or biweekly, usually on Friday or Saturday afternoon or evening, to execute well-planned programs of orations, debates, moderated discussions, skits, and musical entertainment. San Jose's 1900 catalog explained, "The purpose of these societies is to acquaint their members with the customs and practices of deliberative bodies, to give an impetus to literary investigation, and to develop a talent for literary pursuits, public speaking, and extemporaneous discussions." At Greensboro, meetings of the Cornelian and Adelphian societies adhered strictly to Roberts Rules of Order, first published in 1876. J. S. Nasmith, who graduated in 1872 from Platteville, Wisconsin, remembered, "The Philadelphian literary society that met in Prof. Purman's room, was a great help, for the training it gave in writing and speaking, and the quickening of thought that came from the debates." Meetings were occasionally open to the public, and "joint meetings" between two societies were quite common; a public joint meeting at San Jose in 1887 drew over 400 spectators. At the state normal in Greeley, Colorado, the two literary societies were great rivals and competed each spring in oratory, essays, and debate, in the town's Opera House before a large audience.[52]

Both single-sex and coeducational societies flourished throughout the country, allowing both women and men to explore intellectual topics and debate one another. In the South and the East, where social mores were generally traditional, single-sex societies were more common. Florence housed the all-male Dialectical and Lafayette societies from its opening, and female students formed Browning and Dixie in 1889. Browning and Dixie held occasional joint meetings and an annual joint debate, but did not meet with or compete against Dialectical or Lafayette. At San Marcos, students established the Chautauqua and Harris-Blair societies for men and Comenian, the Shakespeare Club, the Idyllic Club, the Pierian Club, and the Every Day Society for women. While these women's and men's groups often exchanged visits, they did not engage one another in debate. In a unique arrangement, New York's normal schools housed branches of statewide single-sex societies. Those for women included Clionian, Agonian, Arethusa, and Alpha Delta, and those for men included Delphic and Philalethean. In many ways, Delphic functioned as a brother society to Clionian, as Philalethean did to Agonian. Unlike southern societies, each pair held joint meetings regularly, and ended the academic year with a public joint meeting. A few normals in the South and East did have coeducational societies. At Troy, Alabama, the Baconian Society became coeducational after just one year as a men-only group, and Pine Bluff's Junior and Senior Societies in the 1880s and 1890s were mixed. At West Liberty, West Virginia, the coeducational Bryant and Irving societies, formed in 1886, were much more popular than earlier single-sex groups, and soon began to hold a public competition during commencement week. Bridgewater's Normal Lyceum and the Erodelphian and Huyghenian societies at Indiana, Pennsylvania were also coeducational.[53]

Gender segregation was less rigid in the Midwest and West, where societies were most likely to be coeducational. Lyceum at Ypsilanti and Wrightonian and Philadelphian at Normal were mixed, as were the Lyceum, Literati, and Belles-Lettres societies at Emporia. When Lyceum members at Ypsilanti in 1870 passed a resolution "That the ladies ought to be allowed to debate; that the interests of the society and its existence depend upon their debating," they in many ways set a precedent for Midwestern and Western societies; a chronicle of Emporia's Literati reported in 1889, "at the present time it is no uncommon occurrence for the debate to be conducted wholly by ladies." Such societies allowed female and male members to participate on a relatively equal basis. Single-sex societies did also appear at midwestern and western institutions, but they generally existed alongside mixed societies. For example, although the Young Men's Normal Debate Society was one of the most active groups at San Jose, several other societies there were coeducational; they were especially protean, as close to two dozen mixed and single-sex societies formed and re-formed among San-Jose students. At Oshkosh, coeducational Lyceum, formed in 1871, was for a time the only society. Male students then formed Protarian, which was later complemented by the Ladies Literary Society. The latter group disappeared in the early 1880s, and shortly thereafter, Protarian was transformed when, as the society journal reported, a "debate was interrupted by the rustling of a few angels who wished to gain entrance . . . a band some twenty-five strong filed in amidst deafening applause." The integrated society was called Phoenix, and, along with Lyceum, remained dominant at Oshkosh, although students did occasionally form additional single-sex groups. While the main societies at Cedar Falls were single-sex and Peru also had a men-only and a women-only society, the coeducational Philomathean Society at Peru was especially prominent in campus life during its "unbroken continuous existence" from 1868 into the 1910s. When all-male Ciceronian and all-female Sapphonian appeared in the shadow of the two coeducational groups at Normal, Illinois in the 1870s and 1880s, they occasionally met jointly, and held an annual contest beginning in 1891. Such joint meetings of single-sex groups and the more popular mixed societies enabled female and male students to engage one another in debate.[54]

Whether they were single-sex or coeducational, literary societies were a lively forum for intellectual exploration. Recognizing that these student groups reinforced curricular aims, administrators and faculty members encouraged them. Some normal schools, especially in the Midwest and West, provided students with the added incentive of academic credit for society work. The 1887 Cedar Falls catalog stated that society work satisfied some rhetorical requirements in the English Course, while an "arrangement" at Emporia allowed "members of the societies to have credit in the regular school record on declamation, essay, and oration rehearsed before members of the faculty." At Tempe in the mid-1890s, members of Zetetic and Hisperion were excused from rhetoric classes. Oregon's normals had a similar policy, as they stipulated: "A student has the choice of performing his rhetoricals before the school or before the society. We prefer that it be done in the society." A few normals went a step further than merely providing incentive; at Kirksville, Missouri through the 1890s and Weatherford, Oklahoma and Las Vegas, New Mexico in the 1900s, joining a society was a requirement. Before society membership at Ellensburg, Washington became optional in 1899, the faculty monitored attendance at society meetings and recorded grades for students' performances. While the faculty was involved in society activities to different degrees at different institutions, the typical structure was management "exclusively by the students, subject to the approval of the faculty," as Mankato, Minnesota reported in 1890.[55]

Regardless of whether society activities had faculty supervision or earned class credit, they enriched academic life in numerous ways. Many societies maintained library collections and reading rooms, which provided additional resources for class and society work, as well as general exposure to literary culture. The trustees at Millersville, Pennsylvania provided the societies with funds to expand their libraries, and the Alcyone Literary Society at Troy collected 450 books for its library at an 1898 fund-raising event. Normal's Wrightonian and Philadelphian donated their libraries to the school's general collection in 1889, enriching intellectual life in a very material way.[56] Literary societies also furthered the goals of the academic curriculum by focusing consistently on refining their members' styles of expression and composition. Students wrote and delivered orations and essays on different subjects from week to week, and debating fostered poise, precision, and accuracy in oral communication. In addition to joint debates between two societies at the same school, it was increasingly common for one normal to challenge another to a debate. After Oswego challenged Oneonta in the early 1900s, for example, Oneonta's three-man team won easily. In the Midwest, statewide and even interstate normal-school debate and oratorical contests attracted large crowds. Yearly Inter-Normal Contests between Normal and Carbondale, Illinois began in 1879. By the late 1890s, the Inter-Normal Oratorical and Declamatory Association of Missouri sponsored yearly contests between the state's three normals for white students, while an oratorical league of the normal schools in Missouri, Kansas, Iowa, Illinois, and Wisconsin organized larger-scale contests. Competitions at individual

Figure 3.1 Students who represented Wisconsin's Oshkosh State Normal School in inter-normal oratorical competition, as pictured in *The Normal Advance*, February 1902. (*The Normal Advance* 8 [Feb. 1902])

Credit: UW Oshkosh Archives. Copyright 2004 UW Board of Regents.

normals determined who competed at the state level, and a state-level contest determined who went on to the inter-state competition. Women readily entered oratorical and occasionally entered debate contests. After Elizabeth Shepard won Oshkosh's contest with her speech on Ulysses S. Grant, 183 supporters accompanied her to the Wisconsin state contest at Stevens Point, cheering, "For Grant when she spoke in electrical tones . . . She made the cold shivers run down our backbones." Orating and debating in both small- and large-scale competitions allowed students gain cultural polish; ironically, women challenged the gender conventions of high society in the process.[57]

Literary-society members also refined their writing skills by producing serial newspapers or magazines. In the 1870s, 1880s, and 1890s, individual societies often produced their own publications. Typical titles were *Normal Thought*, published by the Standard Society at Buffalo, New York, *Normal Ray*, published by the Baconian Literary Society at Troy, and *The Students' Offering*, which was a coordinated effort by all of the literary societies at Cedar Falls. At Cortland, *The Normal News* was published monthly by the Young Men's Debating Club with help from the Ladies' Normal Debating Club until 1892, when it became an all-school publication in which each society had a section. By the late 1890s, most normals had campus-wide student magazines in which literary-society members and activities were prominent. Student editors of *The Crucible* at Greeley wrote in 1897 that they were "interested in providing culture foods for the students." Particularly long-running, Oshkosh's *The Normal Advance* began in 1894 as a bimonthly journal and would evolve into a weekly newspaper by the 1920s. Such publications included essays, reviews, short stories, and poems with both serious and frivolous themes, and news of alumni achievements and campus events—especially the doings of the literary societies. In addition, by the mid-1900s, students at many normals began to publish yearbooks. They, like society and campus-wide publications, featured the activities of the literary societies.[58]

In their publications, orations, and debates, literary-society members further pursued some of the topics covered in the curriculum. In the societies, unlike academic clubs, they were able to dip into different subjects from meeting to meeting. Society members delved into history, for example, with programs on American presidents Washington and Lincoln in celebration of their birthdays in February. In Oshkosh's Lyceum, the female members debated the men on whether "Lincoln was a greater man than Washington," and the women were victorious in their defense of the negative side. A meeting of Brockport's Phothepian Society in 1891 featured addresses on "England and Englishmen," "The Norman Conquest," and "Cromwell," as well as an impersonation of Sir Walter Raleigh. Students also weighed the actions of military heroes and villains, including John Brown, Robert E. Lee, and Benedict Arnold. Leone Spoor spoke on "Napoleon" in Oshkosh's oratorical contest of 1898 and Elizabeth Shepard won the following year's competition with her speech on Grant. Great women of history did not escape the normalites' notice; members of Brockport's Arethusa dressed as Joan of Arc,

Mary, Queen of Scots, and Queen Elizabeth to represent these figures in society discussions, and Geneseo's Agonian debated, "Resolved, that Cleopatra was a better woman than Helen of Troy." Geneseo students also tackled historical subjects as broad as "The Rise and Decline of the English Realm" and "the course of progress from the beginning, down to '77," and as esoteric as "Resolved, that the character of the Puritans was overestimated."[59]

The main focus of the literary societies, of course, was literature, and it was in this area that they most helped students to become conversant with the high culture of the well-educated. Literary societies regularly studied a variety of British and American authors. The Browning clubs at Florence, San Jose, and Oshkosh were named for Elizabeth Barrett Browning, a favorite among their members, just as the Shakespeare(n) societies at Cedar Falls, San Jose, and San Marcos were named for the great playwright. Members of San Jose's Shakespeare Club in 1900–1901 debated the merits of each play they read, and their counterparts at San Marcos in 1904–1905 made *Twelfth Night*, *A Comedy of Errors*, and *A Midsummer Night's Dream* their work for the year. The programs of many other literary societies also focused on the life and work of Shakespeare and other British authors. One of Ellensburg's societies devoted a meeting to "Three Scotch Writers," and Bertha Schuster Beach, Whitewater '83, remembered that the Young Ladies' Literary Society devoted an evening to Oscar Wilde, with "huge sunflowers everywhere, [and] aesthetic dresses that blinded the eyes." At Brockport in 1887, the character of Nicholas Nickelby made an appearance at an Arethusa meeting, and seven years later an Alpha Delta gathering included an oration on Charles Dickens and a play entitled "Pickwick's Trial." Other popular British authors were Jane Austen and Alfred, Lord Tennyson.[60]

Studies of the poetry and prose of John Greenleaf Whittier, Henry Wadsworth Longfellow, Washington Irving, Ralph Waldo Emerson, and Mark Twain, deepened students' familiarity with American literary culture. The program for an 1889 Alpha Delta meeting at Brockport included essays entitled "American Poetry," "Humor in American Poetry," and "The Poet Longfellow." San Marcos' Every Day Society declared Irving its "patron saint," and imitated him in the 1909 yearbook, presenting a supposed "unpublished tale" in which a lost hunter dreamt of "the girls of the land where the sweet peas grow," who were suspiciously similar to Every-Day members. Societies also delved into American regional and vernacular literature. "Pacific Coast Literature" was the subject of a society "round table" discussion at Ellensburg, and Oneonta's societies brought Jack London to campus, where he spoke on "Experiences of a Tramp." Southern literature was a focal point of an Agonian meeting at Geneseo in 1892 and a Dixie meeting at Florence in 1902. Throughout 1906–1907, San Marcos' Comenians focused on Southern literature, while the Idyllics studied "Folk-Lore." At a meeting during the same year, Oshkosh's Phoenix society looked at "Negro Writers and Their Work."[61]

In many literary societies, students explored great literature by performing it. At Oneonta, for example, various societies staged Tennyson's *The Princess* in 1892, Shakespeare's *As You Like It* and *The Merchant of Venice* in 1901, and Dickens' *Tom Pinch* in 1905. Finally, normalites also went beyond the examination of the works of individual authors, to reflect generally on the value and importance of literature. At San Jose, the Allenian society debated, "Resolved, That books have done more good toward the advancement of civilization than newspapers," and *The Normal Index* published a provocative article that declared dime-store novels to be "vile trash" and an article on "The Mission of Poetry as Compared with that of History" by student May F. Blackford. In 1900, Miss Fosmire read her essay, "The Nature of Poetry," for Geneseo's Wendling Club, and a couple of months later, her school-mate Dorothy Covey read "The Influence of Great Writers" in the Arethusa-Gamma Sigma commencement exercises.[62]

Most literary-society meetings featured live music, and a few also included intellectual discussions of the subject. At one meeting of Oshkosh's Phoenix, for example, several musical performances accompanied a talk entitled "Introduction of Music into the United States." The societies' treatment of music, like its approach to literary and historical subjects enabled normal-school students to obtain prestigious cultural knowledge. By the latter part of the nineteenth century, the culturally refined also shared an enthusiasm for travel. In society meetings, normalites accordingly traveled vicariously throughout the United States and the world. In 1876, members of the Normal School Philologian Society at Westfield enjoyed "An Account of his (Mr. Diller's) Geological Vacation . . . up the Valley in the State of New York" in the form of an illustrated lecture. In 1899, "Miss Dopp favored" Oshkosh's Phoenix "society with some of her experiences among the Mormons in Salt Lake City." A decade or so later, a program entitled "Travels in the West" took Florence's Dixie society, figuratively, to Salt Lake City and Yellowstone National Park. Students' vicarious travels also took them to Europe and more foreign locales. In 1880–1881, the Ladies' Literary Society at Oshkosh studied Spain and Germany, and in other years both San Marcos' Comenian and Oshkosh's Phoenix studied China.[63]

As the globe shrank in the eyes of society members, they also studied and debated world diplomacy. In the late 1880s, Cortland's Normal Debating Club weighed the resolution "That England has no right to interfere in the Russo-Turkish war," and Westfield's Philologian society heard a talk on "the state of disturbance" in "the Japan of today." On the eve of the Spanish-American War, Geneseo's Agonians debated, "Resolved that Cuba should be annexed to the United States." In the wake of the war, they debated, "Resolved: That the present American war with the Phillipines [*sic*] is unjust," while at San Marcos the Harris-Blair and Chautauqua societies held a public debate on "the Philippine question." Meanwhile, students from Oshkosh and Stevens Point debated, "Resolved, that the United States immediately own, operate and construct the Nicaragua Canal." In 1907, students from San Marcos debated the Monroe Doctrine with representatives of North Texas State Normal School in Denton. Such discussions must have been especially interesting at institutions with students from the countries under discussion, such as the normal at New Paltz, New York, where young women from Cuba studied in the "Cuban Annex" in the 1890s. Pennsylvania's normal schools had students from Canada, Switzerland, Russia, and even India, as well as a large number of male students from Central and South America and the Caribbean beginning in the 1890s; in just one year, 1906–1907, Bloomsburg hosted 9 Puerto Rican and 29 Cuban students.[64]

On the domestic front, normal students considered many prominent social issues and progressive reforms, which were special concerns of the middle class by the late nineteenth century. In 1902, Castleton's junior class debated whether "High License Local-Option is better than Prohibition in dealing with the Liquor Question." As these debaters sought to curb the social problems that resulted from alcohol abuse, other students tackled poverty and racism. Hulda Felschew entered an oratorical contest at Oshkosh with "Prisoners of Poverty." Also at Oshkosh, Lyceum debated, "Shall the Negro be debarred from voting," and the Alethean society discussed "The Negro Question." In Illinois, the Carbondale-Normal debate in 1880 was over the resolution "That the migration of the African race from the Southern States will promote harmony among the citizens and the prosperity of that section." The social stresses of immigration underlay the previous year's Inter-Normal Contest between the same institutions, as they debated, "*Resolved*, That the United States Government ought to take steps to secure such modification of the treaties with China as would prevent further Chinese immigration." Oshkosh's Phoenix also pondered the exclusion of Chinese and Japanese immigrants. In a similar vein, the Agonians at Geneseo debated, "Resolved, That immigration is not conducive to the welfare of a country," and a joint debate between two societies at San Jose considered the resolution "that the benefits of foreign immigration outweigh its evils." Labor conditions and unionization were also prominent topics. In the mid-1890s, members of Lyceum and Agonian both debated the benefits and liabilities of trade unions, and ten years later, Florence students convened for a Dialectal-Lafayette debate on, "Resolved, That Labor Unions Tend to Promote the General Welfare of a Country."[65]

As progressive campaigns for government regulation of business and political reform appeared in the news, they also showed up as debate topics. In 1896 at Oshkosh, Phoenix and Lyceum squared off over the resolution "that our Railroads should be owned and operated by the Government." The same topic was the subject of the San Jose–Chico debate in California more than a decade later. During the spring of 1901, municipal ownership of utilities was a popular topic—it was the subject of the Agonian-Philalethean joint meeting at Geneseo and the San Jose–Chico debate. The society-sponsored lecture series at Oneonta in the 1890s and 1900s included Robert LaFollette on corporate trusts and Lincoln Steffens on "Graft, What It Is, and What to Do About It." Meanwhile, the 1908 San Marcos–Denton debate focused on the resolution "That Texas should adopt an amendment to the constitution providing for the optional Initiative and Referendum, applicable to all laws." The following year's "triangular debate" among the

Figure 3.2 Members of teams representing the Phoenix and Lyceum literary societies in school-wide debates at Wisconsin's Oshkosh State Normal School, as pictured in the 1900 yearbook. (*The Quiver* [1900], 62)

Credit: UW Oshkosh Archives. Copyright 2004 UW Board of Regents.

normals in Weatherford, Alva, and Edmond, Oklahoma was on the resolution "That Congress should take immediate steps toward the adoption of the Income and Inheritance tax law." Another popular reform issue as well as debate topic was women's suffrage. At an Agonian-Philalethean joint meeting at Geneseo, mixed-sex teams argued for and against, "Resolved, that the right of suffrage should be equally extended to both sexes, the illiterate of both sexes being excluded." After a similar debate among Phoenix's women at Oshkosh, a man in the audience remarked, "that each woman should have a *voter*, but not a vote." A special "suffrage edition" of Oshkosh's *The Normal Advance* included editorials for and against votes for women. In debates such as these on suffrage and other political and social issues, normalites acquainted themselves with topics of great concern to middle-class progressive reformers.[66]

As literary societies' studies of progressive reform, travel, and other topics exposed students to "a higher degree of culture," the structure and setting of society meetings fostered social refinement. Although it occasionally broke down, decorum was of the utmost importance during weekly or biweekly meetings. Between the various speeches and debates, members entertained one another with musical selections. At Oneonta, all of the literary societies by the 1890s had stringed-instrument groups and glee clubs. Many societies throughout the country owned pianos, and put them to good use, such as when a duet played the William Tell Overture between the president's address and an essay at an 1875 meeting of Brockport's Arethusa Society. Students also tried their hands at interior decoration according to high standards of taste. In 1885, the literary society at St. Cloud, Minnesota redecorated the institution's reading room with new carpets and chandeliers. Beginning in the 1880s, normal schools set aside rooms for each individual society or for pairs of societies to share, and students furnished them very carefully. Normal, Illinois' Philadelphian reported in 1882 on the "splendor" of its meeting hall: "The new carpet . . . the richly dyed curtains swinging from the arch, the beautiful scenes of life portrayed by master hands, and the costly chandeliers, cast in a model of rare design, command the admiration of our friends and the respect of our rivals." Societies at most normals had similar rooms by the latter 1890s; at Brockport, they were decorated as parlors, with small libraries as well as pianos. The groups held their regular meetings in these rooms, and hosted dinners and receptions for other societies or faculty members. At Emporia, the Literati Society held a "Dickens social" and a "Shakespearean social." These more social gatherings called for good manners and graciousness, complementing the more intellectual pursuits of the societies.[67]

At state normal schools in the late nineteenth and early twentieth centuries, the extracurricular literary societies and clubs built upon the firm foundation that the required "elementary" curriculum established in academic subjects. Student activities, as well as elective or advanced subjects in the formal curriculum, offered students "a higher degree of culture" through prestigious knowledge and social refinement, or elements of cultural capital. For many normalites, this aspect of their education represented a substantial stretch from their unsophisticated backgrounds. Burtt N. Timbie, who graduated from Bridgewater in 1896, reflected, "These were years of mind and soul awakening. We came to love learning for learning's sake."[68] Their minds awakened, normalites graduated with a store of prestigious knowledge but without certification in the form of a bachelor's degree. Normal schools enabled students to cross some, but not all, of the boundaries associated with social class; in many ways, normalites gained cultural-class standing, but not social-class status. Officially, the normals trained students for teaching, an occupation with poor pay and ambiguous status. Thus, as students prepared in the normals' culturally rich atmosphere to be teachers, they were—apparently unintentionally—socialized for the precarious social standing of teachers.

4

Teacher Education: Breathing "The Ozone of Teaching"

The status of teaching in the professional culture that took hold and dominated American middle-class society by the 1870s was ambiguous. The requisite professional associations, advanced studies, and licensing requirements appeared in the field, yet education leaders and the public were unwilling to grant teachers autonomy or authority. Historians of teacher education traditionally have attributed teachers' problematic professional status mainly to a failure on the part of the normal schools to balance liberal and technical elements in teacher education; according to Merle Borrowman, "an excessively technical concept of the professional sequence" caused the normals to instruct future teachers in little more than "tricks of the trade." Jurgen Herbst argues that by the late nineteenth century normal schools abandoned the cause of professionalizing elementary-school teaching and instead helped administrators to become "the professionals in public education."[1] Such analyses, however, sell short the normal schools of the 1870s, 1880s, 1890s, and 1900s. They prepared students for a vocation with a long history of ignominy, the most recent development in which had been the arrival of large numbers of women. The work of historians who examine the professions and teaching through the lens of gender suggests that larger social structures, norms, and expectations overshadowed anything the normal schools were or were not doing in preparing teachers. As long as society defined "professional" as "educated, altruistic, genteel, and implicitly male," women and anyone from a lower-class background—the majority of teachers—would not gain acceptance as professionals.[2] It was thus not surprising that it was administrators, who were mainly men, who enjoyed unquestioned professional status in education. To overcome teaching's legacy of low status would have been a tall order for any institution; failure in this regard did not mean that the normal schools were unsuccessful in creating a strong professional spirit.

Professor James E. Lough of Oshkosh State Normal School in Wisconsin reflected the normal schools' conception of professionalism when he declared in 1901, "The one who teaches merely because of the personal gain is a tradesman; the one who teaches because he loves, and has a heart and head interest in his vocation, is a professional."[3] State normals had stressed "heart," or emotional commitment to teaching as a special calling, from the beginning. By the latter decades of the nineteenth century, the normals' approach to teacher preparation had matured significantly; "head," as well as hand, balanced the exaltation of teaching. Teaching's technical or "manual" aspects received a great deal of attention in curricula that stressed teaching methods and model schools that provided teaching experience as a valuable form of preparation.[4] In addition—contrary to Borrowman's claims—mental engagement in and liberal reflection on both broad and narrow aspects of education were integral to teacher-education coursework and student initiatives in their literary societies and other activities. Heart, head, and hand worked together at the normals to create a professional culture that was distinctive as well as pervasive; Principal Albert Boyden of Bridgewater State Normal School in Massachusetts referred to "the ozone of teaching which permeates the atmosphere of the Normal School."[5] In their classroom studies in teaching, model-school observation and practice teaching, and self-initiated activities, normalites indeed breathed "the ozone of teaching."

Studies in Teaching

Pedagogical instruction at state normal schools before the 1870s was hardly professionally inspiring. Because the field of education was so new, lessons in how to teach consisted generally of a series of lectures by the principal that offered a lot more inspiration for the heart than helpful training of the hand or engagement of the head in the teaching enterprise. As the object method began to take hold as a theory of teaching, it was immature theoretically and subject occasionally to ridicule. The tide was turning by the 1870s, however, as state normals incorporated more sophisticated topics and approaches in their teacher-training curricula. Well-qualified instructors were crucial to increasing curricular breadth and depth, and thus fostering a professional spirit.

Normal-school faculties between the 1870s and the 1900s reflected the balance of hand, head, and heart that guided their institutions' approach to teacher education. True to the normals' philosophical commitment to practical training, experience in the field of education was a predominant feature of faculty members' backgrounds. Many instructors, especially women, had taught as well as attended normal school before their appointment to the faculty. Jennie Hayden, who served on the faculty at the state normal in Farmington, Maine from 1872 to 1881, had graduated from Farmington in 1867 and then taught for five years in the public schools. Viola Boddie taught both before and after attending Peabody Normal College in Nashville, and then joined the faculty at the new state normal in Greensboro, North Carolina in 1892. After graduating from Emporia, Kansas in 1869, Martha Spencer taught in no fewer than six different towns and served as superintendent of education in one before returning to Emporia to join the faculty in 1883. Other instructors had extensive practical experience without having trained formally in a teacher-education institution. For example, before joining the faculty at Ypsilanti, Michigan in 1896, Charles Oliver Hoyt taught in a rural school, was principal of a high school in a small town, then served as superintendent of schools in larger towns, including Lansing.[6]

An 1889 publication of the Kansas State Normal School in Emporia applauded instructors' experience in the field, stating that Spencer's "training and her familiarity with public-school work admirably fitted her for the position." Similarly, Minnie Curtiss "had won a reputation as a teacher in Carmel, N.Y., and in Adams, Mass., which made her a desirable acquisition to the faculty," and Lizzie Stephenson's experience as a teacher and county superintendent gave her "a practical knowledge of the needs of teachers in our public schools." The profile of Dorman Kelly stated, "His career in teaching was a valuable preparation for a teacher in a normal school," before listing the many teaching and administrative positions he had occupied in the schools of his native Indiana. At Cedar Falls, Iowa in 1900, President Homer Seerley stated, "Our board of trustees have made it almost a custom to select practical teachers who have been in charge of school systems and who know what good teachers are . . . rather than young men who have just finished graduate courses at higher institutions." Under Seerley between 1886 and 1897, 22 of the 37 instructors hired had experience in the public schools, 5 as superintendents. Hands-on experience in the education system was thus an important qualification for teaching at Cedar Falls and other state normal schools.[7]

While very prominent, practical experience was by no means the only defining characteristic of members of normal-school faculties; many instructors were also active scholars and thinkers in education as well as other fields. Even if President Seerley did not seek them, holders of academic degrees were present on normal faculties. At New York State's normal schools in the 1870s and early 1880s, 23 percent of instructors held collegiate degrees, including many master's and a small number of doctorates. More than half of Seerley's hires at Cedar Falls between 1886 and the turn of the century held bachelor's degrees, a few had earned MA's, and one had a Ph.D. At Winona, Minnesota in the late 1900s, almost half of the members of the faculty possessed collegiate degrees; 14 percent had master's and 5 percent had doctorates.[8]

Perhaps more representative of many instructors' academic orientation were their efforts to educate themselves during their tenure on the faculty. Many devoted leaves of absence, summers, evenings, or weekends, to language studies in Europe, scientific studies at American universities, or classes in various subjects. In the early 1880s, Mary Lee and Mary Sheldon, both of whom had bachelor's degrees and were active members of the faculty at Oswego, New York, spent two years traveling in Europe and studying at the University of Cambridge. Emporia's Dorman Kelly "spent his summers in travel and study. The summer of 1886 he passed at Harvard, taking special [scientific] work." While she taught at Farmville, Virginia from 1883 to 1893, Celeste Parrish spent her Saturdays studying with a tutor at Hampden-Sydney, a men's college nearby. Several other women on the Farmville faculty were "special students" at the all-male University of Virginia. While teaching at

Ypsilanti in the late 1890s and 1900s, Abigail Pearce, Bertha Buell, and Estelle Downing earned bachelor's degrees at the University of Michigan as well as master's degrees at Michigan, Radcliffe College, and the University of California, Berkeley, respectively. Faculty members enriched their knowledge in a variety of fields, but education was, not surprisingly, the most popular. Emporia's Martha Spencer "spent part of the summer of 1887 in a careful study of physical training, at Harvard," and Ypsilanti's Charles Hoyt used a leave of absence to earn a Ph.D. in 1903 at the hub of the Herbartian movement in education, the University of Jena in Germany. Closer to home, it was quite common for normal-school instructors to undertake further studies in education at the University of Chicago, Teachers College of Columbia University, the University of Wisconsin, and Stanford University.[9]

For many faculty members as well as principals, publishing was an outlet for scholarly endeavors. While some published in other academic fields, the majority who wrote published books and articles on education; the Midwest alone provides voluminous examples. Five years after earning his doctorate, Hoyt published *Studies in Modern Education*. Seerley focused on the problems of rural schools in *The Country School* and many articles in Iowa's education journals, while Emporia's President Albert Reynolds Taylor wrote a book on child study. Many instructors tapped into their teaching experiences as well as academic knowledge of teaching methods to write textbooks for normals and the public schools. Hoyt's colleague Wilbur P. Bowen published *A Teachers Course in Physical Training* in 1898. At Cedar Falls, Annie E. McGovern wrote *A Manual of Primary Methods* and compiled a children's story and song book, while David Sands Wright published *A Drill Book in English Grammar, Geometrical Outlines*, and *Bible Study Outlines*. At Normal, Illinois in the late 1870s and 1880s, President Edwin C. Hewett wrote pedagogy and arithmetic texts while faculty member Martha Haynie published texts entitled *Syntax and Analysis* and *Etymology*; Buel Colton, who joined the faculty in the late 1880s, published textbooks in zoology and physiology. Among the books Albert Salisbury wrote during his principalship at Whitewater, Wisconsin from 1885 to 1912 were *The Theory of Teaching and Elementary Psychology* and *School Management*. In addition to publishing mathematics textbooks, Ypsilanti's Charles Fritz Roy Bellows published a journal entitled *The School* in the mid-1870s. Principals and faculty members throughout the country wrote and edited publications ranging from the *State Normal Exponent*, a quarterly education magazine founded at Troy, Alabama in 1893, to a series of bulletins and "exercise books" for California's teachers published at the state normal in San Francisco from its opening in 1899.[10]

Finally, faculty members' involvement in various professional associations reflected their mental engagement in teaching, as well as their emotional commitment to the field. At the institutional level, faculties met to discuss various educational matters. In regular meetings at Kirksville, Missouri in the 1870s, the faculty addressed "Success in Class Management," "Problems of Successful Teaching," and other issues. After publishing *The School*, Bellows helped to found the Pedagogical Society at Ypsilanti, with the goal, he said, of "promoting professional enthusiasm in the faculty as a whole, and thereby conserving the professional welfare of the school at large." Especially active in the 1890s, the society discussed papers by faculty members, many of which were subsequently published. Perhaps inspired by Ypsilanti, a faculty club appeared at nearby Kalamazoo as soon as it opened in 1904. At the state and national levels, many faculty members (not to mention principals) were active in educational and academic organizations. Throughout his tenure at Cedar Falls, Wright was highly involved in the Iowa State Teachers' Association, serving as president in 1904. His colleague Melvin Arey was the secretary of the Educational Council of the ISTA for an entire quarter century, as well as president of the Iowa Academy of Science in 1905 and fellow of the American Academy for the Advancement of Science. Ypsilanti instructor Edwin Strong served terms as president of the Michigan Academy of Science, Teachers Association, Schoolmaster's Club, and Association of High School Principals, and was director of the National Education Association from 1892 to 1894. His colleague Estelle Downing chaired the International Relations Committee of the National Council of Teachers of English. Meanwhile, Farmville faculty were active in organizations including the Virginia State Teachers' Association, the Southern Education Association, and the National Education Association. They helped one another prepare presentations for meetings, and those who attended reported on the proceedings to the rest of the faculty.[11]

With extensive experience in the field and a high level of involvement in scholarly pursuits and educational associations, many normal-school faculty members clearly were devoted to their work. It follows that the 1889 Emporia publication had high praise for several instructors' heart-felt commitment. Martha Spencer, for example, was held up as "the shining example of the devoted, the untiring, the inspiring Christian teacher. She lives in her work, and it lives in her." Another instructor was "a very thorough teacher—a true disciple of normal methods." Lizzie Stephenson's teaching and administrative experience was balanced by "a most earnest, conscientious, self-sacrificing spirit, [which] makes her a valuable teacher." And Lillian Hoxie had it all: "She is a woman of brilliant intellect, wonderful energy, and marked enthusiasm. This, with a ready fund of knowledge, rare powers of expression, and a thorough love of teaching, make her unusually popular in her profession."[12] Faculty members like Hoxie embodied their institutions' vision of professional commitment through one's hand, head, and heart. As instructors guided students through the teacher-education curriculum, this notion of professionalism shaped the normal schools' "ozone of teaching."

Throughout the last three decades of the nineteenth century and the first decade of the twentieth, studies in teaching at state normal schools emphasized methods and classroom management, and thus focused on the practical aspects of the teaching enterprise. Some normal schools, especially early in this period, stipulated simply that methods were to be covered in the regular academic curriculum. At Florence, Alabama in the 1870s, catalogs stated that within academic subjects, "In addition to a thorough knowledge of each subject embraced in the course, students will receive instruction as to the best methods of imparting information." Throughout this period, the assumption at normal schools generally was that instruction in academic classes would pay some attention to methods. Most normal schools did not rely on this approach exclusively, however, and also offered separate classes in teaching methods. Such classes had names like Principles and Methods of Instruction (at Providence, Rhode Island in the early 1870s), Methods of Teaching (in Oregon in the early 1880s and Vermont in the late 1890s), Primary Methods (at San Jose, California in the late 1890s and early 1900s), or simply and most commonly, Methods. At Oshkosh in the early 1870s, Theory and Art of Teaching included "Specific Methods of teaching each Elementary Branch"; by the late 1870s and early 1880s, Art of Teaching stood alone, and focused around a "careful examination of specific methods of teaching the different elementary branches."[13]

A small number of state normal schools, especially by the late nineteenth century, organized classes on the methods of teaching particular subjects. The course of study at Washington's normal schools in the 1890s included Reading Methods, Arithmetic Methods, and Grammar Methods. By this time, the teacher-education curriculum at Florence had expanded to include Methods in Arithmetic, Methods in Language, Methods in Geography, and a general methods course. In the early 1900s, the English Course at Geneseo, New York required Music II and Methods, Physical Geography and Methods, Drawing II and Methods, Methods in Number and Grammar, and Methods in Science and Geography, as well as a broader methods class called General and Primary Methods. Geneseo's Classical Course required the same classes, with the addition of Arithmetic and Methods. In fact, most institutions expected all students to take the same methods classes, even if different courses of study were intended to prepare students for different teaching positions. At Cedar Falls, which had differentiated courses of study from the 1870s, the first suggestion of specialized methods instruction did not appear in a catalog until the mid-1880s, and the first actual specialized instruction, for kindergarten teaching, did not begin until 1893. While more common by the 1890s and 1900s, this practice remained unusual. Winona, Minnesota added a Kindergarten Training Department as early as the 1880s, and Ellensburg, Washington inaugurated specialized training for kindergarten work in the 1890s. The Kindergarten Course that began at Geneseo in 1906 included specialized study in elementary science and nature study. Meanwhile, Kalamazoo's Rural Life and Education Department prepared teachers for Michigan's one-room schools. Thus, while a small number of students was able to specialize in areas such as kindergarten or rural-school teaching, the assumption underlying methods instruction from the 1870s through the 1900s was that all students would benefit from the same classes.[14]

Catalog descriptions revealed how methods classes covered the technical or manual elements of teaching. The 1893 Ypsilanti catalog explained that instruction in mathematics methods aimed to "show what has been and is considered the proper sphere of arithmetic and methods of teaching the subject," and "to show the best methods, in the light of present criticism, of presenting typical chapters in the various grades." "General work in method" at Florence in 1899–1900 included "the aim of education and of instruction, the principles of self activity and unity in instruction, a study of analysis, synthesis, induction and deduction, formal steps of instruction, and an inductive study of questioning as a means of awakening and directing mental activity." San Jose's catalog descriptions of various methods classes in the early 1900s spelled out the "specific principles" of teaching the subjects, suggesting that the classes centered on these practical approaches. Primary Number Methods instilled five principles: "(*a*) Teach numbers at first concretely—objectively; (*b*) Things and clear ideas before words or symbols; (*c*) Proceed from particulars to the general; [and] (*d*) In primary grades, teach processes, not rules." Similarly, Primary Language Methods emphasized three basic principles: "(*a*) With young children language is learned by imitation; (*b*) The habits formed in language are more important than the knowledge gained; [and] (*c*) The study of the *science* of language belongs to the years of judgment." Meanwhile, the 1904–1905 catalog of the Southwest Texas State Normal School in San Marcos explained that Primary Methods would include "the study and application of the principles on which approved methods of teaching in the primary grades are founded," and "teach normal methods. The purpose will also be to make known to students the means, devices, and appliances used and approved by primary teachers in our best modern schools."[15]

Instruction in classroom management and school structures accompanied instruction in methods, so that normalites became familiar with the practical elements of organizing classes and schools as well as teaching. As early as 1872, Theory and Art of Teaching at Oshkosh included not only methods, but also topics ranging from "Organization and Classification of Schools," including "Warming, Ventilation and care of Schoolroom," to "School Government—its *moral* aspect," including "Power, how exercised" and "Causes of disobedience." Beginning in the late 1870s, a separate class called School Management covered these topics. At San Jose in late 1870s and 1880s, professional instruction in the middle year consisted of "Methods of teaching, and instruction in classifying, grading and managing schools"; by the mid-1880s, senior-year studies included school law. In the late 1880s, School Organization and Laws at Farmington and School Economy and Management at Emporia covered the same territory. The latter class focused on six elements of the structure of a graded school and the teacher's position within it: "School organization," "Employment," "Government," "Physical conditions," "Moral culture," and "Supervision."[16]

School Management at Oshkosh remained unchanged throughout the 1890s, while normal schools from Pennsylvania to Washington offered a class with the same title. San Marcos' 1904–1905 catalog stated, "In School Management, the aim will be to prepare students to understand and overcome the difficulties encountered in organizing, grading, and conducting schools. Special attention will be given to all questions relating to discipline. The grading of schools and courses of study will be carefully studied." At San Jose by the turn of the century, School Economics covered this subject matter; the catalog summarized:

> The aim . . . is to familiarize students with the methods of procedure required by law to be used in organizing and conducting a school. Attention is given to the formal matter of making out school reports and to the plans for keeping school registers. The subject of professional ethics is dealt with in several lectures. The educational system of the State is discussed and compared with that of one or more typical States of the Union. The practical application of pedagogical theory to school work also receives attention.

Meanwhile, Florence's curriculum also covered the ins and outs of the organization and management of schools; the catalog explained, "The object of school economy and school law is to make the student intelligent regarding plans and arrangements, objects and requisites of public schools; best ways of applying the resources of a school; correct ideas with regard to teachers' motives, qualifications (physical, mental, moral, professional), rights and duties; school organizations and government." Practical instruction in education thus included a thorough treatment of the structures of schools.[17]

While the main focus of school management and methods classes was practical or manual, they often included a mental or liberal-education element as well. Although the San Marcos and Florence catalogs did not explain exactly what understanding the "difficulties" of running a school meant or what constituted intelligence regarding the functioning of a public school, their course descriptions implied thoughtful, multidimensional consideration of the issues. The administration at Florence also emphasized that work in the normal department went far beyond imparting tricks of the trade: "The work of this department requires, as do all others, both time and earnest effort. It is not, as many have supposed, a cunningly arranged set of devices, plans or rules whereby one may, in a short while, easily become an expert teacher, regardless of previous preparation or development. Those who enter upon it with such an idea will be disappointed." Similarly, Clarence F. Carroll, principal at New Britain, Connecticut from 1881 to 1894, suggested that instruction in methods would prepare teachers to determine the best approach for a given class rather than provide them with a script to follow: "The term 'methods' as commonly understood, had become utterly offensive to all good teachers. Teachers known to possess skill are employed, and they are allowed the greatest freedom. Certain results are exacted but a *set method* has never been sought or tolerated in any department." While catalog descriptions of methods classes stressed technical elements of teaching, they often also stated that students would learn to investigate and evaluate different approaches to teaching a subject. Math methods at Ypsilanti, for example, included "Especial attention . . . to the bibliography of the subject, the Normal Library being quite complete in works of reference." Primary Methods at San Jose made use of "New and valuable educational journals and writings on experimental work in the primary grades, together with the recommendations of the 'Committee on Rural Schools' and of the Council of Education in the California Teachers' Association," while the San Marcos catalog stated that instruction in Primary Methods would aim "to appeal to the intelligence, to stimulate thoughtful inquiry, to direct systematic investigation, [and] to discover sound principles."[18]

Normal schools had long endeavored to "stimulate thoughtful inquiry" and "discover sound principles" of teaching and learning, but the Pestalozzian object approach had failed to provide a useful theoretical framework. Normals continued into the 1890s to offer instruction in object teaching: Oshkosh's Theory and Art of Teaching included "objects" and Oneonta, New York offered Methods of Giving Object Lessons. But the object method's underlying philosophy became more important than its practical application, as the Pestalozzian interest in the development of children's minds planted the seed of psychological studies at the normals.[19] Beginning in the 1870s, the developing field of educational psychology provided both a theoretical base for principles of teaching and substantive questions for ongoing inquiry into the learning process. Bridgewater's Principal Albert Boyden reflected:

> The teacher as an educator must know *what* the different mental powers are, the *order* of their development, and *how* they are called into right activity. He must know the *different kinds* of knowledge, the *order* of their acquisition, and the *method* of acquisition. The principles of education are derived from the study of the mind. The methods of teaching and training are determined by these principles.

And Whitewater's Principal Salisbury asserted, "practical and scientific study of the phenomena and development of the child-mind, [is] the true material of the teacher's art."[20] Acting on such convictions, state normal schools throughout the country offered instruction in psychology. As it matured and expanded at the normals in the 1870s, 1880s, 1890s, and 1900s, educational psychology enhanced practical instruction and fostered normalites' "head interest" in their future vocation by engaging them in intellectual discussions and investigations of the teaching and learning process.

After arising within philosophy of education, psychology established itself as a field of study at the normal schools during the 1870s and 1880s. Mental and Moral Philosophy at Bridgewater in the late 1860s focused increasingly on "the general development of the mental powers," while Philosophy of Education at Winona in 1870 focused on topics including: "Nervous mechanism. The senses. Sensation. Observation. Memory. Imagination. Reasoning." Senior-year work at the Michigan State Normal School in 1868–1869 included "Philosophy of Education based upon a knowledge of Psychological powers, analysis of such powers, Minute and careful analysis of the Mental acts concerned in each." Three years later, the Ypsilanti catalog no longer linked psychology to philosophy, as it explained that all courses of study covered "the elementary principles of Psychology," to which the Full English Course added "a thorough knowledge of Psychology, with its applications to the work of education in respect both to principals and methods." By the late 1870s, the principal at Westfield, Massachusetts, John Dickinson was writing a book entitled *Psychology for the Normal Schools*, Psychology was a separate class at Normal, Illinois, and seniors at San Jose studied "Laws of mental growth and development." In the 1880s, state normals from Castleton, Vermont, to Troy, Alabama, to Cedar Falls, Iowa included a psychology class in their teacher-education program, while Mental Philosophy at the normal schools in Pennsylvania and Oregon reflected psychology's fading connection to philosophy. Students at the African American state normal in Pine Bluff, Arkansas studied or wrote Essays in Mental Development, and at Emporia, Kansas in the late 1880s, Mental Science focused on "the facts of the student's own consciousness," including "the facts of observation" and "the order and laws of the development of the faculties." Albert Boyden's son Arthur, who also served as principal at Bridgewater, observed, "During the eighties the psychology became more analytical and specific."[21]

By the 1890s and 1900s, specific observation and analysis were indeed prominent in normal-school instruction in psychology, which was required at most state normals as a foundation for methods and further study of teaching and learning. The 1893 Ypsilanti catalog asserted that "instruction in Psychology must precede any such special instruction in methods of teaching" because the "truth is recognized that the art of teaching must be based upon the science of education, and that the science of education has its ultimate basis in the science of mind." Similarly, the 1898–1899 Florence catalog stated, "the Normal School realizing that education is the development, the drawing out of the mental powers, seeks to give teachers a knowledge of the nature of the mind, and the laws of its development, before they attempt to deal with the minds of the children." Again recognizing psychology as "a foundation for all educational theory and method," Florence's catalog the next year also emphasized that students themselves conducted research in the area: "The work is made as fully inductive as possible, students finding material for their work in observation, experiment and literature." The empirical nature of psychological studies at Oshkosh was also clear, as a catalog explained:

> The method adapted to develop clearness, interest and constructive power, will advantageously pursue the following plan: (*a*) Use texts to excite and direct introspective attitudes merely, not for the purpose of memorizing another's doctrine. (*b*) Observe mental activity critically and with effort at systematic record. This may be done through memory of experiences or by determinate seeking for manifestations in others. (*c*) The tentative construction of mental theory confirming or modifying existing ones, as the case may be. . . . The phase

of observation and making of records should extend over as much time as possible and not be limited to the brief time formally prescribed for the branch in our curriculum, since in this field, necessarily involving attendant queries and inferences, the mind is brought to self-examination most effectively.

At San Jose at the turn of the century, students read the work of William James and utilized the "splendid psychological library of the school." In addition, according to the catalog, "demonstrative and experimental work, as conducted in modern psychological laboratories" was "extensively employed to classify and emphasize mental processes," while each student was "expected to specialize along some line of children's development, and to embody results in a paper" reflecting "the nature of the investigation."[22]

Investigation into children's mental development led to two overlapping movements at the normal schools beginning in the 1890s: Herbartianism and child study. Based on the teachings of German philosopher and educator Johann Friedrich Herbart (1776–1841), Herbartianism systematized psychological theories into concrete approaches to teaching based on engaging and fostering children's interests. At Illinois State Normal University, Principal John W. Cook and three professors with German Ph.D.'s, Charles McMurry, Frank McMurry, and Charles DeGarmo, worked to spread Herbart's philosophy. In *How to Conduct a Recitation* and *Elements of General Method Based on Principles of Herbart*, published in the early 1890s, Charles McMurry spelled out the five formal steps that became the heart of scientific methods at American normal schools. In their methods classes, normalites in the 1890s and 1900s learned to plan and execute lessons through "preparation, presentation, association and comparison, generalization or abstraction, and practical application." According to historian Charles Harper, the "great contribution of the five formal steps was to focus attention on the possibility of a general method that would come from careful analysis of an act of thought. 'We are now prepared,' said Charles McMurry, 'to inquire into the mind's method of approach to any and all subjects.' "[23]

While Normal, Illinois was the epicenter of the movement, Herbartianism guided instruction and inquiry at state normal schools throughout the country. In New York, for example, Herbartian theories were popular at Oneonta, Potsdam, and even Oswego, once the home of object teaching. Margaret K. Smith graduated from Oswego in 1883, studied in Germany from 1885 to 1887, and then returned to her alma mater as "Teacher of Exact Philosophy as founded by Herbart." She, along with instructors Grant Karr and Mary E. Laing, both of whom had studied at Jena, Germany, researched and spread the new theories at Oswego. Like object teaching, Herbartianism ran the risk of becoming overly formulaic. Normal's Cook reflected, "That Herbartianism becomes dogmatic is the real danger." But, he asserted, "If we start with the determination that the child shall have a clear intellectual life all of the time and that he shall not be sacrificed to the gods of stupidity by immersing him in formal work, the whole trouble will end." Although the trouble did not end as normal schools at times oversimplified the five formal steps, Herbartianism helped to infuse teaching methods with an element of psychological theory. While these developments had a huge impact on practical preparation for teaching, they also encouraged further empirical research in educational psychology.[24]

Herbartianism and educational psychology in general centered on the child. Arthur Boyden at Bridgewater consequently observed that the normal schools of the 1890s and 1900s found "that the subject matter was not the center of its activities, it was the child waiting to be taught. The laws of child growth must be known, the varying capacities of children's minds must be perceived, and the equipment and instruction must be adjusted to these varying capacities." Initiated by psychologist G. Stanley Hall after he became president of Clark University in Worcester, Massachusetts in 1889, the child-study movement emphasized direct observation as the route to unraveling the laws of child growth. Working closely with Hall, principal E. Harlow Russell at the state normal in Worcester infused psychology classes with child-study methodology, requiring students to undertake extensive observations of the ordinary behavior of children and record their findings with great precision. While Hall was delighted to have the records compiled by class after class of Worcester students for analysis, Russell celebrated child study's ability to "awaken curiosity concerning the phenomena of child-nature, excite intelligent sympathy with children and contribute to skill in discipline and instruction."[25]

At many normal schools in addition to Worcester, child study played an important role in educating both the hand and the head of the future teacher. Hall was the graduation speaker at Willimantic, Connecticut in 1894, and Primary Methods at San Jose in 1900 required "child-study on the part of the students" because "a knowledge of children, their interests and capabilities, must be the basis of all true method work." At Ellensburg, after studying Elementary Psychology in their second year, fourth-year students studied Applied Psychology and were encouraged to observe children at school and at play. They were to note their own direct observations on white paper, their childhood memories on yellow paper, and information gained through reviewing the literature on green paper, after which all students' observations could be analyzed as a whole. After Will Monroe, a pioneer in child study, joined the Westfield faculty in 1896, he guided students as they observed and gathered data in the practice school, then wrote up their results in papers, some of which were published. While such research was usually part of general psychology classes, some normal schools offered separate classes in child study. Vermont's normal schools, as well as Ypsilanti and Emporia, all offered Child Study by the turn of the century; even President Cook at Normal introduced a class in the subject. Thus, "the child waiting to be taught" was a prominent subject of study at state normals.[26]

Educational studies at normal schools in the late nineteenth and early twentieth centuries extended beyond psychology and methods. Throughout this period, most state normals offered instruction in history and philosophy of education, usually in the advanced course, or third or fourth year. It was not unusual for the two subjects to appear in the title of a single class: in the late 1880s, course catalogs at Farmington, Maine and Tempe, Arizona listed History and Philosophy of Education, and in the 1890s, Fairmont, West Virginia used the same class title, while Brockport, New York offered Philosophy and History of Education. However, the most common title for a class covering this territory was simply History of Education, which appeared in catalogs from Providence to Pine Bluff and from Oshkosh to San Jose. As psychology broke away from philosophy, classes in the latter subject alone were rare, although Pennsylvania's normal schools offered Philosophy of Education in the 1890s and 1900s. Classes in educational history and philosophy rounded out students' liberal education in the field while also enhancing their practical training and emotional commitment to education.[27]

The most significant contribution of these classes was encouraging students to think liberally, or broadly, about the field of education. History of Education at Emporia, according to the 1889 catalog, focused on "the study of the educational systems and methods of ancient, medieval and modern nations." Through "careful inquiry into the lives and educational theories of all the great writers and teachers, together with a critical survey of the development of systems and methods that now find general acceptance, as well as full discussion of current educational questions," Emporia students were to expand their knowledge of past approaches to education and perhaps view contemporary issues in a new light. Similarly, History of Education at Florence in the late 1890s intended "to give the student a clear view of the general trend and progress of education; to help them reason from effects to causes; to aid them in finding the significant principles in great educational systems; to make some acquaintance with great educational reformers, and to appreciate in some measure the forces that are operative in our own educational era." Educational history at San Jose also shared the aim of broadening students' thinking about their future field of work; instruction in 1900 covered "the history of civilization . . . in its educational aspects, with special reference to the typical factors in the development of the social ideas which have been directive of the life, as well as the education, of nations, and with the express end of understanding the typical features of the existing system of educational practice and theory."[28]

History and philosophy classes often engaged students in a centuries-old dialogue among leaders and philosophers about the purpose and meaning of education. While Emporia began with "inquiry into the lives and educational theories of all the great writers and teachers," Florence acquainted students "with great educational reformers." San Jose sought in late 1870s and early 1880s to do the same through what catalogs called "Reading of educational works." The 1900 catalog went into more detail, explaining that the investigation of education in "the history of civilization" included "the educational theories of representative men of ancient and modern times." In the class, these theories were "studied comparatively, and as stages in the development and conscious appreciation of the basic principles and ideas of education." Listed as "aids" to the course were almost two dozen books, including

Socratic dialogues and works by Comenius, Locke, Rousseau, Mann, and Barnard, to be used as "sources of original investigation" on which students would report to one another. Class discussion, then, would be a continuation of ongoing debates about broad questions in education. Oneonta students in the early 1890s engaged in the same dialogue through writing graduation theses on historical developments in education or the philosophies of history's great educators, such as Froebel or Pestalozzi. *The Oneonta Herald* published the essays of the class of 1890. Meanwhile, instruction in history of education at Greensboro focused on educational reformers. Students read critically Rousseau's *Emile*, Pestalozzi's *Leonard and Gertrude*, and other such works in order to understand and analyze educational theories.[29]

In describing History of Education, the 1897–1898 Oshkosh catalog referred to "the consideration of educational theory in its development through racial ideas and exigencies" as "the philosophic aspects of history." Indeed, it was often difficult to distinguish where history ended and philosophy began in these normal-school classes. The few classes that normals offered strictly in philosophy of education furthered liberal exposure to the field. At Westfield in the 1870s, principal John Dickinson confronted students in Moral Philosophy with questions about the legal, ethical, and moral dimensions of relationships among children, parents, teachers, and the state, while Philosophy of Education at Emporia in the late 1880s examined "the nature, form, and limits of education; its physical, intellectual, and moral elements." Philosophical discussions were a popular approach to fostering a broader interest in and understanding of education.[30]

While focused mainly on liberal exploration, instruction in history and philosophy of education reflected normal schools' practical orientation and emotional commitment to teaching as well. The 1898–1899 Florence catalog explained how history was practical: "The successful teacher studies the experiments of teachers of the past, notes the successful ones and those that have proved failures. . . . Some of the so called 'new methods' are as old as Plato and Socrates and Aristotle." Teachers were thus responsible for understanding "the most careful and painstaking experiments of the past." The 1904–1905 San Marcos catalog stated simply, "In the study of the History of Education the controlling purpose will be to find, in the experience of the past, light for the future teacher's guidance." These institutions recognized the potential practical results of classes that encouraged students to think broadly about their future vocation.[31]

Also woven into descriptions of history and philosophy instruction was a thread of the "spirit of consecration" that had been so strong at the early normal schools. The 1900 San Jose catalog explained that through their investigations and discussions, students were "to discover something of the evolution of modern school systems and methods of instruction, and, by the grand and noble examples, the self-sacrificing labors of the biographical characters discussed, to encourage and inspire the young teacher to a more devoted effort in behalf of children." In other words, studying the great teachers of the past would inspire normalites to view teaching as a calling. A sense of consecration also arose from an almost millennial notion of progress in some normals' approaches to the history of education. Just as a Florence catalog mentioned the "general trend and progress of education," a sense of faith in progress infused descriptions of history of education at Ypsilanti. In 1880, the Elementary Course included the study of "Progressive development of improved methods of elementary teaching, illustrated by reference to the lives, labors, and principles of the great leaders in educational reform and progress." In 1893 in History of Education, students still studied "the great leaders." "In this way," the catalog explained, "the relation of the present to the past is clearly shown and the direction of real progress is discovered."[32] Studies in teaching at state normal schools in the 1870s, 1880s, 1890s, and 1900s maintained an emotional commitment to teaching as a calling. Unlike in earlier years, however, attention to the hand as well as the head overshadowed the focus on the heart. Instruction in teaching methods, school management, educational psychology, and history and philosophy of education, had a practical bent and also fostered mental engagement in the field of education. Observation and practice teaching in the model school would further strengthen normalites' heads, hearts, and, especially, hands.

Figure 4.1 Students practicing a model lesson at Geneseo State Normal School in New York, ca. 1900.

Credit: Livingston County Historian, Mount Morris, New York.

Observation and Practice Teaching

Classroom instruction in education was just one aspect of the state normals' formal teacher-training program; observation of experienced teachers and practice in conducting actual lessons were central components of normal education. In the mid-nineteenth century, most institutions had been less than successful at providing these experiences for normalites. By the 1870s, however, the model, practice, or training school was, in the words of normal-school historians, the "center and the core of school life" or the "body and soul of teacher training." The presence of the model school reminded students of the seriousness of their undertaking. Lottie Matthis, who graduated from San Jose in 1884, remembered, "From the beginning of our Normal course much attention had been given to methods of teaching; and our attentive faculties were stimulated by the well known fact that the time would come when we should be given an opportunity for putting theory into practice, as pupil teachers in the Training Department." While the impending pupil–teacher experience inspired whole-hearted commitment, model schools fostered professionalism mainly through knowledge gained by practical experience. The 1886–1887 Florence catalog explained,

> The training of teachers would be very incomplete, if it should stop with academic and professional work. As a Medical College must have its hospital and daily clinic, so must a training school for teachers provide them an opportunity for testing their professional knowledge. This is offered in the Model School in connection with the Normal College. Here the students are assigned regular classes, and are required to exemplify the most approved methods of instruction and government. . . . Their work is neither experiment, nor is it observation, but actual teaching, just as in the hospital the practitioners give actual medical or surgical treatment to their patients.

Similarly, the 1901–1902 Tempe catalog noted that "to enable one to arrive at a thorough understanding of educational theories, an opportunity for observing their practical application, and for assisting in it, is essential. To supply the opportunity a training school is provided." Matthis remembered that at San Jose, when "the news came at last that Senior A was to go down stairs" to teach in the model school, "the stoutest heart quaked, for well we knew how much depended upon this test of our ability in school management." Observation and practice teaching provided the practical experience that was essential to the normal schools' approach to professional education.[33]

For Matthis and other normalites in the late nineteenth and early twentieth centuries, the training school was a ubiquitous presence in daily life, as it often occupied a central location on campus. At Framingham, Massachusetts, an enlargement of Normal Hall in 1870 made room for a one-room model school, which, a student remembered, "was fitted up like other town school-rooms, with desks, blackboards, etc. and a platform and teacher's table in front." It grew during 1880s and 1890s to occupy most of the first floor of the new May Hall. In addition to small recitation rooms, the practice school, as it came to be known, had a principal's office and large rooms for the primary, intermediate, and grammar departments. The town of Edinboro, Pennsylvania by the 1870s located a public elementary school at the local state normal. By the late 1880s, this practice school occupied 12 classrooms in 2 buildings. When the town built a separate school building a decade later, the model school shrank but was still a presence on the lower floor of Normal Hall. Upon the opening of the state normal at Troy, Alabama in 1887, the catalog announced that a local public school "has been placed under the same management as the Normal School, that it may be used as a *practice school*. . . . The practice teaching in this school will form an important part of our training." As at Edinboro, this public/training school grew until the public school moved away from the normal, at which point the normal school organized its own model school. Other southern state normals, such as Florence in the 1870s and early 1880s and the African American state normal in Pine Bluff, Arkansas throughout most of the late nineteenth century, had preparatory departments that doubled as training schools. By the late nineteenth century, some normal schools were able to build separate campus buildings to house their training schools. Completed in 1893, San Jose's was connected to the main building by an enclosed corridor that the normal students dubbed the "Bridge of Sighs."[34]

Even when a model school was not physically located on a normal-school campus for the entire period from the 1870s through the 1900s, observation and practice teaching were prominent in normalites' daily lives. The state normals in Castleton, Johnson, and Randolph, Vermont used the villages' public schools as model schools throughout these decades; even though the model schools were not on campus, the small size of the villages and close proximity of the buildings assured easy access. Principal Albert Boyden described a similar arrangement at Bridgewater, Massachusetts in 1880:

> An arrangement has been made with the Town of Bridgewater, by which one of the schools of the town is to become a school of observation and practice for the normal school. This arrangement will furnish an opportunity to the members of the senior class to observe a good school with reference to organization, its course of studies, its methods of teaching, and its discipline. In addition to this, the class will have an opportunity to put to the test under the direction of skilled teachers, the principles they have learned in the normal school.

After utilizing the town school throughout the 1880s, Boyden and his staff delighted in the establishment of an on-campus training school in 1891.[35]

Policy decisions regarding model schools were often driven by the availability of resources. For most of the 1870s, Westfield students observed in the local Green District School, but this arrangement ended in 1879 when financial cutbacks prohibited the normal from paying the required fee to the school district. It was not until the normal was able in 1892 to build a new building, the first floor of which housed a training school, that Westfield students again had access to common-school classes. The Normal School of Arizona sent students to Tempe's public schools for observation and practice teaching until it was able in 1895 to open its campus model school. Milwaukee, Wisconsin opened with a model school in 1885, but turned also to the local schools in the early 1890s because the model school was unable to grow apace with increasing normal-school enrollments. Greeley, Colorado opened in 1891 without the resources for an on-campus model school, but teacher educators there were so dissatisfied with sending students to the local schools that they found the means to open their training school within only a year. Other normal schools utilized both the local public schools and the campus model school more intentionally and successfully. When the state normal in Buffalo, New York opened in 1871, it had an active practice school; "In addition to this," a school bulletin announced, "the schools of the city will be open at proper times and under proper regulations as 'Schools of Observation' and as such will be visited by normal pupils." At New Britain, beginning in the 1880s, students observed and taught first in the campus training school, and then in a public school in New Britain or a neighboring town; Adelaide Pender '88 remembered, "After a few weeks of training in the New Britain Model School we were sent either to Bristol or Newington."[36]

In campus model schools and local public schools, Pender and other normalites apprenticed with experienced teachers called supervising or, most commonly, critic teachers because they, as the Buffalo bulletin explained, were "critics of teaching" as well as the "permanent teachers" of training-school classes. Critic teachers in campus model schools were members of the normal-school faculty; some worked exclusively in the training school while others also taught methods or other pedagogy classes. Arrangements for critic teachers in the public schools varied; in the graded city school that was Oswego's practice school, these instructors were employed by the school board but selected by and responsible to the normal school. Critic teachers taught model lessons, reviewed student teachers' lesson plans, and supervised student teaching. In other words, they facilitated experiential learning, which came through both observation and practice.[37]

Students at most normal schools spent a significant amount of time observing the teaching of model-school pupils by critic teachers as well as advanced normalites. The 1898–1899 Florence catalog described a rather unstructured approach to observation: "The practice rooms are open at all hours to students of all grades, thus affording each opportunities to observe the Methods of expert teachers." Most normal schools, however, included observations at specific points in the teacher-training course. Some teacher-training programs required observations from the beginning, while others saved them until later in the course of study. In the 1880s and early 1890s, seniors at the Rhode Island Normal School spent a lot of time in Providence's schools, both observing and teaching. After the institution established its own model school in 1893, principal Willliam E. Wilson explained that not only would "practice in the training school . . . run along parallel with the studies of the last year," but "all students will observe the work in all grades in the model rooms." At this point, classroom observations became central to junior-year studies at Providence. At the turn of the century, Oneonta's seniors spent ten weeks observing critic teachers' lessons for two periods each day, and then taught one or two classes while also still observing one model class each day.[38]

Observation in the model schools was hardly a passive process; students actively evaluated what they witnessed, and discussions of their evaluations enhanced classroom studies and the teaching of the observers as well as the observed. To Boyden at Bridgewater, the primary purpose of the model school was "to train the normal students in observing and teaching children." Thus, observations enhanced instruction in child study and child-study methodology guided observations: "The normal students have a definite course in practical child study, under the careful direction of Miss Hicks, the principal of the Model School, and make reports on their study," focusing on "the school as a whole and the details of school work in the different grades," in addition to "the individual child." At Greeley, juniors worked in small groups with critic teachers to observe and discuss one or two recitations each week. The 1897–1898 Oshkosh catalog elaborated on a similar procedure: after students observed in the model school or a nearby city school, each wrote a long "account of the work done, setting forth the excellence and defects of class instruction and general school management." In discussions of their accounts, students answered "such questions on the reported lesson as may be asked by the class or director," justifying "criticisms, whether favorable or the reverse." Finally, Matthis remembered her classmates' enthusiasm for observing her teaching at San Jose: "Of critics there was no lack, for in each class room was stationed an able body of Senior A's and B's, each individual armed with a pencil and open note book, all ready for an attack upon order, neatness, manner, or discipline."[39]

Throughout this period, practice teaching was the capstone of the formal teacher-training program, allowing students to grow professionally by nourishing their craft. During rare periods when an institution, such as Westfield between 1879 and 1892, did not have a campus training school or access to a local public school, students' only practice came through teaching mock lessons to their classmates. At Pine Bluff, where scarce resources did not allow for critic teachers, advanced students simply assisted the principal with his teaching in both the preparatory and normal departments. Often maligned in earlier decades, "mutual instruction" was more likely in these years to be a supplement to, rather than a substitute for, practice in teaching actual children. When the Iowa State Normal School had difficulty attracting pupils to its training school because it was a bit distant from the center of Cedar Falls, other normalites and preparatory-department students served as additional pupils for student teachers. Principal James C. Gilchrist reported in 1885, "we do not depend on the model school alone for the opportunity to train students how to teach. Practice classes are formed from the Normal students themselves, giving very excellent results. The regular students of the Normal are employed as occasion may offer to afford training in the advanced parts of the common school studies." When Cedar Falls' model school was closed from 1886 to 1892, this system allowed the normal to continue providing practice-teaching experience. At Bridgewater during the 1880s, mutual instruction supplemented practice teaching in the local public school. Teaching her classmates had a significant impact on Sarah Dixon '85, who reflected, "Few can estimate the value of this experience for after life. To mount the platform and brace feet, mind and heart for this difficult task called continually for a large measure of self-control and personal power."[40]

Working with real school children was equally, if not more, valuable. Planning, executing, and reflecting upon actual lessons was challenging and instructive. The Florence administration explained that student teachers were "required to exemplify the most approved methods of instruction and government. They are not allowed to imitate blindly what they may have seen, nor are they allowed to follow out without thought the methods they have been taught." Critic teachers worked with practice teachers at each step to be sure their work was thoughtful and well-informed. The 1898–1899 Florence catalog described the process: "Each [student] teacher has a class for whose teaching he is held responsible. He must plan the work for each day's recitation, must submit the plan to a competent critic teacher and must teach the class under the supervision of this critic." Similarly, a training school connected with the Rhode Island Normal School reported in the late 1900s:

> Each pupil teacher has kept daily lesson plans. Every evening, before leaving the school, the day's work has been criticised and the next day's lessons discussed and written out carefully in the plan book. In this way the work has been very definite . . . All teaching experiences have been based on the Herbartian steps and from observation and suggestions these lessons have been carefully prepared and skillfully taught. The aim has been to impress upon the minds of the teachers that careful, pedagogical preparation of the lesson is a necessity for skillful teaching.

While these evening meetings encouraged some self-reflection, student teachers at Worcester, Massachusetts and Normal, Illinois were required to document their reflections in diaries of their teaching experiences. At Normal in the early 1870s, president Richard Edwards stipulated that students' diaries should describe each day's work, including "the difficulties that have been encountered, the methods by which they have been surmounted, the various expedients resorted to for securing an interest in the lesson, and the modes of reviewing and thus rendering permanent the acquisitions of the pupils." These diaries, according to Edwards, were "useful in many ways. They turn the thoughts of the young teacher to his methods and details of his teaching, and prevent him from doing any thing thoughtlessly."[41]

Learning from student teaching was of course not limited to self-reflection, as critic teachers and fellow normalites scrutinized practice teachers' lessons. New Britain's Adelaide Pender described the intensity of being observed: "As I stumbled through that lesson, my classmates sat around the edges of the room taking notes fiendishly (I thought) and frantically. I felt like a small fly being drawn into a spider's web." While this experience was anxiety-inducing for Pender, such processes allowed fellow students and the critic teacher to provide each student teacher with extensive and thorough feedback on her or his performance. At Florence, after each teaching experience the student met with the critic teacher; at that meeting, the catalog explained, "his work is criticized, and the particulars in which he succeeded and those in which he failed are pointed out." Each student teacher at San Jose was "constantly guided, aided, and encouraged in his work by wholesome advice, direction, criticism, and commendation." Troy's director of the model school evaluated student teaching on a form that listed the following categories: "Power to control," "Power to interest," "Skill in preparation of lesson," "Skill in questioning," "Skill in illustrating and explaining," "Judgment in assigning lessons," "Voice," "Manner in classroom," and "Care of blackboard." The feedback Pender received was less bureaucratic but equally thorough: "criticism concerned my method, my preparation, manner, personal appearance, anything. EVERYTHING! . . . I gave a lesson for that superior teacher, Miss Page, one day, and she pulled me to pieces everywhere."[42]

Analysis of teaching experiences continued in seminar discussions in which students digested feedback on their own performances and learned from others' mistakes and successes. At Normal, Edwards and the practice teachers met twice each week to read from and discuss their diaries, and Saturday meetings at Worcester allowed student teachers to go over the week's events with the faculty. Discussions of student teaching at Edinboro, according to the 1881 catalog, were "held twice weekly or oftener, at which the excellencies and defects are pointed out that all may profit thereby." At Oshkosh after the turn of the century, practice teachers, critic teachers, and model-school supervisors held a "general meeting" late every Wednesday afternoon to discuss the teaching experience. After investigating New York's normal schools in the late 1870s, a state legislative committee reported that such discussions were thorough and instructive:

> usually near the close of the day, these pupils who have been acting as teachers, as well as those who have been "observing," meet, and one of the normal critics who has been watching their work, reviews it, pointing out and explaining not only the errors, but the excellencies observed. In addition the pupil-teachers question and are questioned as to what is well or ill done, and suggestions are made by way of improvement. The committee were very deeply impressed with the great value of this exercise . . . excellencies are ascertained and developed, and professional zeal is awakened.[43]

As they reflected in seminar discussions upon their student-teaching experiences, normalites drew upon their extensive involvement in the training school. Whether on campus or off, it had been a large presence in their normal-school education. They had observed the teaching of more advanced students and critic teachers, worked closely with critic teachers to plan their own lessons, and finally executed those lessons under the watchful eyes of other students and teachers. These discussions were also an opportunity to integrate the experiential knowledge gained in the model school with the technical and liberal knowledge acquired in classes. The normal schools' multifaceted approach to teacher preparation justified Florence's boast, "We have not graduated a student who has not exemplified an *aptness*" for teaching.[44] As the formal teacher-training program of state normal schools assured graduates' aptness for their field, extracurricular activities further awakened their "professional zeal."

STUDENT ACTIVITIES

Student-initiated extracurricular activities complemented classroom instruction and model-school experiential learning, as normalites in the late nineteenth and early twentieth centuries undertook all sorts of group efforts to enhance their preparation for teaching. Even class cheers at Ellensburg, Washington reflected the pervasiveness of the "ozone of teaching"; senior "yells" included:

> Plato, Rousseau, Abelard!
> Harris, Parker, Mann, Barnard!
> Pestalozzi, Froebel, Rein,
> W.S.N.S. Ninety Nine!

and

> Methods, Teaching, Observation!
> Pedagogy, Education!
> Profs and Pedagogues we're numbered!
> W.S.N.S. Nineteen Hundred!

As explained in chapter 3, normal-school students involved themselves extensively in academic clubs, literary societies, and publications. The most prominent focus and outcome of these intellectual activities was the acquisition of prestigious knowledge and polished writing and speech, or cultural capital. When normalites explained the benefits of society membership, they drew connections between cultural refinement and teaching skill. Excellent examples of this appeared in a yearbook published in 1905 by the students at Geneseo, New York. The page devoted to the all-female Clionian Society began by describing the group as "a literary organization which aims to give to its members a broader culture than is afforded by class room exercises and routine," and ended by stating, "The fraternity spirit among our chapters is a very helpful one and our girls, wherever they have gone, have been successful teachers and a credit, not only to the Clionian Fraternity, but to the teaching profession." In his description of "The Benefits of Society" membership in general, student Andrew Gilman wrote:

> the freedom and ease of expression and the improved compositions of the society member fully justifies the belief that this work is essential to a successful

> school. . . . Many of our students come from localities where they have never had the advantages of social life. They feel awkward and out of place in refined society. They lack the polished and graceful manner of the society member. To such this training is of great value, for what success can one expect as a teacher, or in any profession, if he is not able to meet others with ease.[45]

These descriptions suggest that normalites had a broad view of the abilities necessary for good teaching. While their extracurricular work cultivated these general skills, it also examined teaching and education. A few student journals focused almost exclusively on education and some clubs formed around educational issues; in addition, the majority of clubs, literary societies, and publications tackled educational issues at least occasionally. Student-initiated activities not only enhanced the technical or manual function of the state normal schools, but also allowed students to explore liberal or mental issues, and to identify emotionally with the education profession.

Normalites throughout the country used student activities to broaden and deepen their knowledge of methods and basic teaching skills. In the North, the student editors of *The Normal Advance* at Oshkosh, Wisconsin published various elementary-level lesson plans and supplementary reading lists, written either by students or instructors. "Nature Lessons," by Stella H. Jillson in 1894 described helpful activities for fourth graders and articles on "Literature and Supplementary Reading for the Schools" and "Memory Work in Language Training" in 1898 contained various suggestions for lesson planning in reading and language arts. In March 1902, the editors announced a "New Addition to Paper": "The Staff purposes [*sic*] to make the Normal Advance more of an educational paper than it has been heretofore. . . . beginning with this number, several pages of each issue will hereafter be devoted to the discussion of some educational subject. The articles will be contributed by members of the faculty." That issue included "Spring Nature Study," by H. N. Goddard, and subsequent editions of the *Advance* featured "Suggestive Lesson-Plan for 'Paradise Lost' " by Ellen F. P. Peake, "Rapidity in Calculation," an article about teaching math by Emily F. Webster, and "A Bibliography for Teachers of Drawing" by Henry Emmett. In the East and West, Castleton, Vermont's *The Normal Student* and San Jose, California's *The Normal Index* and *The Normal Pennant* published similar articles, with titles such as "The Value of Nature Study" and "Literature in the First Four Grades." In 1885, *The Normal Index* reprinted pieces from *The School Journal*, and in the early 1900s under editors-in-chief Daisy P. Rudd and Jennie Jones, *The Normal Student* included a "Practice School Department" that published exemplary work by elementary students and lesson outlines by student teachers.[46]

In the South, the literary societies at Florence, Alabama in the late 1880s launched *The Normal Gem*, "A Monthly Journal, Devoted to the Interests of Education in Alabama." The April 1, 1889 issue included short, practical articles on "Methods of Recitations" and "Objects of Recitations." At a 1902 meeting of Florence's all-female Dixie Society, the affirmative side won the debate on "Resolved that Mother Goose Rymes [*sic*] should be taught in the school room," as members of the all-male Dialectical and Lafayette societies observed. At Southwest Texas State Normal School in San Marcos, the all-female Comenian Society reported in the 1905 yearbook that it worked to "cultivate some of the great love which Comenius had for little children." That year, it focused on "The Child in Literature and Art," explaining, "we are not a social club, but a band of teachers with a purpose in view in belonging to this Society. . . . to prepare ourselves a little better (realizing that every little counts), for the great field of work lying out before us." Six years later, the yearbook reported:

> The purpose of the Comenian Society during the session of 1910–'11 has been to give to its members helpful suggestions for making the children's hour in the school room pleasant and profitable. . . . Our programs have been largely devoted to story telling and dramatization of stories that all children should know and love. In our choice of stories, we have been careful to select those which are not only interesting but profitable to children. . . . Each girl whose privilege it has been to tell a story assumed the role of teacher while the other members of the society became an enthusiastic class of young children. Among the stories which have been most skillfully presented were "The Three Bears," "Life of Robert E. Lee," and "The Miraculous Pitcher."

Activities such as these reinforced and deepened normalites' studies of teaching methods.[47]

Extracurricular work enabled students to add dimension to other elements of their technical education as well. *The Normal Index* at San Jose published articles on everything from "Moral Training and How Far the Teacher is Responsible" to "Manual Training." The student author of the latter outlined arguments in favor of manual-training instruction, including "eye training," character formation, and the development of powers of observation through clay modeling. An 1886 article on California's schools that began, "The public school system of California has attained a high standard of excellence," extended students' understanding of school structures while stoking their state pride. Published in February, 1890, "The Presentative and Representative Faculties of the Human Mind" declared, "It is highly important that we, as teachers, should have some insight into the workings of the human mind." While this article and others like it reinforced and extended what students were learning in educational psychology, students at Ypsilanti, Michigan formed the Child Study Club to explore further this important subfield. Similarly, students at Kalamazoo, Michigan founded the Rural Sociology Seminar as a forum for further study of issues in rural education. The founding members stated their intention to explore "the means by which the rural school can contribute to the general progress of the community in which it is situated." At the February 1905 meeting of the club, students read papers on "The Early History of Farm Organizations," "Biographical Sketch of Justin S. Morrill," and "The Relation between Nature Study and Elementary Agriculture."[48]

Self-education also took students beyond technical aspects of their training. As discussions in the Child Study Club and Rural Sociology Seminar added depth to members' practical knowledge, they at the same time fostered liberal thinking about these areas of education; understanding the history of farm organizations and the author of the Land Grant Act helped future rural-school teachers to situate and evaluate their work in a larger context. In other clubs, society activities, and publications, students further enhanced the liberal side of their education, especially in the history and philosophy of education. At Oshkosh, for example, students taking History of Education in the fall of 1902 formed their own "pedagogical seminar," which met on Thursday afternoons to pursue "a special study of schools in the United States." In New York, Geneseo's *The Normalian* published an article on Horace Mann, and Plattsburgh's *The Cardinal* quoted Pestalozzi and Froebel. In 1891, San Jose's *The Normal Index* announced, "the last descendant of [Johann] Henry Pestalozzi, Karl Pestalozzi died a few weeks ago in Zurich. . . . although the blood descendants of this greatest of all modern reformers are dead, his educational children will live to the end of time. No name is held in more affectionate remembrance than that of Henry Pestalozzi." As students explored further the work of educators and philosophers introduced in class, they felt strong connections with their ideas.[49]

Students used their activities to explore and refine their opinions on all sorts of educational topics that were probably mentioned but not discussed as philosophical issues in class. Many debates and articles focused on the relative merits and purposes of various fields of study. When Florence's Lafayette Society in 1881 debated, "Resolved that the practical part of an education is more beneficial to society than the Literary part," the society's president was more convinced of the "literary" side of the argument and thus decided for the negative team. At Geneseo, a Philalethean Society debate in 1892 had the opposite outcome; the affirmative side convinced the judges that, "the study of Latin and Greek does not repay the time spent on them." The sister society to Philalethean, Agonian debated, "Resolved that higher education is not of practical value to business men." And an 1889 article in San Jose's *The Normal Index* posed the question, "Does a High School Education Fit One for the Practical Duties of Life?" and answered, "Since High Schools give an education one-sided and far above the needs of most people . . . we can say surely, that they do not."[50]

Other debates and articles covered classroom procedures and district and state regulation. At Oshkosh, in 1879 the Ladies Literary Society debated, "Resolved, that the marking system is detrimental to the mental growth of students." Two and a half decades later, the year-end Lyceum–Philakean Debate tackled the resolution "That a uniform system of text-books, to be selected by a text-book commission, should be used in the common schools of the State." *The Normal Advance* reported:

> The debate was opened by Ellen McDonald, who said that the present plan was not a system, that it did not provide for competent judges in selecting text-books.

> She also argued that in order for County Superintendents to do their best work a uniform system of text-books was necessary. Adolph Ruehl said that Wisconsin should adopt a system which is found efficient in other States. He then proposed such a general plan.

Compulsory education was a very popular debate topic. In 1872, the question for "Inter-Society Contests" at Normal, Illinois was, "*Resolved*, That it would not be wise for the State of Illinois to pass a law compelling all persons between the ages of seven and sixteen years, not otherwise well instructed, to attend school for at least four months each year, or for an equivalent amount of time." Ten years later, Lafayette at Florence debated whether "education in a republic should be compulsory," and in 1891 when Geneseo's Philalethean debated whether "a compulsory educational law should be passed in the State of New York and enforced by state authorities," the team representing the negative side was victorious. Finally, in 1906, the affirmative side convinced the judges at San Marcos "That Texas should have compulsory school attendance laws." As normalites debated educational policy and the purposes of instruction, they familiarized themselves with issues that were tangential to their formal studies yet important to their future occupation.[51]

Another issue unlikely to be addressed in the formal normal-school curriculum but very present in the schools and society was gender. Not surprisingly, student societies and publications considered all sorts of philosophical questions about the relative treatment of the sexes in education. The men in the Lafayette Society revealed themselves to be progressive thinkers when the negative side won an 1880 debate on "Resolve [*sic*] that the two sexes should not be educated together" and an 1882 debate on whether "the Mental capacities of the sexes are equal" was "decided in favor of the affirmative." Florence's *The Normal Gem* in 1891 ran a first-page article entitled, "Should Boys and Girls Receive Equal Advantages?" The author, "A Member of the Third Year," presented the argument that they should "receive the same mental training," in part because "There is so much more real pleasure in this life to an educated person." The author's opinion was clear in the title of a similar story in San Jose's *The Normal Index*: "Why Women Should be Educated." In the mid-1890s, Geneseo's Agonian debated whether "women make better school commissioners than men" and "men make better teachers than women." The school-wide debate topic at Oshkosh in the fall of 1902 was, "Resolved, that co-education as it now exists in the secondary schools and colleges is more desirable than any education which might be obtained in separate schools for the sexes." The women who argued each side of the question must have reflected upon their own experiences at the normal school.[52]

Students' considerations of educational topics also led them to broad philosophical questions about the importance of education for, and impact of education on, the individual and society at large. Agonian at Geneseo debated whether "Nature has a greater influence than Education on the formation of character," and the Young Men's Normal Debate Society at

San Jose squared off against the local high-school team over "*Resolved*: That a man's success in life depends more upon his education than upon his natural ability." Arguing the negative side, the San Jose team was victorious. Rather than natural ability, the importance of wealth in relation to education was in question when Florence's Lafayette held a debate on "Resolve [*sic*] that money has a greater influence on man than education." The Ladies' Literary Society at Oshkosh, meanwhile, tackled religion: a meeting in April, 1879 featured a debate on, "Resolved That the school teacher exerts a greater influence for good than the minister," and a meeting the following November included a debate on, "Resolved that Education has done more to civilize the world than religion." Two decades later, Sarah C. Walsh wrote an article on "Social Significance of Education" for *The Normal Advance*. Through discussions of these rather intangible issues, normalites encouraged one another to define their own educational philosophies.[53]

As their self-initiated activities took students' liberal education far beyond the formal curriculum, they viewed their field through a wide-angle lens. Their investigations even included schooling in other countries and education outside the realm of public schooling in the United States. Many publications had sections similar to *The Normal Advance*'s "Educational Notes," which included short quotations or facts, many of which were about other countries. For example, this section in the November–December 1896 issue stated, "Russia has promised to establish a public school system for Siberia"; and "Prussian rural schools have on an average forty-two weeks of school. Rural schools in the United States average not quite sixteen weeks." In 1890, *The Normal Index* reported on a lecture at the normal on "Education in China" by the pastor of the First Presbyterian Church in San Jose; the article concluded, "Mr. Minton told much more that is intensely interesting, as well as instructive, about these comparatively unknown people." In 1906, *Echoes from the Geneseo Normal* familiarized readers with the work of "Geneseo Teachers in the Orient." Two years later, *The Normal Advance* reported on "A New School" in Switzerland, which offered, among other things, manual and outdoor activities in afternoon. The author commented that the school "shows an idea of education rather different from what we ordinarily think of as the only kind. . . . We should welcome any suggestion which will help us to become conscious of the many problems involved in teaching and to know something of what is being done in places outside of our own school in the effort to settle some of them." Looking "outside of our own school" also meant studying other approaches to education in the United States. Students Mary Morgan and Asa Royce entered Oshkosh's Oratorical Contests in 1897 with speeches on Booker T. Washington, and senior E. A. Loew entertained the Phoenix Society in 1901 with a lecture on the educational system of the Jesuits. A decade later, Florence's Dixie Club studied Jane Addams's work at Hull House in Chicago.[54]

This wide-ranging examination of education also included political issues, as students used their activities to follow developments in state governance of public education and the activities of teachers' associations. Student publications discussed education bills under consideration in their state legislatures. For example, *The Normal Index* in February, 1889 included a detailed editorial about upcoming legislative decisions on education. The article explained, "The present session of our State Legislature bids fair to be a memorable one in the annals of educational reform, there being now under consideration several important measures tending in that direction," and then urged the legislature to "strengthen and render fully operative that clause in the school law which prohibits the dismissal of teachers in city schools without cause, duly proved" and pass "the bill aiming to make the High School a part of the state educational system." The editorial also supported the "more far-reaching" proposal of "giving women the privilege of voting at all municipal elections, and for all school officers, city, county, and state," asking rhetorically, "Why should this privilege not be given?" The *Index* staff clearly was well-informed on educational politics in California, and invited other students to pay attention as well. Students at Oshkosh weighed in on the politics surrounding school-superintendent positions in the early 1900s: Philakean debated Lyceum on "the present method of electing the county superintendent of schools" and the Oratorical Association debated "the Proposed Amendment to the Constitution of the State of Wisconsin, relating to the Superintendent of Public Instruction." In a Faculty Prize Contest Oration published in *The Normal Advance*, B. C. Streeter argued for government interference in education, stating "Popular education is one of the strongest agents of a wise paternalism." In a similar vein, when *The Oneontan* at Oneonta, New York reported on Joseph Mayer Rice's investigations of urban schools, the editorial staff called for action: "It is to be hoped that his sharp criticism will sting our citizens out of their self-satisfied indifference to the evils of the present system."[55]

An interest in teacher associations and activism accompanied political awareness. Student journalists consistently covered the activities of teachers' associations in their states. At San Jose, *The Normal Index* reported on meetings of the State Teachers' Association and a county institute, and *The Normal Pennant* included "Gems From the N.E.A. Speakers in Los Angeles." Plattsburgh's *The Cardinal* likewise reported on a meeting of the National Education Association in San Francisco. A similar story in *The Normal Advance* commented, "At the December meeting of the Wisconsin Teachers' Association, our Normal was well represented. Fifteen members of the faculty, many alumni, and several under-graduate students were present. At one of the hotels a large table was filled by Oshkosh students; it made one think of one of our class recitations." Such reports brought normal students into the fold of professional education associations. As teachers unions began to form by the turn of the century, student publications followed them as well. In October, 1900, Geneseo's *The Normalian* included a note on the organization of "A National Teachers' Union." In March, 1902, *The Normal Advance* carried a three-paragraph report on a school-wide talk on "the struggle which started in Chicago between the teachers and the Board of Education for an increase in salary." The article described the Chicago Teachers Federation and the efforts of Margaret Haley and Catherine Goggin, and observed, "The schools of Chicago were being deprived of much needed funds." While normalites were not necessarily educational activists, they certainly made themselves aware of union activity, the work of other teachers' associations, and educational politics.[56]

As they strengthened their technical and liberal preparation for teaching through self-initiated activities, normal students also enhanced their emotional commitment to the field. The "spirit of consecration" was clearly present in student publications and society meetings. *The Normal Index* declared, "The sphere of a teacher's usefulness and influence is not bounded by the walls of the school room," while student editors of *The Crucible* at Greeley, Colorado referred to fellow students as "co-workers in the cause of education" and the *Normal Outlook* at Ellensburg, Washington dedicated itself "to the cause of true education." The first volume of Castleton's *The Normal Student* referred reverently to "the spirit of the teacher" and "the character of the teacher," while the first edition of San Marcos' *The Pedagogue* declared on behalf of graduating seniors, "I intend to represent the best there is in a teacher's life and work, and bring to my profession the nobleness and dignity of a high calling. . . . I shall remember my training and stick to my work and try to bring my little part of the world a step closer to knowledge and progress." At Oshkosh, it was in a 1901 speech for the Phoenix Society entitled "Need of a Professional Spirit," that Professor Lough declared, "the one who teaches because he loves, and has a heart and head interest in his vocation, is a professional." A decade and a half earlier, the same society had asserted, "We have good reason to believe that at no distant day teaching will be recognized as a profession." While the Phoenix society and other students may have been a bit too optimistic, their self-initiated investigations of teaching methods and the ramifications of various educational philosophies and policies insured that emotional commitment underscored rather than overpowered the technical and liberal aspects of their preparation for teaching.[57]

As Principal Boyden observed at Bridgewater, an "ozone of teaching" permeated state normal schools in the late nineteenth and early twentieth centuries. The formal education curriculum, training school, and student activities provided in-depth technical and experiential training, a broad liberal background in the field, and emotional inspiration. This approach incorporated each student's hand, head, and heart into the preparation process. One of Boyden's students, Sarah Dixon recalled that readying all of these aspects of herself to conduct a practice lesson "called continually for a large measure of self-control and personal power," or professionalism. In another decade and a different state, Carrie Houghston included the following stanza in her "Psalm of Normal Life" at San Marcos:

> For in the teachers' field of battle,
> In the busy work of life,
> We of hardships must not prattle,
> But be *Normals* in the strife![58]

Houghston and Dixon's normal-school education enabled them to view themselves as professionals during a time when society was reluctant to grant this status to women (and people from lower social classes). As the "ozone of teaching" implicitly taught normalites that they were professionals, the state normal schools' overall climate, or public sphere, also subtly challenged views of class and gender in nineteenth-century middle-class society.

5

"Noble" Men and "Not Necessarily Bloomer Women": The Public Sphere, Gender Attitudes, and Life Choices

The men and women who attended state normal schools between the 1870s and the 1900s shared an extraordinary experience. As these institutions accommodated students' nontraditional backgrounds, normalites enjoyed a lively intellectual life and the opportunity to develop a professional spirit. Another important dimension of normal schools was the public sphere that extended beyond the formal curriculum and academic and pre-professional student activities; students became involved in the community life of the local towns and created their own social sphere on campus. In the process, they socialized themselves for participation in middle-class society through observation, emulation, and the formation of extensive webs of personal connections, or social capital.[1]

Like intellectual life and preparation for teaching, the normals' public sphere initiated students into middle-class life while also violating one of its central tenets, separate gender spheres. Normal-school culture seemed to adhere to the rural and lower-class view—necessitated by practical considerations—that gender roles were somewhat flexible. Among the middle and upper classes, the Victorian notion that the public realm was men's domain and the private, domestic realm was women's domain, remained strong into the late nineteenth century. Even as it became more acceptable by the turn of the twentieth century for women to be visible in public, the socially proper approach was through separate women's organizations, which grew and multiplied in urban areas and on college campuses. College women were excited to socialize and participate in campus life, but they did so mainly with other women as their social outlook remained quite conventional.[2] At state normal schools, female and male students interacted freely and shared leadership in the public sphere, implicitly challenging Victorian gender norms for both sexes. As a result, many women students formed and acted upon a fundamental belief in autonomy for women, although they usually stopped short of identifying themselves as feminists. One instructor captured this ideology when she described the "new woman" as "a college woman, but not necessarily a bloomer woman, for she is distinguished not by what she wears, but by what she wants and what she does."[3] Sharing the public sphere with these women, men students became aware of the uniqueness of their experience as well as their own gender privilege in the larger society. In a poem entitled, " 'The Noble Nineteen' or the Boys of S.N.S.," one male student acknowledged their rareness while suggesting that he and his male classmates did not wish to capitalize on their privileged status:

> And we will be your friends, true;
> The best and noblest kind.
> The Normal boys, although few,
> Their like is hard to find.[4]

Although he did not necessarily intend to, the author also alluded to social class in his use of the concept of nobility. Following graduation, a degree of class mobility, as well as flexibility regarding gender roles, characterized male and female students' personal and career paths. Indeed, movement into the middle class and simultaneous disregard for its gender prescriptions defined the public sphere, gender attitudes, and graduates' life choices at state normal schools.

THE PUBLIC SPHERE

Involvement in two interlocking public spheres prepared students at state normal schools for middle-class life, as the structure of school life acquainted them with the larger communities in which the institutions were located and provided them with a safe smaller space in which to create their own social world. Only occasionally did students receive direct instruction in the ways of the middle class. Women students at Whitewater, Wisconsin in the 1870s attended weekly meetings at which the wife of principal Oliver Arey discussed proper hygiene and manners, and evaluated students' reports on when they slept, studied, and went visiting, as well as how they bathed and cared for their teeth. At Ypsilanti, Michigan in the 1880s, 1890s, and 1900s, instructor Julia Ann King held Friday "Conversations" for students and other members of the faculty dealing with conduct, social conventions, and religion. King's colleague Bertha M. Buell later observed that these gatherings "inspired 'a feeling that one's life among neighbors shapes the world community.' " Throughout the late nineteenth and early twentieth centuries, most normals held a required all-school assembly or chapel meeting every morning. These assemblies usually included announcements, some sort of devotional service, and a talk by the principal that occasionally dealt with behavior—or, essentially, how to uphold middle-class standards. At St. Cloud, Minnesota in the late 1870s, the chapel talks by principal David L. Kiehle, a Presbyterian minister, were basically sermons on student behavior. Students at Bridgewater, Massachusetts in the 1870s, 1880s, and 1890s remembered principal Albert Boyden's chapel talks as "helpful." Lottie F. Graves '93 explained,

"Mr. Boyden used to speak on many subjects which he thought would be for our good, even in the fall announcing when it was time to put on our winter flannels," and Mary Ellen Clapp '96 reflected that the talks "created ambition in his students."[5]

Rules for student conduct were a means of indirect instruction in middle-class mores. All state normals published rules of one sort or another. Some, like Fairmont, West Virginia in the mid-1870s, trusted students to make their own decisions because "students who will not govern themselves cannot hope as teachers to govern others." More commonly, though, normal-school catalogs listed detailed instructions for when students should study, exercise, sleep, and receive visitors, with a middle-class flavor. Sarah A. Dixon (Bridgewater '85) remembered, "Naturally there were rules and regulations . . . The one requiring the greatest courage to keep was the silent study hour from 7:00 to 8:30 P.M. . . . All must attend church on Sunday, must walk or exercise in the open air an hour each day, must refrain from visiting other rooms at certain prescribed hours." The 1871–1872 catalog for Oshkosh, Wisconsin stated that school rules were intended to cultivate "habits of self-control" and "the value and right use of time." In the late 1870s and 1880s, Oshkosh's catalogs directed students not only to avoid "indulgences known to be injurious," including liquor, tobacco, profane language, gambling, and dancing, but also to always be courteous: "Let it be your daily resolve to cast all your influence for the common good," the catalogs stated. In the interest of "simplicity and economy," the trustees of California's state normal schools in 1885 adopted a resolution "That the Faculties of the Normal Schools be requested to discountenance, by advice and counsel, all extravagance, such as expensive dress, the making of costly presents, and other things of like nature among the pupils." Even Fairmont's catalogs by the 1890s listed rules for attendance, visiting hours, "Care of clothing, books and person," and "Polite and respectful deportment," as well as the prohibition of "All defacement of the walls, seats, desks or other property of the schools."[6]

Victorian gender prescriptions were woven through normal-school regulations. While ubiquitous rules restricting visiting and mandating "exercise in the open air" aimed in part to preserve women students' reputation and femininity, some normal schools regulated male–female relations more directly. Men and women often had separate study halls and sat in separate sections of the classroom; in Whitewater's chapel, they sat on different sides of the central aisle. Dixon explained that at Bridgewater, "Our relations to the young men were carefully guarded. I believe we could converse with them for a short time after supper in the reading room, but if we wished to accompany one of them to a public place of amusement our escort must get a special permit." Lewis H. Clark, an 1879 Whitewater graduate, remembered that any man who "desired to call upon a lady" had to first ask for Mrs. Arey's permission. "Imagine how long a bashful young man would hesitate before he faced the chances of two refusals, that of the lady, and that of Mrs. Arey," Clark reflected. If he had attended normal school in Pennsylvania, Clark would not even have had an opportunity to face refusal, as rules for relations between male and female students were unusually strict at that state's normals. Not only were men and women forbidden from visiting, riding, or walking together, but, as Kutztown's catalogs stated, it was "expected that the ladies and gentlemen of the Institution will treat each other with politeness, but no conversation between them will be allowed in the lecture rooms, or in the halls." Male and female students in Pennsylvania used separate entrances and stairways. At Shippensburg, students referred to the black line painted on the chapel floor to separate male from female territory, as the "deadline." Although students would interpret gender rules quite liberally in their own activities, campus regulations familiarized them with all sorts of middle-class sensibilities.[7]

Town Life

Rules and occasional instruction in manners structured students' involvement in life in town and on campus, yet remained in the background while their myriad activities in the two interlocking public spheres served as important instruments of socialization into the middle class. In the wider sphere, students had considerable contact with society in the thriving cities, towns, and villages that housed normal schools. The latter decades of the nineteenth century witnessed tremendous urban growth—by 1900, 38 cities had populations over 100,000 and 40 percent of the population lived in urban areas—fueled by huge advances in transportation, communication, and mechanization. These changes also facilitated the ascendancy of the urban middle class in American society. In large and small urban communities, including those that housed state normal schools, middle-class citizens enacted the mores that differentiated them from the lower classes. They pursued high culture in performance halls, grand hotels, libraries, and museums, and formed social capital in community associations, churches, and their neighborhoods.[8] Still primarily concerned with family and home, urban middle-class women were becoming "recognizably modern," as they utilized public transportation, enjoyed access to cultural venues, and participated in women's organizations.[9] For both female and male students—most of whose backgrounds were neither urban nor middle class—the community that surrounded the normal school was new and exciting, and they eagerly took advantage of opportunities to become involved in and imitate many urban middle-class activities and associations. As they observed middle-class mores firsthand and became personally acquainted with local citizens, normalites began to assimilate into urban society; they then enacted in their own campus public sphere what they were learning in town, albeit with a very liberal interpretation of the proper sphere for a modern woman.

The largest American cities, of course, had an abundance of cultural venues and associations. Fewer than a dozen state normal schools were located in cities whose populations approached or exceeded 100,000 in 1870 and 200,000 by 1900, such as Providence, Baltimore, Buffalo, and Milwaukee. Still, one-fifth of all state normals were located in sizable urban areas. In addition to this handful of large cities, close to two dozen others whose populations exceeded 10,000 in the 1870s and 20,000 or 30,000 in the 1900s, such as New Britain, Connecticut, Petersburg, Virginia, and Terre Haute, Indiana, housed state normal schools. Oshkosh, Wisconsin and San Jose, California exemplified to a tee these growing modern cities. After a decade of rapid growth, Oshkosh in 1870 was Wisconsin's second-largest city, with eight hotels, 33 lumber companies, flour mills, grain elevators, steamboat and railroad service, a short stretch of paved street, and a population of 12,633, one-third of whom were foreign-born. Students who attended Oshkosh State Normal School, even in the early years, enjoyed the opera house and gas lighting on some streets. When the California State Normal School relocated to San Jose in 1871, the city already contained a stately county court house, foundries, flour mills and woolen factories, three banks, three benevolent societies, several educational institutions, and a population of nearly 10,000. Transportation was relatively convenient on two railroad lines, a "horse-car line," and broad, macadamized streets. Even San Jose's earliest students had access to five newspapers, Brohaska's Opera House, and the Musical Hall. Both cities continued to grow at a rapid pace: in 1890, San Jose had more than 18,000 and Oshkosh more than 22,000 residents, and in 1910, San Jose approached, and Oshkosh exceeded, a population of 30,000.[10]

Half of all state normal schools were located in towns with populations generally between 2,500 and 5,000 in the 1870s and 1880s and between 3,000 and 10,000 in the 1890s and 1900s. While not as large as Oshkosh and San Jose, these communities were also bustling, modernizing towns. Brockport, New York, for example, was a trading center on the Erie Canal and the railroad. By the mid-1890s, steamboats appeared alongside mule-drawn canal boats, and the railroad depot was busy with hourly departures and arrivals. The village housed flour mills, a canning company, and various manufacturing companies; the Moore and Shafer Shoe Company employed 400 workers. Brockport had many retail establishments, banks, hotels, restaurants, saloons, and a candy store. Citizens were active in temperance and missionary societies, read two local newspapers, learned how to cure their ailments from traveling "medicine shows," and enjoyed performances at Ward's Opera House, which opened in 1877. While Brockport did not have public transportation, many other towns of this size, including Westfield, Massachusetts and Normal, Illinois, added trolley lines during this period. Two southern towns that typified the small urban communities that commonly housed state normals were Florence, Alabama, which had a population of over 6,000 by 1890, and San Marcos, Texas, whose population surpassed 2,500 in the early 1900s, when the normal opened there. The site of the new normal was called Chautauqua Hill due to its history of hosting myriad intellectual and recreational activities for San Marcos' citizens. In the 1900s, San Marcos boasted more than thirty mercantile establishments, railroad service, an opera house, and "well paved and macadamized streets," while Florence's normal-school bulletin was able to mention the development of many "of the conveniences of the larger cities . . . such as electric lights, gas, good streets, an abundance of pure water, and a street car system."[11]

Most of the towns that housed state normal schools for African American students, such as Frankfort, Kentucky and Fayetteville, North Carolina, were similar in size to Brockport and Florence. With a population of over 2,000 in 1870, nearly 9,000 in 1890, and over 15,000 in 1910, Pine Bluff, Arkansas exhibited the high level of activity and change that characterized modernizing cities of this period. In addition, it was a thriving center for Arkansas' black population, which gave students of the Branch Normal College a special and encouraging taste of urban life. Greater than 4,500 in 1890 and 6,000 in 1910, Pine Bluff's black population in itself was large enough to constitute a bustling city. In Pine Bluff, the Colored State Fair was an annual event beginning in 1881, a black newspaper called *Echo* began publication in 1900, and the state's first black-owned bank opened in 1902. D. B. Gaines observed in an 1898 publication entitled *Racial Possibilities as Indicated by the Negroes of Arkansas*: "The city and vicinity of Pine Bluff have many men of color who stand as evidence of racial progress and possibilities. Probably no county in the state surpasses Jefferson, of which Pine Bluff is the county-seat and chief city, in wealth and educated colored citizens." These citizens, Gaines continued, "are doing a large and profitable business. It is clearly demonstrated that the Negro is capable of engaging in the higher pursuits of life." Black culture thrived in Pine Bluff.[12]

Other communities that housed state normal schools were not quite large enough to be frenetic urban areas, but were still busy villages. Over one-fourth of state normals were located in towns whose population approached but never exceeded 2,500 between the 1870s and the 1900s. Despite their small size, established villages such as Geneseo, New York and Castleton, Vermont exposed normal students to a thriving civic life. As early as 1870, Geneseo housed a variety of stores, three hotels, two livery stables, and a machine shop that manufactured lathes and sewing machines. The village also had railroad service, a newspaper, masonic organizations, a library and reading room, and boasted "good side and cross walks, beautiful street shade trees, public parks, water works and gas, and its tasty and substantial residences." Castleton had a wide main street that could "challenge a comparison with any other in New England." A "substantial" town hall and two hotels were in operation by the early 1880s, and ten stores thrived while an "electric road" served the town by the 1900s. Castleton's electric road and Geneseo's railway service made the bigger cities of Rutland and Rochester, respectively, easily accessible for residents as well as normalites.[13]

Many of the towns that housed state normal schools in newly settled frontier areas had a different ambience than Castleton and Geneseo, yet still introduced students to civic life. During the Territorial Normal School's early years in the late 1880s, for example, Tempe, Arizona had fewer than 1,000 residents, but two hotels and a public hall, as well as a river ferry, a grain mill, seven stores, three restaurants, and four saloons. Weatherford, Oklahoma was a typical frontier settlement on the railroad with dirt streets, unpainted frame buildings, and not many more than 1,000 residents when the normal opened there in 1903. The town was also known for its saloons, gambling, and brothels, but its citizens soon banded together to clean up this image; by 1910, Weatherford was a calmer, reform-minded community of over 2,000.[14]

Day-to-day life for normal-school students involved much contact with public life in these villages, towns, and cities. Most important, a majority of normalites lived in town rather than on campus because most normal schools did not have the resources necessary to build and maintain adequate dormitory space. Very few normals had the means to enforce regulations such as Shippensburg's declaration in the 1870s that "No student will be allowed to board outside of the school building unless with parents or near relatives, except by special permission of the Board of Trustees." Those normal schools that did have residence halls usually had a limited number of rooms and prioritized housing female students: institutions including Castleton, Brockport, Geneseo, Pine Bluff, and St. Cloud had dorm space at one time or another, but were not able to offer rooms to male students, or even all of the women. Castleton's catalogs urged women to live on campus, but also mentioned that rooms were available with "private families." Pine Bluff had a small residence hall for women beginning in 1889, but, as the Arkansas State Superintendent of Public Instruction reported in 1906, male as well as many female students throughout this period lived "with white and negro families of the town." After the all-female state normal school opened in 1892 in Greensboro, North Carolina, most, but not all students lived in the dormitory; of the 223 students who attended in 1892–1893, 143 resided on campus, 10 lived at home with their families, and 70 resided in private homes in town. Not only were the town dwellers at Greensboro and other institutions situated in an urban setting, but they were also in a position to link their dorm-dwelling classmates to town life.[15]

At the majority of state normals in the late nineteenth and early twentieth centuries, all students lived in town, either with families or in boarding houses. Arriving students at Oneonta, New York in the 1890s received a list of 144 private homes on 44 different streets that the principal had approved as residences; most shared a room or even a bed in one of these homes. Such residential arrangements allowed normalites to experience urban middle-class family life first-hand. When Fairmont opened in 1873, the catalog promised that students would "find comfortable homes in good society." The 1898–1899 Florence catalog explained, "There are no dormitories for students. They secure homes with cultivated, refined, discreet people, who can guide them in their amusements, advise them when necessary, and give them a pleasant social life." Many women students at Weatherford defrayed the costs of room and board by cooking and cleaning for their hosts, who still treated them as members of the family. Students from the local area usually continued to live with their own families, yet attending the normal enabled them to have new experiences with urban life. For example, normalite Adelaide Pender's family lived "in the country" near Southington, Connecticut, where, she explained, "Social life was outside my scheme of things." Riding the train to New Britain each day enabled her to join a "merry party" of other commuting women students, some of whom "flirted outrageously with the brakemen and conductors or any other masculine who met them half way."[16]

Normalites who lived in various types of boarding houses were removed from family life, yet enjoyed their close proximity to other aspects of middle-class urban society. At Cheney, Washington, students lived in town, mainly in group homes named for the house mother, who owned the home and provided room and board and acted as an authority figure. Meanwhile, at Ellensburg, Washington, a local physician in 1896 remodeled a three-story brick building near the county court house; for the next decade and a half, female students lived on the upper floors and various businesses occupied the first floor. At Oswego, New York, normalites lived in private homes, in boardinghouses called "clubs" because it was common for groups of students to rent several rooms and to share responsibilities for chores such as food service, or at a former hotel called the Welland. Privately owned, the Welland housed 100 women students who enjoyed its indoor toilets and bathtubs in the 1870s and 1880s and piano and telephone in the 1900s.[17] Whether they lived in privately run dormitories like the Welland, smaller boardinghouses, or family homes, students were not beyond the reach of school regulations, as prescriptions for evening study hours and restrictions on social visits and absences applied to off-campus residences.[18] Even with the help of boardinghouse keepers, however, it was difficult for principals and faculty members to monitor all students. In 1910, principal Henry T. Burr at Willimantic, Connecticut complained that some landladies took "no responsibility whatever" for enforcing school regulations, and that several "young men" were boarding in houses where women also lived. Although many normal-school leaders shared Burr's frustrations and Pine Bluff principal Isaac Fisher's concern that students who did not live on campus might experience "unrestrained liberty and freedom in the city," living off-campus more commonly exposed them to the polite society of the urban middle class.[19]

In addition to the location of their residences, students' involvement in local churches helped to immerse them in the town's public sphere. During the late nineteenth and early twentieth centuries, most normal schools required, or at least strongly recommended, that all students attend church services in town. Rules for "The Sabbath" at the Eastern Maine State Normal School in Castine were typical: "Proper observance of the entire day is expected. Pupils are required to attend public worship at the church which they or their parents may select. There are at present three churches in town, Unitarian, Methodist, and Congregationalist." At Shippensburg, Sunday-morning services at a local church were required, and Sunday-evening services were optional yet popular, in part because students who attended were able to stay in town until 10 P.M. A student at the Western Kentucky State Normal School in the late 1900s, Gordon Wilson attended Sunday school and the morning and evening services every Sunday at the State Street Methodist Church in Bowling Green. Wilson and other students were much more than members of the audience at local churches. Many were active participants in Sunday-school classes. A Bible-study class at the Presbyterian church in Geneseo in the late 1890s and early 1900s attracted between 75 and 100 students each year. Pine Bluff students taught Sunday school in the town's three black churches; Principal Joseph Carter Corbin was proud to report in 1877 that high Sunday-school attendance was the result of the students' excellent teaching. And Wilson became the regular teacher of a religious-education class for girls at the Broadway Methodist Church in Bowling Green.[20]

Students' involvement in church life also had a social side. *The Normalian* at Geneseo reported in 1901, "Soon after the opening of school in September each church gives a reception to the students of its denomination, thus affording them an opportunity to become acquainted with each other and with the pastor and others of the church." Then, throughout the academic year, Sunday-school teachers would reach out to students, "invite them to their homes, encourage the holding of class suppers, and in other ways show them the pleasant attentions of friendship." The popular class at Geneseo's Presbyterian church ended each year with a "graduate banquet." The 1900 banquet, according to junior Edith Tammany, was held "at the residence of the class's patron on Main street. Covers were laid for some forty guests. After a very delicious repast," the guests enjoyed toasts and responses. Church socials were popular among normalites; Wilson certainly enjoyed them in Bowling Green. San Jose students were attracted to inter-church events; *The Normal Index* reported in 1887: "The 'Bazaar of the Centuries' and the lunches, both given at the California Theatre under the auspices of the ladies of the various Churches, were well patronized by the Normal students." When the Florence and San Marcos catalogs boasted in the early 1910s that, "The Christian people of the city assist in throwing around the students the best possible religious influences," they celebrated the web of social ties, as well as religious education, that students gained through their involvement in church life.[21]

Outside of church, normalites were involved in various community events, meetings, and causes. Some gatherings were purely social. Pine Bluff's black community hosted a yearly picnic with games and good food for graduating normal students, and townspeople in San Marcos hosted a "Citizens Barbecue" for students at Southwest Texas State Normal School. Community clubs and organizations brought members of the middle class together for socializing as well as self-improvement, and often welcomed students. In 1902, for example, Castleton students formed connections with local citizens while studying Germany and the Philippines in the Good Literature Club, and enjoyed a reception in their honor by the local Young People's Christian Endeavor Society. Other events through which students became involved in the local community centered on entertainment. When a circus train made yearly stops in Chico, California in the 1890s, normal students and residents alike were drawn to what an instructor termed "the mysterious charm of circus domestic life." At Greeley, Colorado, meanwhile, students and citizens together enjoyed the "road shows" that stopped to perform in the Opera

House. And at Christmas time in the early twentieth century, Willimantic students sang carols outside the hospital and the home for the aged.[22]

When politicians and other important visitors came to town, normalites witnessed first-hand the outpouring of pomp and civic pride, in addition to learning about the political system. After U.S. President William H. Harrison went to San Jose in 1891, *The Normal Index* reported that normal students, including "young ladies attired in the national colors" lined his route through town. Geneseo students in 1900 heard Governor Franklin D. Roosevelt speak at their court house, and Westfield students were among the 10,000 people who assembled in the town's Pare Square in 1902 to hear a speech by President Theodore Roosevelt. Students also formed connections with local citizens through involvement in various political and civic causes. Beginning in 1877, Geneseo's Delphic and Clionian societies donated funds to the village's reading rooms. After disastrous fires during the 1900s, Pine Bluff students donated the money they had raised to purchase an organ to the homeless and helped the owners of a gutted lumber yard salvage materials. In Bowling Green, Gordon Wilson joined local efforts to prohibit liquor sales, and was deeply disappointed when the "dry" campaign was unsuccessful in the polls. Students' contributions to local causes were an extension of their involvement in middle-class urban life. An 1889 publication of the state normal in Emporia, Kansas summarized, "The students are allied to the citizens by ties better than those of dependence: They are friends who are kindly welcomed as boarders in the homes of the city, and as workers in the church, in the Sunday school, and in every worthy enterprise." By living in town residences and attending church and public events, normalites came into contact with middle-class mores in the bustling, modernizing towns of the late nineteenth and early twentieth centuries. Unlike the rural areas where most of them had grown up, the communities in which state normal schools were located enabled students to learn the ways of middle-class society and tap into social capital. Because students generally were temporary residents of these towns, the actual connections they formed with local citizens were less important than the first-hand understanding they gained of how to build and use social capital. At the same time, women students' experiences in these expansive surroundings increased their confidence as actors in the public sphere.[23]

Campus Life

The normal campus functioned as a safe space in which women could exercise their budding self-assuredness and all students could enact their new sense of middle-class social life. Political scientist Robert Putnam observes that the years between 1870 and 1920 marked "a veritable 'boom' in association building."[24] Normalites not only replicated the association building they witnessed in the larger society, but also enabled female students to shine in every type of formal and informal activity. The resulting campus life was remarkably lively; one observer wrote that San Jose State Normal exuded "enthusiasm and mutual confidence."[25] Through involvement in all sorts of events and organizations, normalites both deepened their understanding of social capital and built strong ties with other students, faculty, and administrators; they created a microcosm of middle-class society that was largely devoid of separate-spheres ideology.

At the most formal level, daily "chapel" and occasional ceremonies punctuated the normal schools' public sphere. From Cortland, New York in the 1870s to Weatherford, Oklahoma in the 1900s, required daily assemblies or morning exercises in the chapel brought together the whole normal-school community. At Whitewater, the faculty sat on the front platform and students sat in the audience, arranged by seniority, from seniors in the front to model-school pupils in the back, and sex, with females on one side of the aisle and males on the other. In addition to announcements and talks by the principal, these gatherings often included hymns and prayers; a visiting committee reported in the early 1870s that Whitewater's assemblies "were deeply impressive; we have seldom seen a religious gathering, even in churches, more devout" and inspiring of "healthful moral and religious influence," which would "assist in making successful teachers and good citizens." At Charleston, Illinois in the early 1900s, Principal Livingston Chester Lord said that morning exercises were "for family prayers: to sing a hymn, to ask for your daily bread, to hear some words of wisdom from the Book." He added that the Bible was read "as literature," and chapel was not for proselytism. Instructors and visiting lecturers also addressed the Charleston community during morning assemblies. Chapel programs at Buffalo were even more varied, including worship services, addresses by visiting speakers, dramatic performances, and speaking contests. Gordon Wilson's biographer reported that chapel meetings at Bowling Green "provided an unsurpassed means of communication and contributed to a sense of unity" on campus.[26]

Like daily chapel, celebrations of institutional milestones and yearly graduations were organized and executed by the principal and the faculty. These ceremonies also unified the campus community, as well as infused it with a sense of history and grandeur. When students, faculty, and honored guests assembled to listen to speeches commemorating Castleton's one-hundredth anniversary in 1887, and Geneseo's and Oshkosh's twenty-fifth anniversaries in 1896, students realized that they were part of a growing tradition. Dedications of new buildings reminded students that their school's history was ongoing, and they were part of it. When Bridgewater celebrated a new building in 1891, the *Boston Herald* reported:

> The dedication of the new normal school building in this old town this morning was the occasion for one of the most notable gatherings ever seen here. Many of the distinguished educators of the State came down by the early trains to congratulate Principal Boyden on the completion of the grand edifice . . . every seat, numbering more than eight hundred, was occupied.

Speakers at the grand event included Boyden, a general, and a judge. Two years later, a similar dedication ceremony at Fairmont featured speeches by an ex-governor, the mayor, a professor, and senior Annie Linn. While such one-time celebrations brought the campus community together to mark milestones in the life of the institution, elaborate yearly ceremonies honored graduates for what they had achieved. Usually spanning several days, graduation exercises often included a baccalaureate sermon by a local minister, academic speeches by all graduates or student representatives of the senior class, and speeches by the principal and perhaps a visiting dignitary, as well as the ceremonial conferring of degrees. Formal events—Cora A. Brown (Oswego '73) remembered that all of the women in her class "wore black grenadine over black silk" while the men "wore their Sunday best"—these ceremonies also showcased normalites' assimilation into middle-class society.[27]

To the extent that the principals and guest speakers at the most formal ceremonies were usually male, these events reflected the Victorian notion that the public sphere belonged to men. This was not true of student-organized celebrations, however. Normalites created their own tradition of more- and less-formal events in which women and men were equally active and visible. The most formal of these events highlighted women students' place in the campus public sphere and celebrated social capital, as students displayed publicly their bonds to one another. The capstone of each year's social as well as intellectual life, graduation festivities included "class days." Fairly widespread by the 1890s, class-day exercises varied somewhat from school to school. Generally, students of both sexes took the stage to read the class history and prophesy, and make other speeches that illustrated their social networks. Oshkosh's class days in the late 1880s and early 1890s included humorous "botanical" and "zoological" analyses of the seniors, and Castleton's in the late 1890s included addresses to the juniors that offered advice as well as gentle gibes. In 1909, Castleton's class day consisted of "The Celebrated Case of the Class of 1909," a mock trial in which women played the judge, the sheriff, and lawyers. Instead of organizing class days, Geneseo's all-female and all-male literary societies joined forces to stage anniversary exercises that were, according to one graduate, "quite as important as the graduating exercises. Dignified and impressive, with essays and orations varied with music, they drew [large] audiences." By the turn of the century, Geneseo's societies added plays to their programs. At Florence and Pine Bluff, students staged both class-day and anniversary exercises.[28]

Other less-formal events also made a practice of affirming and encouraging the growth of social capital, as well as women's prominence in public. From week to week, campus life was alive with student-initiated events. At all sorts of socials and receptions, students enjoyed the opportunity to interact in a semi-structured public setting. "Normal Socials" in the boarding-hall parlor were a regular Saturday-evening occurrence at Westfield in the late 1870s and 1880s. Students and faculty members enjoyed games, music, and even dancing—with the approval of principal Joseph Scott: " 'Don't dance if your parents disapprove,' he would say, 'but if I can help you have a better time, I'll dance too.' " Socials were end-of-the-term events at Cortland,

Figure 5.1 Cover illustration on program for Anniversary Exercises of the Arethusa and Gamma Sigma literary societies at Geneseo State Normal School in New York, June 21, 1900. *Credit:* College Archives, Milne Library, SUNY Geneseo.

where the local paper reported in January 1883:

> The closing term sociable of the Normal School was held in the chapel on Saturday evening last. A large number of students and their friends were present, and a most pleasant time was enjoyed until the gong struck 10 o'clock. During the evening Miss Clara Whittimore and Mr. C. C. Roberts gave readings, and vocal music was furnished. A comical burlesque of the "Engle Clock" was given by a few students which caused uproarious laughter.

Class receptions were also fixtures on many schools' social calendars. In 1898, the annual fall reception that Oshkosh's seniors hosted for the juniors took the form of a "Japanese party," and in 1899 it was a Halloween party. The juniors hosted a reception for the seniors each spring until 1900, when they began to take them on a boat ride instead. Oshkosh's class of 1903 gave an all-school reception at which, the student magazine reported, "all the different counties represented in the class were divided into seven groups and for each group a booth was erected . . . For a program a speaker from each booth described the especial charms of the counties represented by his or her booth." At San Jose, the section of the senior class called A's regularly hosted receptions for the senior B's, and vice versa. For one such gathering in 1887, the senior A's decorated the normal basement with fans, lanterns, and flowers to create a "Chinese" atmosphere.[29]

Excursions and picnics were additional events through which normalites strengthened their social ties. San Jose students ventured to Alum Rock, a nearby landform. When a group of juniors went there in 1887, *The Normal Index* explained, "the day was pleasantly spent in gazing at the falls and other wonders, in dancing, and in disposing of the many good things brought along for lunch." During the late 1880s, San Jose Professor C. W. Childs and his wife hosted class picnics at their home. The agenda for the visit of one group of seniors included: "1st, Geologizing; 2nd, Melonizing; 3rd, Luncheon; 4th, Concert; 5th, Storming of the [grape] vine." Pennsylvania's climate allowed for annual class sleigh rides at Indiana beginning in the 1890s, while Oshkosh's setting on Lake Winnebago encouraged student groups to take boat trips. In 1899, the glee club sailed to nearby Neenah, and, beginning in 1900, the junior–senior boat ride across the lake to Calumet Harbor included refreshments and dancing. The Philalethean and Agonian societies at Geneseo held an annual picnic at a beach on nearby Conesus Lake during the 1900s, and San Marcos seniors had a yearly picnic at Jacob's Well, a natural fountain. At the normal school in Richmond, Kentucky, meanwhile, trips to "the mountains" meant leaving on the midnight train, and hiking from the Berea station in time to watch the sunrise over the peaks; after cooking and enjoying bacon and coffee together, students hiked until it was time to catch the noon train for Richmond. Hiking, camping, and having picnics in the desert were all popular among Tempe students in the late 1900s.[30]

While all-school ceremonies, socials, and excursions encouraged and celebrated sociability or informal associations among members of normal-school communities, students also interacted and built social ties through all sorts of smaller groups. For many normalites, the others with whom they lived and/or ate their meals constituted a primary social network. Those who lived on campus became close to their room- and dorm-mates, and communal meals brought groups of students together around dining-hall and boardinghouse tables. At Huntington, West Virginia and St. Cloud, Minnesota, only women lived in the dormitory, but both men and women took their meals in the dining hall. Michael Dignam (Westfield '82) remembered a similar situation: "We had our meals together in a large hall," where Principal Scott sat at the head of one table, women students sat along the sides, and instructors and men students filled in the ends of tables. The dining hall at Slippery Rock, Pennsylvania at the turn of the century accommodated 400, including President Albert E. Maltby and his wife Harriet, who dined at a corner table. Maltby was proud to report that the hall was "a model of artistic taste and modern convenience." He even purchased a Victrola, so that "while the students and faculty dine they also listen to fine strains of music"; these were inspiring conditions under which to form social

Figure 5.2 Oshkosh State Normal School students on the porch of a Lincoln Avenue home near the normal school, 1902. The home was a rooming house where some of the students lived. Photo by Effie Howlett.

Credit: Courtesy of the Oshkosh Public Museum.

ties. A 1907 account of the Illinois State Normal University described a different approach: "during the last thirty years the majority of students have boarded in clubs. . . . The number of students in each club has varied from ten to fifty, the average being perhaps twenty." These clubs adhered to President John W. Cook's belief that "the presence of young men and young women at the same table" would be "advantageous to both." Bowling Green's Gordon Wilson enjoyed his boardinghouse companions—both male and female—tremendously. He ate slowly "to lengthen out the club-like atmosphere," played along when Nettie Depp used a dictionary to tease him about the way he pronounced words, and agreed with Corbett McKenney that their evening discussions deserved the title, the "Cherry Hill School of Philosophy."[31]

While Wilson's philosophical discussions with his boardinghouse mates remained unstructured, the popularity of all sorts of student organizations reflected normalites' propensity for the formal association building so popular among members of the urban middle class during this period. As explained in chapter 3, literary societies were the most widespread and long-lasting associations at the normal schools; virtually all students participated. Shorter-lived academic clubs were an additional forum for structured interaction, as were ever-popular religious associations. What began in 1871 as informal prayer meetings of Isaac Eddy Brown and six other men students at Normal, Illinois' Presbyterian church, became one of the first campus chapters of the Young Men's Christian Association (YMCA) in February 1872. The following November, Brown's sister Lida invited other women students to her room for prayer and hymns, and the group soon grew into a chapter of the Young Women's Christian Association (YWCA). Students at Emporia organized their longstanding weekly prayer meeting as a YMCA in the early 1880s, but withdrew from the state organization when it refused to approve their plan to include women students as members; the coeducational Young People's Society of Christian Endeavor later thrived at Emporia. The Students' Christian Union of Cortland Normal School, founded in 1891, aimed "to promote an earnest Christian life among students, to increase their mutual acquaintance, and make them more useful in service to God." From Shippensburg to Ellensburg, though, YM- and YWCAs were the most popular religious organizations from the 1890s into the twentieth century. They met, sometimes jointly, for devotional services, held Bible-study classes, and undertook service projects on campus. Kirksville, Missouri's YWCA and YMCA, organized in 1895 and 1896 respectively, welcomed new students at the railroad station and assisted them with housing and enrollment.[32]

In addition to enriching religious and intellectual life, Christian organizations, academic clubs, and literary societies brought students together for structured social interaction, serving as important vehicles for the formation of social capital. Friendships grew over refreshments before and after regular meetings and during special social events, such as literary societies' parties celebrating Washington's birthday and Halloween. At Buffalo in the 1890s and 1900s, members of the coeducational Normal Literary Society enjoyed banquets, game parties, and even dances, before which they taught one another the proper steps. Students also danced at social gatherings of Oswego's literary societies and Normal Christian Association, but men danced with other men and women with other women, as was the rule. For the Cornelian and Adelphian societies at the all-female state normal in Greensboro, North Carolina, the initiation banquet was a highpoint of the year; society members decorated the dining halls, wrote souvenir menus, organized a receiving line, and shared lively toasts after the meal. The YWCA at Normal, Illinois held teas, and the Germanistiche Gesellschaft at San Marcos, Texas organized four or five socials per year at which members sang German songs and played German games. Student groups also occasionally invited nonmembers to socialize at their events. YWCAs regularly offered socials to welcome new students to campus and encourage them to get to know one another. At one such reception at San Jose, a leader announced at regular intervals new conversation topics including "the last book I read," the weather, and "women's sphere." Literary societies' public debates and "joint meetings" were all-school gatherings, and Geneseo's all-female societies held yearly all-school "fairs" with decorated booths in the gymnasium. For Thanksgiving in 1898, Oshkosh's Christian societies transformed the gymnasium into a barn, where "Grandpa and Grandma Normal" hosted a dinner.[33]

The many gatherings and events at which students formed social networks and learned the benefits of association building usually also involved intellectual or artistic performances that put both women and men students in the public spotlight. The notion of a separate female sphere was absent as all students took the stage in literary societies' public rhetoricals, oratory contests, and debates, and student musicians of both sexes often performed between the academic speeches. Lighter in tone, many receptions and socials included live music and skits. At the "Chinese" reception at San Jose, the senior A's presented a "humorous recitation," a skit, and performances by the senior orchestra and a quartette. Class receptions at Oshkosh invariably included "farces," and at the 1907 fair, according to *The Normal Advance*, "the cowboy girls beat any on the stage." A fund-raising "entertainment" at Castleton likewise included two farces and vocal selections. With other members of their classes and societies, women frequently took the stage as actresses in dramatic productions, such as Sheridan's "The Rivals," by the Oshkosh class of 1899 and "Madame's Boarding House," performed by Florence's Browning Society in 1909. The German clubs at Oshkosh and San Marcos presented plays with titles like "Ein Kopf" and "Nein." Geneseo's literary societies were especially enthusiastic about staging dramatic productions; together, sister and brother societies presented various plays throughout the year and, by the late 1890s, as part of commencement week. In 1904, Pine Bluff students performed "Milestones of Progress," a play by president Isaac Fisher about the evolution of the black race, in Pine Bluff and nearby Fort Smith.[34]

Dramatic associations per se were not established in most normal schools until the 1910s, probably because before that time the literary societies and classes were so enthusiastic about staging productions. Musical organizations, on the other hand, began to appear during the 1870s and 1880s. While student musicians performed within societies and clubs as soloists and in small groups, separate organizations gave them the chance to perform in larger orchestras and choirs. By the late 1880s, many normal schools had orchestras. The Geneseo orchestra played at all sorts of functions, both on campus and in town, and the Normal Voluntary Band was active at Kirksville beginning in the early 1890s. Mandolin Clubs were popular by the 1900s; Oshkosh's had six mandolins, three guitars, and a piano. Vocal performance groups included mixed choirs and women's and men's glee clubs. Organized in 1875, the coeducational Song Circle thrived at Brockport for more than two decades, and even performed in a nearby town in 1896. By the turn of the century, the Normal Chorus was the main vocal group at Brockport; members practiced weekly, sang at morning chapel, and staged a midwinter concert and spring recital. Choral groups at Cedar Falls, Iowa in the 1880s and 1890s included the Minnesingers and the Troubadours for men and the Cecilians, the Euterpeans, and the Bel Cantos for women; the mixed Choral Society performed at commencement and staged a May Festival beginning in 1903. Students at Normal, Illinois in 1898–1899 organized separate

Figure 5.3 Principal Joseph C. Corbin (second row, center) with members of the college band, Branch Normal College, Pine Bluff, Arkansas, ca. 1881.

Credit: University Museum & Cultural Center, University of Arkansas-Pine Bluff.

men's and women's glee clubs, and at San Marcos during the 1900s, women sang in the Schubert Club and men sang in the glee club. Membership in these many musical groups offered another opportunity for building social capital, as they also further called into question the notion of a separate women's sphere. It is hardly surprising that San Jose's *The Normal Pennant* in 1909 described the "highest type" of woman as: "no longer the pale, listless individual who spends her time making fancy laces, shedding tears, and doctoring sick headaches, but rather the strong, robust, Grecian type who is able to stand on both feet and look the world in the face."[35]

The Normal Pennant described the "highest type" of woman specifically to praise the sport of basketball for increasing women's strength and robustness. Physical activity, which had a social component and put women in the spotlight, was a visible part of normal life in the late nineteenth and early twentieth centuries. Most normal schools required physical exercise, although their expectations were at times rather vague. For example, the 1888 San Jose catalog simply stated that students were to partake in "at least one hour a day" of "vigorous exercise . . . regularly by daylight, and in the open air." In most cases, this meant walking. Adelaide Pender remembered how this type of exercise at New Britain lent itself to forming social bonds:

> Another form of social life was afforded by walking arm and arm around the edges of the assembly hall. The first time I tried this promenade I was sure the

> distance was half a mile, but, the oftener I did the promenade, the hike seemed too short. Before school, during intermissions, the noon period, and often after school, the students walked and walked, exchanging school gossip, discussing lessons, groaning over assignments . . .

Most normal schools required some sort of calisthenics or, as the 1895 Tempe catalog put it, "marches and movements." An 1889 history of the normal at Emporia, Kansas stated that calisthenics was "a regular exercise" throughout the 1870s and 1880s. The institution added a basement gymnasium in 1880, and throughout the decade obtained "apparatus" including "chest weights, walking-rings, ladders, etc." All Emporia students were "required to engage in the daily exercises in light gymnastics," and time in the gym was available for "pupils who may wish to use the heavier apparatus." The 1890–1891 Florence catalog explained, "As a means of developing the muscular tissue of the body and cultivating a proper use of the same, certain actions in free gymnastics are taught. Drills are with wands, clubs and dumb-bells for the same purpose." Normal schools occasionally subscribed to particular methods of calisthenic instruction; under the Del Sarte system of rhythmic gestures at Oswego, students "took calisthenics in the auditorium," where, one graduate remembered, the instructor and the pianist "put us through our paces. We stood between the seats or marched through the rows, while counting lightly in time with the music." Twenty women students from Cortland demonstrated such light gymnastics in a performance drill at an 1894 teachers institute in nearby Homer, New York.[36]

Exercise requirements generally did not make distinctions between men and women. At Huntington, the gymnasium erected in 1875 was reserved for male students, and instead of using the gym apparatus, women took daily required walks as a group and enjoyed use of the croquet grounds on campus. This was unusual, however; if anything, institutions were more likely to restrict the times of day when each sex was allowed to use the gymnasium, or perhaps to require additional military drills of the men. In the mid-1880s, Castleton's catalog stated, "Lessons in calisthenics are given in connection with the reading classes. Also, once a week, the young men are drilled in military tactics, including the manual of arms." Similarly, at Cedar Falls between 1892 and 1903, all male students partook in required military drill for three hours per week, with supplies and an instructor from the local armory. Required military drill for Tempe's male students began in 1898; the Tempe Normal School Cadet Corps was soon very visible at parades and other events, and guarded President William McKinley during his visit to Arizona in 1901. Although it was not a requirement for them, women students occasionally organized their own military drills. At West Liberty, West Virginia, the "broom brigade" consisted of women who conducted military drills with brooms, and even performed during commencement in 1887 and 1888. The women at San Jose also organized military-drill companies, and even took on the men in a competition. A student publication reported in March 1892, "the long delayed competitive drill between Companies

A [composed of women] and B [composed of men] took place in front of the Normal building, before a large and interested audience." Company A was victorious.[37]

This competitive military drill presaged athletic games' importance as student activities at state normal schools. Male students took the lead in organizing competitive sports, beginning with fledgling baseball teams. As early as the late 1860s and early 1870s, Platteville, Wisconsin, as well as Whitewater and Oshkosh, fielded teams for intramural competition or games against area teams. At Brockport in 1873, the "Normal Nine" took on the "Village Nine," and the normal team won the four-hour game that was finally called on account of darkness. Baseball hardly dominated student life, however, as teams waxed and waned according to variations in interest and the number of men students.[38]

When football and basketball teams appeared, beginning in the late 1880s and mid-1890s respectively, male students continued to struggle to keep their athletic teams afloat. The men at Brockport organized a football team in 1889 and began yearly contests against the University of Rochester in 1894, while the men at nearby Geneseo did not form a football team until 1896 and then had trouble maintaining it from year to year. In Wisconsin, most of the normal schools had football teams by 1895 and basketball teams by the turn of the century. Football was quite popular among the men at Whitewater, even with a 106 to zero loss to the University of Wisconsin, yet they continually sought more support from the women; the student newspaper in 1904 offered the following "Advice to the girls": "Uphold the Foot Ball Team and you will contribute indirectly to 'the attraction that keeps the boys in the Normal.' " Tempe's football team managed in 1899 to defeat the University of Arizona, but when men students at Greeley decided in 1901 to withdraw from an intercollegiate league, the student paper reported that the decision was a sensible one because, "In the place of attempting to cope with the colleges that could pick out a team from many men they will play class against class." Ben Van Oot (Oswego '05) described the less-than-ideal conditions under which he played on his institution's first men's basketball team: "The school had no money for uniforms or equipment, so we provided our own. . . . The gym where we played was adequate as far as floor area was concerned, but the ceiling was low and was supported by six iron posts which interfered considerably with our playing." Competition in football and basketball remained primarily intramural, with occasional inter-school match-ups, often as sideshows at inter-normal debates or oratorical contests. Playing on these teams surely allowed male students to build important social bonds with their teammates, but the small numbers of men and limited funding for athletics prevented men's sports from capturing the public spotlight as they did at colleges and universities during the same period.[39]

Although men took the lead, women soon had their own teams and tournaments. Basketball especially was the territory of female normalites; by the late 1890s, it was a full-blown craze that placed women in the normal-school limelight. A poem published in the student magazine at Oshkosh,

"A Valentine to a Basketball Girl," captured the strong bonds that basketball created and the enthusiasm with which Oshkosh women took to the courts. The author, presumably a male student, lamented:

> Will you be my valentine?
> Or, alas! have I a rival?
> If it's the game of basket ball
> Then all my hopes I stifle. . . .

By January 1897, the women at Oshkosh had formed four basketball teams, and intense intramural competition was underway. By late fall of 1898, the women of San Jose also had four basketball teams, and two years later, they formed the Normal Girls' Athletic Association with the intention of having "the best basketball team in the state, or even on the Coast." Meanwhile, the Girls' Athletic Association at Geneseo arranged for games among teams representing the different all-female literary societies; 90 women joined the association in 1901–1902. A women's basketball team called the Gypsies was San Marcos' very first athletic team in 1903, and by 1906 it entertained challenges from the Nymph and the Topsy teams. Women's basketball was underway at Ellensburg, Washington by 1895, when the local newspaper reported that "the ladies are getting to be splendid players . . . it is a very exciting game." A decade and a half later, according to Ellensburg's

Figure 5.4 Women's basketball team, State Normal School at Ellensburg, Washington, 1904.
Credit: Central Washington University Campus Photograph Collection, Washington State Archives, Central Branch, Ellensburg, Washington.

Normal Outlook, it appeared that "the whole school—barring a few exceptions—was divided into teams, and that some of these were playing continuously—nights and Sundays excepted."[40]

Normal women also took on inter-school competition in basketball. In 1900, Millersville, Pennsylvania played a team from Miss Stohr's School in nearby Lancaster, and San Jose played a team from Palo Alto High School. The following year, the women's team at Kirksville was the first basketball team to represent that institution against another school when Kirksville lost to the American School of Osteopathy by 25 points. After the women of St. Cloud took on nearby Brainerd High School, the normal newspaper reported that the team "nobly maintained the reputation of the Normal School . . . The Brainerd team was strong at guard and center but rather weak at securing goals. For the Normal team the guarding of Miss Langvick prevented any scoring by Brainerd . . . The match resulted 9–0 in favor of the Normals." In 1904, the women of San Jose traveled as far as Reno, Nevada for a game, and the following year they played teams from high schools, Stanford University, Mills College, and the University of California at Berkeley. Geneseo and Fredonia, New York met for two women's basketball games in 1904–1905, and a women's team from Florence defeated the Patton School in 1910.[41]

Women's basketball was not only popular among the participants; it also drew many spectators. The Kirksville-American School of Osteopathy game had a sizable audience, and Geneseo's *The Normalian* reported in 1902 that an inter-society game on a Saturday evening drew a "large crowd." The crowds became so large at Oshkosh in 1897 that for a short period, the players barred spectators, much to the dismay of the male students, one of whom lamented in "A Stroke of Fate":

> They did make plays and no mistake,
> Those girls in blue and yellow;
> And to spur on the lusty crowd
> We cheered them to a fellow.
>
> Now, never a boy meant any harm
> In cheering on the players;
> But several of the girls one day
> Objected to us cheerers.
>
> Our cheering made them play too hard
> Was all they had against us;
> But ever, ever since that day
> The girls will not admit us.

The men's appeals were eventually successful, because the 1898 yearbook stated: "Members of the foot-ball team are often found watching the game from some good safe position," and a later game between two class teams at Oshkosh drew "an audience of about three hundred invited guests."[42]

Even at the height of women's basketball's popularity, female and male normalites diversified their endeavors and even initiated some coeducational ventures in athletics. Normal-school men began to compete in track and tennis, and women also added tennis and various other sports to their repertoire. Cortland had tennis "grounds" as early as 1888, and students organized a lawn tennis association in 1897. Lawn tennis was also very popular among students at Normal, Illinois in the early 1890s; both women and men set up nets wherever they found level ground on campus. San Jose's tennis club advertised, "It is not an exclusive club, quite to the contrary, for every young lady of the Normal is cordially invited to join." The women at San Jose also took up fencing, while some at Geneseo tried "indoor baseball." San Diego's location on the water was ideal for boating; school catalogs in the early 1900s boasted that, "almost daily," crews of women students could be seen "rowing on the still waters of the bay." Oshkosh women played competitive hockey beginning in 1906. Meanwhile, women at Greeley competed in sports ranging from target shooting to golf at their annual field day, and by 1911, the members of San Marcos' five women's societies made a tradition of a yearly social event and track meet, the "famous annual Hill and River meeting."[43]

While athletic activities increased women's visibility in the public sphere, they generally allowed them to form social ties only with other women, as men formed ties with other men. Still, although women and men never met on the basketball court, they did participate together in some non-contact sports. For example, men and women occasionally held joint tennis tournaments at Oshkosh. The rules for an 1897 tournament had "the ladies contesting for a '97 racquet and the gentlemen for a similar prize." *The Normal Advance* further explained, "The final contest will be held commencement week, and then the winners of each prize will play, the award being a two-pound box of candy." San Jose and San Marcos had coeducational tennis clubs by the late 1900s. Oshkosh students formed a coeducational bicycling club, which came to be the "Pedagogical Pedagogues Bicycle Club," in 1896–1897; club members rode together in the late afternoons when the weather was good for "wheeling." San Jose's commencement festivities in 1901 included a baseball game between mixed faculty and student teams, and in 1903 featured a baseball game between the men of the faculty and the women of the senior class. Two years later, San Jose held a spring "field day" in which coeducational class teams competed in a chariot race, tennis, hurdles, a sack race, and a "100 yard rope-jumping dash." These coeducational athletic endeavors epitomized men and women sharing the public sphere, and served as another avenue through which men and women built social capital.[44]

Female and male normal-school students not only created a social world in which women played a visible and active role, but they also shared leadership responsibilities in campus life. Of course women headed all-female literary societies, YWCA chapters, and sports teams. This brought them recognition among their peers as well as greater visibility on campus, as they made welcoming speeches at receptions, planned events with other societies, challenged other teams to basketball games, and represented their organizations at end-of-the-year festivities. Geneseo's *The Normalian* reported: "To be elected final president of one of the literary societies at the Normal is one of the highest honors that can be conferred upon a student."[45]

Outside all-female associations, women were part of the leadership structure of coeducational clubs and societies, and of their classes and student government associations. Men served as organization presidents in numbers disproportionately greater than their representation on campus, but women did occasionally serve as presidents, and often occupied other offices. Oshkosh's senior or graduating classes between 1888 and 1907 had 18 male and 2 female presidents. In the 14 classes that had vice presidents, 8 were women and 6 were men. Women usually, but not always, were class secretaries, and men usually, but not always, were class treasurers. The constitution of the campus-wide self-government association formed at Oshkosh in 1896 stipulated that each class would be represented by one woman and one man. When Cortland instituted a new student-government system in 1901, all students had an equal vote in electing members of the legislative, executive, and judicial branches. The first elected chief justice was Grace PerLee, a woman who took her position very seriously; she later remembered, "In order to learn court procedure I consulted a local lawyer, David VanHoesen, and attended some court sessions in the old court house." When San Jose's student body in 1898 elected Harriet Quilty its first president, *The Normal Pennant* remarked: "Miss Quilty needs no further introduction to our students, her great abilities as a leader are known from Juniors to Seniors." In the 29 semesters that followed, 19 more student-body presidents were female and 10 were male. At San Jose, Cortland and other normal schools, Quilty, PerLee, and other women organized and presided over all sorts of student-government and society meetings and events, heightening the visibility of women in campus life.[46]

At state normal schools, students both learned the value of and began to accumulate social capital by participating in the public spheres of their schools' host cities and of campus life. They lived with town residents, were active in churches, and took part in other local organizations and causes, while they also participated in campus gatherings, outings, societies, and athletic teams. At the same time, the Victorian ideology of separate spheres seemed very distant as students interacted without a significant gender hierarchy. Even in the heart of the South, Florence's administration in 1884 endorsed the school's coeducational social environment:

> This Institution was one of the first in the South to open its doors to young women and young men, and place them on the same footing. For ten years they have been associated in our halls and class-rooms. The experiment has been successful beyond the expectation of its friends and advocates, and its success has done much toward furnishing a solution of this vexed question.[47]

The spirit of "enthusiasm and mutual confidence" that pervaded San Jose and other normal campuses reflected the comfort that students, regardless of their gender, felt in public and in a middle-class environment. Students were so secure in public interactions that they occasionally engaged in good-natured teasing, even regarding gender.

Gender Attitudes

Outside the normal-school campus where students created a comfortable public sphere with few gender boundaries, the late nineteenth and early twentieth centuries were a time of consternation in the popular press, politics, and middle-class society in general regarding proper sex roles. National developments including the entrance of females into colleges and universities, the movement for women's suffrage, and the rise of the unmarried career woman, caused great concern for the survival of middle-class femininity. In addition, the shift from small-scale entrepreneurial capitalism to industrial wage-labor capitalism, the notion that the country no longer had a frontier in need of taming by "rough riders," and the specter of the independent woman created urgent fear that middle-class masculinity was endangered. Thus, according to historians David Tyack and Elisabeth Hansot, "femininity and masculinity were hardly to be taken for granted but rather required conscious buttressing," and an obvious place to start was in the pubic schools. Educational policy-makers were less worried about feminists in the teaching ranks than they were concerned that the preponderance of women teachers was sapping the virility of the nation's young men. One solution to the gender crisis, therefore, was more men teachers; the middle class wanted to bring more of the rough-and-ready masculinity of Theodore Roosevelt into the schools. Female domination of the teaching field, however, raised questions about the manliness of men who sought entry. At the same time, teaching allowed women a degree of independence even though it remained within the boundaries of middle-class femininity.[48] As normalites both shaped and were shaped by the campus public sphere as well as their experiences in the classroom, they could not escape the gender issues confronting the middle class. Yet, perhaps due to a combination of their rural backgrounds and the distinctive gender characteristics of the profession for which they officially were preparing, both female and male students at state normal schools were not prone to express extreme views on feminism or masculinity; rather than fretting excessively about gender, their words—and later deeds—simply supported individual autonomy.

Female Students and Feminism

Women students at state normal schools enjoyed full access to and visibility in the public sphere, as well as rich opportunities for intellectual and professional development. In the process, they (and their male counterparts) absorbed an expansive vision of women's capabilities. They tended, however, to stop short of full support for women's rights; female students as a whole were ambivalent about the subject. Most protested that they were not radicals, and yet they often seemed to hold and act upon a fundamental belief in autonomy for women. In her study of Troy Female Seminary in the mid-nineteenth century, Anne Firor Scott found that Emma Willard was *both* "a prime exemplar of 'true womanhood' " and "a thoroughgoing feminist." The typical female normalite was unconcerned about similar contradictions in her belief system; she weighed the issues one at a time, and often, but not always, supported rights for the individual. As a longtime instructor at Oshkosh Normal in Wisconsin, Rose Swart explained in 1895, the normal-school woman was "not necessarily a bloomer woman, for she is distinguished not by what she wears, but by what she wants and what she does." Scott further explained that Willard and the students she mentored exhibited "the ambivalence which is common when values are in the process of change. Indeed the most effective purveyors of new values were often those who had some attachment to the old, and therefore were not so frightening." Even though they did not identify themselves as feminists, normal-school women likely helped to widen the possibilities for their sex. According to Nancy Cott, during the 1910s the first wave of the women's movement gave way to "modern feminism," which, unlike the earlier "woman movement," stressed individual rights above all else. But female normal-school graduates' ambivalent gender attitudes—as well as the quietly radical choices they would make about their lives following normal school—suggest that "modern" feminist notions of female autonomy existed as an unexamined undercurrent long before the second decade of the twentieth century.[49]

Normal-school students were certainly aware of feminist social currents. Among their instructors were a few outspoken supporters of women's rights, as many faculties included a "bloomer woman" or two. Mary V. Lee, M.D., who taught at Oswego, New York from 1874 to 1892, was an adamant supporter of dress reform. "I weep to think how the body is repressed," she said. She sold patterns for "sensible dresses" to women students, and shortened her own skirts by four inches in order to avoid "gathering up dust and germs." Lee also refused to wear a corset, arguing, "God gave me a proper skeleton to support my flesh and to protect my organs. Why, therefore, should I impose over this one an artificial skeleton?" On the faculty at Normal, Illinois, June Rose Colby, the first woman to earn a Ph.D. from the University of Michigan, was an outspoken feminist who insisted that her skirts have pockets, arguing that women had as much use for them as men did. Julia Flisch at Milledgeville, Georgia and Estelle Downing at Ypsilanti, Michigan spoke out and wrote articles supporting suffrage and other opportunities for women; in addition, Flisch took a public stand on the Spanish–American War and Downing promoted world peace. It was not uncommon for married women to teach at normal schools; two of the first five women employed at Indiana, Pennsylvania were married. Lucy Aldrich Osband, who taught natural sciences at Ypsilanti in the 1880s and 1890s, was married to another faculty member, William M. Osband. After he bought a controlling interest in the local newspaper, both Osbands contributed regularly to the editorial page. Married women and outspoken feminists were in the minority on normal-school faculties, yet their visibility on many campuses exposed students to feminist ideas.[50]

In their literary societies, graduation speeches, and publications, normalites explored fully women's social position, rights, and lifework options. Suffrage was popular among the progressive-era issues and reforms tackled by the literary societies, and even frivolous-sounding debate topics such as "Resolved, that women do more courting than men," at an Agonian-Philalethean joint meeting at Geneseo, New York, and "Resolved that women should be given the right to propose," at a meeting of Oshkosh's coeducational Phoenix society, reflected thoughtful consideration of women's social position. Elloie Powers at Florence, Alabama addressed "Girls, and What Girls Can Do" in her graduation speech, and Addie Washington at Pine Bluff, Arkansas presented "Fair Play for Women" in a commencement-week "Junior Class Program." Between 1878 and 1891, graduation speeches by women at Castleton, Vermont included "Woman's Work," "Woman's Rights," "Woman's Opportunities," and "Work of Women in the Nineteenth Century." During the 1890s, members of Agonian at Geneseo discussed women's options in professions such as medicine, and debated, "Resolved that womans [*sic*] chief aim should be marriage." The all-female Sapphonian Society at Normal, Illinois, debated resolutions including "the maiden lady can do more to benefit humanity than the married lady," and "woman should compete with man in all possible vocations."[51]

Students also came into contact with pathfinding women of their day. At Oshkosh, the Ladies Literary Society featured discussions of the work of Elizabeth Cady Stanton and Julia Ward Howe; and Bertha Hansen entered the yearly oratorical contest with a speech on temperance reformer Frances Willard. When the Every Day Society at San Marcos, Texas focused on "the Great Women of the Day," they studied several women writers, Grace Dodge, Ella Flagg Young, Jane Addams, and Queen Wilhelmina of Holland. Lecture series often featured prominent women, including Mrs. Ballington Booth of the Salvation Army, Florence Kelley of the New York Consumers League, Mrs. Platt-Decker of the National Federation of Women's Clubs, and Jane Addams. When Mary A. Livermore, a social reformer who supported suffrage and temperance, spoke on "What shall we do with our daughters?" at Geneseo in the late 1870s, she argued against raising girls "with the idea that they are to be ladies with nothing to do but sit and wait for a husband." Many of the topics students explored in their myriad activities reinforced Livermore's message.[52]

Despite their exposure to feminist ideology, most normal-school women did not seem to identify with the general movement for rights. Normalites generally did not march in suffrage parades, lobby for property rights, or wear bloomers (except perhaps on the basketball court). When considering abstract radical ideas, they were as likely to take a conservative as a liberal stance. When Geneseo's Sarah E. Youngs considered "Woman in Politics" in *The Normalian*, she concluded that a woman "makes her best appearance as a politician . . . within the sphere were God made her supreme—the household and social life." An 1889 article on dress reform in San Jose's *The Normal Index* urged women to dress in a "becoming, modest and sensible" manner. Three years later, a female student-editor of the same publication concluded her reflection on, "Which is Preferable For Women, Public or Private Life," with the unequivocal statement, "there is no doubt that private life is woman's proper sphere." The editor and most of her classmates were reluctant to count themselves in the woman's-rights camp, even though in their campus activities they violated her dictum on almost a daily basis.[53]

While most female normalites did not consider themselves radicals, they fairly consistently supported certain basic individual rights for women. Crucial to female autonomy was the option of not marrying, and they trumpeted their right to exercise that option. Describing "The Bachelor Maid," San Jose's Lucille Potter made grand pronouncements of equal rights: "She does not ask privileges or consideration above that accorded to her brother, but she demands recognition that shall give her equal opportunities for life, liberty, and the pursuit of happiness. . . . We are treading a new path." Most states during this period required that women teachers remain single; thus, the prospect of "bachelor"-hood was undoubtedly familiar to normalites. Unfazed, San Marcos' class of 1905 made light of the requirement in their "Normal Commandments," which included: "Thou shalt not commit matrimony." After one of the members of the class got married, the yearbook joked: "Some of the Seniors were horrified, and would have as her epitaph, 'She that loveth housekeeping and a husband more than us is not worthy to bear our name.' " Several years later, when San Marcos' Pierian society debated whether marriage should be a goal for women, the negative side won with its defense of "single blessedness." The prospect of unmarried life was hardly daunting to normal-school women.[54]

Although their own attendance at normal school was not controversial, female normal students supported unfailingly women's right to education at all levels. San Jose's *The Normal Index* in 1886 shared the good news of a study by *Good Health* on "Health of College Girls," which did "not indicate that hard and continuous study is detrimental to young women." In *The Normal Gem* five years later, a student at Florence defended higher education for women with equal-rights language: "The theory that woman could not receive the same mental training as men has long since been exploded . . . To be an equal sympathetic companion for man, woman must be educated. . . . To be able to maintain herself and not have to be dependent on some one else, woman must be educated." The article went on to describe the achievements of Mary Somerville in mathematics and Maria Mitchell in astronomy. And, in a 1900 article entitled "Higher Education of Women," Geneseo senior Mabel Deyo made a similar argument with more poetic imagery:

> How can the diamond reveal its lustre unless it is taken from the earth and polished? How can the marble stand forth in all its beauty unless it is taken from the quarry and fashioned by the hand of the artist . . . One great hindrance

to the education of a girl, even in these modern times, is the refusal to admit her to the best schools on equal footing with boys.[55]

When they considered it in terms of individual rights, many normal-school women also supported suffrage. Rhetorical questions on the editorial pages of *The Normal Index* and *The Normal Pennant* at San Jose revealed the editors' opinions on the subject. In 1889, regarding "giving women the privileges of voting in all municipal elections, and for all school officers, city, county, and state," one asked, "Why should this privilege not be given?" Ten years later, another commented: "Governor Gage calmly pocketed the bill intended to allow women to vote for school trustees. Women are qualified to teach, but not to help elect people that choose teachers. Queer logic, isn't it?" In her "Salutatory and Oration" at Oshkosh in 1895, Evelyn Griffin reflected, "There is a remedy offered for reform in politics, perhaps you will tell me it is a wild idea of reform, but it must come, and come through equal suffrage." Griffin then concluded, "When the history of this century is written, even the most conservative historian will give a chapter to 'The Rise of Women.' " A few years later, the headline in junior-class news at Oshkosh was: "An Up-To-Date Move: Suffrage gets a boom—Miss Bridgman becomes [class] president." The yearbook reported, "She is not a 'new woman' or a 'bloomer girl,' yet she believes in woman's rights in so far as they pertain to the ability to govern." The Equal Suffrage Club appeared at the Milwaukee State Normal School in 1901 and had 62 members by 1908. A 1911 history of the institution reported that the club had grown to over 100 members, as "Coming events cast their shadows before."[56]

Regardless of whether they publicly identified themselves as "bloomer women," female normalites helped the cause of women's rights by advocating female autonomy in politics, education, and the decision of whether to marry. Swart said that the "New Woman" was also distinguished by "what she wants and what she does." Many normal-school students' wants for the future—as well as the decisions that graduates would make about what to do with their lives—were based on a fundamental belief in their rights as individuals. In 1897, Nellie Patton's valedictory at Florence included a lofty vision of her classmates' futures:

> Members of this graduating class will form a part of the future citizens of this country. Some of us will go on from here with our intellectual development in college. . . . Others will continue their growth and development in the business of the world—learning to take their places as real women of business, but gaining rather than losing that refinement and charm that makes the true gentlewoman. We will be an influence in politics and the solution of all social problems. . . .[57]

Although somewhat facetious, class prophesies also displayed normalites' support for female autonomy, as well as the comfort they felt in addressing and perhaps joking about gender issues. Graduating seniors expected great things from one another, including academic achievements. Anna Lehner of

Figure 5.5 Students on fire escape of Sutton Hall at Indiana State Normal School in Pennsylvania, ca. 1905.

Credit: Special Collections, Indiana University of Pennsylvania.

Oshkosh's class of 1885 was to be "teaching in a Western College. She is a strong intellectual woman and has done much to advance education in the West." Seniors at Castleton in the early 1900s expected Miss Gibbs to study at Harvard, Jennie Jones to finish college with "high honors," and Sophie Norwood to earn a Ph.D. at Vassar. And, as a widow, Hannah Smith of San Marcos' class of 1904 would publish her studies, "the nucleus of a new psychology." Projections of professional, and even political, success were legion. Oshkosh's classes of 1885 and 1896, Castleton's classes of 1902 and 1903, and San Marcos' class of 1904 were to produce many successful teachers, a couple of principals, and a state superintendent of schools, as well as women writers, editors, lecturers, nurses, a doctor, and a business agent. An Oshkosh woman was to become "the first Governess of the State of Wis.," a Castleton woman would be the governor of Idaho, and a San Marcos woman was to become "a full-fledged congresswoman gifted with customary oratorical accomplishments inherent in womankind." Expectations were not bound by geographical barriers, as these classes were to send missionaries to India and "the icy regions of the Northern light," and teachers to the Philippines and "the Klondike." In addition, classmates pronounced that several women would become outspoken advocates of women's rights. To a certain extent, the students making such predictions were probably teasing one another; still, the prophesies demonstrated that they understood the choices awaiting them and were comfortable addressing women's rights. Although they were ambivalent regarding the larger women's movement, normal-school women believed in autonomy for their sex.[58]

Masculinity and Male Students

Just as gender norms provided fuel for prophesies of female graduates' futures, normalites engaged in verbal play regarding proper male attributes. Occasionally, the humor in pieces for campus magazines hinged on the norms of masculinity, revealing the authors' understanding of society's expectations for men and their assumption that others in the normal-school community would find the gender references amusing. For example, at San Jose in 1885, Mary E. Lynch, one of the editors of *The Normal Index* used loss of virility to convey the nervousness and frustration of taking exams when she wrote that if the exam-taker were "one of the sterner sex, his face lengthens, his eyes sink back, and his cheeks that once glowed with manly strength become pale and sunken." Seven years later, the following mock advertisement appeared in the same publication:

> Wanted—A husband that's healthy, wealthy and wise,
> Who'll take care of baby, bake all the pies,
> Never growl if dinner is not on time,
> But will smile a smile that's very sublime,
> And get it himself.

While the image of a husband who would take care of the baby and get his own dinner was funny because it violated gender norms, the "ad" was also a commentary on those norms.[59]

Women students occasionally joked that their male classmates either lacked or possessed manly qualities. *The Normal Index* published a poem entitled "The Normal Dudes," which began:

They have dudes at the Normal they say,
Who compare well with those of the day.
Oh, don't they look sweet,
As they walk down the street,
All dressed in bright colors so gay.

Looking "sweet" and wearing "bright colors" hardly suggested masculinity. The teasing tone continued as the poem described individual students:

There's Walsh with his numerous blushes and curls,
The well-known favorite of all pretty girls.
And Hughes, the blonde Junior B,
Who's just as sweet as sweet can be.

The author attributed typically feminine qualities to Walsh and Hughes in order to evoke affectionate laughter. When Oshkosh's junior class elected a woman president in 1895, *The Normal Advance* invoked norms of masculinity to add a humorous element to the story. The *Advance* reported that, despite being outnumbered four to one, the junior men had nominated one of their own for president, resolving "to make a manly struggle and bear the defeat, if it came, bravely." That they did, and the class' "bonds of unity" remained intact. Two years later, when Oshkosh's class of 1899 introduced itself in the *Advance*, the 93 female members seized the opportunity to present a tongue-in-cheek description of the 32 male members: "There are fewer hairy faces in our class than in the others. This is doubtless due to the fact that our men have alto or tenor voices and it is only 'basses who can sprout beards.'" These normalites understood gender norms, and had fun with them; rather than fretting about whether the Oshkosh men of '99 or San Jose's "Normal Dudes" were sufficiently masculine to foster virility in the nation's schoolboys, they used gender as a source of affectionate humor.[60]

While students were seemingly unconcerned about the manliness of the nation's future teachers, normal-school principals were quite concerned. Especially as the proportion of female students rose throughout the late nineteenth and early twentieth centuries, they sought to increase the number of male students, and to create a more masculine environment for them. Florence's principal observed in the late 1880s, "The demand for young men as teachers far exceeds the supply. . . . Some steps should be taken to induce young men to take normal training." As one inducement, San Jose's professor C. W. Childs in 1887 invited "a number of the boys" to visit his ranch, probably to help them feel at home at the normal. Two decades later, San Jose's male students and faculty members formed "The Men's Club," the purpose of which was "to promote and cultivate a more united feeling and a better acquaintance with each other" through banquets and other gatherings. More than anything else, however, principals looked to men's athletics to attract and retain male students. Childs, who had taken over the principalship at San Jose, reported in 1891, "It seems desirable to encourage all manly sports, and especially to foster the interest now shown in military drill." Childs' successors further encouraged men's athletics; in 1900, James McNaughton celebrated "the fairness and manliness that have characterized the performance of our students in the public football games this year." Meanwhile, the principals of Wisconsin's normal schools gained control of the formerly student-run men's athletic teams in an effort to increase virility on campus.[61]

Of course, men's sports did not thrive like women's basketball at the normal schools during this period. Still, male students were more visible on debate teams and in leadership positions than their numbers would predict, probably due to their status as the dominant gender outside the normal school and a sought-after minority inside its walls. Grace PerLee, who graduated from Cortland, New York in 1901, remembered, "Boys were so sadly few as compared with girls, as is usual in normal schools, that their opinions of themselves were badly inflated." Castleton's *The Normal Student* in 1903 included a poetic image of a male normalite that was hardly complimentary:

> He wasn't a king, though he held his head
> As high as the haughtiest king,
> And he stalked about with a lordly air
> And a proud and pompous swing.
> He wasn't a king or an emperor,
> Though he built on the royal plan;
> There were forty girls at Normal Hall
> And he was the only man.[62]

Although PerLee's observation definitely suggests that the men thought too highly of themselves, the Castleton poem could have been a warning rather than an observation. In fact, there is no other evidence that the men sought to capitalize on their status or assert their male privilege. Instead, men students endured much teasing by the women with good humor and seemingly without worrying about their masculinity.

The line "And he was the only man," was in keeping with the gentle humourous jabs at male students, most likely written by female students, that appeared in student publications and served as a reminder that the men were following a path unusual for their sex and that they were outnumbered. In the spring of 1886, San Jose's *The Normal Index* reported, tongue-in-cheek, that "The Middle A's, having lost four of their boys this term, are thinking seriously of adopting some plan by which the few remaining gentlemen will be induced to remain at school. The Middle A girls say it is so convenient to have the boys sharpen their pencils." The following fall, the *Index* mentioned, "It has been found that our young men, being so scarce, are very conspicuous, and therefore make excellent targets for bean-bags." Ribbing continued unremittingly in the *Index*, where the following poem appeared in 1891:

> Queer boys there are as we all know,
> In books, on land, on sea;
> But the boys that to the Normal go,
> Are the queerest boys to me.

Ten years later, Castleton's *The Normal Student* noted, " 'The Juniors are informed that they must go to the Training School as women and not as children': there are three young gentlemen in the class. What about them?" While most of the verbal jabs emphasized male students' "queer" or unusual position, some were gentle reminders of their privileged status or assertions that the women did not plan to kowtow to them. The first senior class at San Marcos described itself in the yearbook as "twenty-six young women, each of whom, after solemn deliberation had firmly given up all hopes of matrimony, and eight young men who had given up no hopes at all." At San Jose, *The Normal Index* reported in 1888, "the young men of the school are but a small minority and are, according to the young ladies, by no means the center of social and intellectual development," and *The Normal Pennant* reported in 1899, "tennis belongs mostly to the girls. Of course they are pleased to welcome the Normal boys on courts for a game, but they do not care to have the boys monopolize the courts."[63]

Instead of becoming defensive about their decision to attend normal school or asserting that they deserved special treatment, the male students seem to have played along with the women's ribbing. At Castleton, Harley S. Goodwin wrote in "Social Life Of The Boys At The Normal," that a "careful father" remarked, " 'I am kinder wishin' my boy could come to this school but I don't hardly want to send him 'cause there's so many girls here I'm 'fraid they'll spile him and he'll never be good for nothin' again.' " Goodwin went on to complain facetiously that the seats reserved for "boys" in chapel were directly in front of the principal, that the women expected them to sharpen their pencils, and that, "When Friday night comes around, the girls, of course, expect that we are going to come to the Normal and visit each one personally not less than two or three hours." Men also took teasing about their masculinity in stride, or at least did not make noteworthy efforts to prove their manliness. Instead, Clarence G. Fasset at San Jose wrote in " 'The Noble Nineteen' or the Boys of S.N.S.," that normal-school men were distinctive because they were true, noble friends:

> Now we are less than twenty;
> A fact that's sad but true.
> Of girls there are plenty;
> Of boys there're only a few. . . .
> We're not, as some would think us,
> Almost extinct. Nay! Nay!
> Nor yet are we degenerate,
> As we've heard others say. . . .

And we will be your friends, true;
The best and noblest kind.
The Normal boys, although few,
Their like is hard to find.

Fasset, Goodwin and their male classmates seemed comfortable with their minority status and their masculinity.[64]

At the end of its junior year, Castleton's class of 1903 passed a resolution regarding "the attitude of the fair sex toward the boys": "we will consider it a kindness and for their welfare to treat them as ordinary human beings." The sentiment, is seems, was mutual. Like the women, male normalites gained considerable exposure to the cause and concerns of the women's movement of the late nineteenth and early twentieth centuries. Also like the women, they seemed to be somewhat ambivalent, yet to support autonomy for women in many areas. After listening to a debate on women's suffrage at a meeting of Oshkosh's Phoenix in 1895, one man in the audience "remarked that each woman should have a *voter*, but not a vote." Other normal-school men, however, worked alongside female students on editorial boards and debate teams as they supported individual rights for women. In addition, in closed sessions of Florence's all-male Lafayette Society during the 1880s, the negative side won the debate on, "Resolve [*sic*] that the two sexes should not be educated together," and the affirmative team won the debate on, "Resolve [*sic*] that the mental capacities of the sexes are equal." Similarly, the affirmative team convinced the judges that "the sexes are mentally equal" in an 1899 debate among members of Geneseo's all-male Philalethean Society. Also in 1899, Edgar J. Doudna of Platteville State Normal School entered the Inter-Normal Oratorical Contest in Wisconsin with a speech on Frances Willard; however passionate he was about Willard, he lost to Elizabeth Shepard of Oshkosh. Doudna and other male students seemed to learn from their unique opportunity to develop respect for women as intelligent beings and legitimate actors in the public sphere. Isaac Rogers's quip in his "History of the Junior Class" at Florence in 1913, "There is little need for me to speak much of them, for co-eds speak for themselves," was a lighthearted admission that women were indeed "ordinary human beings." Many would go on to make life choices that reflected their sense of individual autonomy.[65]

LIFE CHOICES

Attending state normal school between the 1870s and the 1900s acquainted students with many of the accouterments of the middle class. Through their studies and activities, normalites absorbed high culture, adopted a professional outlook—albeit in teaching, a field with ambiguous standing as a profession—and created community ties through formal associations and informal sociability. Students learned in many ways to think and act like members of the middle class and thus enjoyed a sort of cultural-class mobility. Still, the normal-school experience was no guarantee of socioeconomic mobility or full middle-class status. As Christopher Jencks and David Riesman have documented, "the distribution of power, prestige, or even self-respect" is fluid; "a farmowner's son is still in the middle of the heap if be becomes a school teacher, even though this means he has nominally risen into the 'professional class.' "[66] Nevertheless, an examination of the career and personal paths of normal-school graduates reveals that many did enjoy a degree of class mobility as their life choices took them far from their provincial backgrounds. Normal school helped to open their eyes to possibilities and thus served as a launchpad to further opportunities. For male graduates, the succession of experiences that began with normal school often led to middle-class, or very occasionally higher, status. Some class mobility was also an element of female graduates' life paths, which, more significantly, often also involved the exercise of a good deal of autonomy.[67]

Men

Normalites generally signed a pledge as they began their studies to teach once they finished, and thus graduated with an obligation to teach. Most men fulfilled this commitment, but did not make teaching a life-long career. The average tenure in teaching for men who graduated from Bridgewater, Massachusetts during the administration of Albert Boyden between 1860 and 1906 was ten years. Of the men who graduated from San Jose, California in the 1870s, 1880s, 1890s, and 1900s, 81 percent taught following normal school. Close to 90 percent of male graduates of Oshkosh, Wisconsin between its opening in 1871 and 1900 worked as teachers or school administrators, logging an average of between six and seven years by 1906. Among male graduates of the advanced course, 91 percent worked in the schools for just over seven years on average, and among male graduates of the elementary course, 84 percent worked in the schools for just over five years on average.[68] After graduating from Farmington, Maine in 1873, Enos F. Floyd taught in California until 1888; he reported in 1889 that he had "taught 496 weeks" and "been a member of the Board of Education of Calaveras Co." since 1876. Orrey H. Hoag, who graduated from Geneseo, New York in 1884, also had a long teaching career for a normal-school man. He taught for four years before attending Geneseo, and then taught in four different districts between 1884 and his death in 1912. Hoag's classmate Herbert E. Millholen was more typical, however; he taught and served as principal at Geneseo's village school for just one year before leaving the field of education.[69]

Men tended to move quickly from teaching into school administration, which had more status among middle-class professions. Millholen enjoyed the title of principal during his one year in the field, and in the 1890s Hoag served briefly as principal and then superintendent but made the unusual move of sticking with teaching. Floyd's classmate at Farmington, Ezra F. Elliot taught for several years and in 1886 became Superintendent of Schools in Polk County, Minnesota. Similarly, Thomas B. Price (Florence '78) taught for several years in Alabama before holding a succession of principalships in Alabama and Texas. James B. Cunningham began teaching in Birmingham after his graduation from Florence in 1886, and by the late 1890s was principal of the city's largest public school and had contributed several articles to publications including *Educational Exchange* and *Southern Educator*. By 1889, several male graduates of the State Normal School of Kansas in Emporia had blazed a trail from teaching into administration: Charles M. Light '75 taught in five different towns and then served for four years as a county superintendent before assuming the superintendency of schools in Chanute; T. W. Conway '79 taught for two years before becoming superintendent of schools in Independence, where he also served as president of the State Teachers Association in 1887; and Benjamin M. Ausherman '84 taught at Humboldt High School for two years, was principal in Greenburg for three years, and then became superintendent of a boarding school for Native Americans. More than 20 percent of male San Jose graduates of the 1870s and 1880s gained experience in administration, and more than 40 percent of 1890s and 1900s graduates did so. Oshkosh did not differentiate between teaching and administration in tabulations of graduates' tenure in education, but it is likely that a large proportion of men's time in the field was in administration. For example, Edwin G. Beardmore '96 taught the seventh and eighth grades for only a year before moving on to a succession of principalships.[70]

By the late 1890s, it was also quite common for men to step immediately into administrative positions upon their graduation from normal school. John W. Calnan (Oshkosh '98) served as principal at Plover, Wisconsin and then at schools in Oshkosh between his graduation and 1905. He moved on to city superintendencies in Wisconsin and then Berthold, North Dakota from 1905 into the 1910s. Following in his footsteps were J. Elden Beckler '00, Frederick William Oldenburg '02, George Gustav Price '04, Walter P. Hagman '05, and many male graduates of other normals. Four men of the class of 1891 at Oneonta, New York became principals immediately upon graduation; one of them was Wilbur Lynch, who oversaw 5 teachers and 300 pupils. Frank H. Sincerbeaux graduated from Oneonta two years later and stepped immediately into a position as a public-school principal. While Sincerbeaux would only remain in school administration for five years, other male graduates had long careers and achieved high administrative positions. Lynch went on to become superintendent of schools in Amsterdam, New York, where a high school would bear his name. Three men who graduated from Kirksville, Missouri occupied the state superintendency in Missouri, and two of them went on to be normal-school presidents: John R. Kirk '78 became president of Kirksville in 1899 and W. T. Carrington '78 became president at neighboring Springfield in 1906. An 1882 graduate of Westfield, Massachusetts, Marcus White taught in Massachusetts and Pennsylvania before becoming principal of New Britain, Connecticut in 1894 and remaining in the position for 35 years. George William Brock (Florence '93) held three high-school principalships in Alabama before becoming the superintendent of Opelika schools in 1902. In 1907, he joined the faculty of the State Normal

School in Livingston, and in 1910 he assumed the presidency of that institution.[71]

Some normal-school men who did not fulfill their obligation to work in the schools or who taught for a short time, elected to pursue occupations that were clearly outside the realm of the professions of the urban middle class. An 1889 listing of the current occupations of 117 men who graduated from Emporia between its opening and 1885 included 37 farmers and/or stock-raisers, a mail carrier, a "quarryman," a "blacksmith and wagon maker," and even a gold prospector. Wilbur L. Strickland (Westfield '72) taught for six years after graduation, but by 1889 was a "carpenter, joiner, agent for farm machinery, etc." A "sketch" written the same year of Frank C. Foster (Farmington '79) summarized: "After graduating, he finished out a term of school, teaching three weeks. He then learned the trade of machinist which he worked at five years, then came home to take charge of the farm, which he has since carried on." Everett C. Dow (Farmington '77) also returned to farming, but his "sketch" indicated that he remained involved in middle-class associations: "Taught 100 weeks . . . Is now farming and managing a milk route. Is superintendent of the M. E. Sunday School in his village, and a member of the S. S. Committee in his town." John G. Voss (Oshkosh '98) served as a county school superintendent for ten years, then became a "breeder of pure-bred Holstein-Fresian, cattle"; while the latter occupation was more removed from middle-class professional life than the former, Voss did serve as secretary of the Wisconsin Holstein Breeders' Association and he represented the state in the National Holstein Association. Although Voss and Dow maintained middle-class mores in their work and personal lives, they and a number of other normal-school men did not pursue professions with middle-class status.[72]

The occupational paths of the majority of male graduates, however, were squarely within the bounds of middle-class professionalism: while some normal-school men rose within the field of education, others established careers in academia, law, medicine, and other fields. In many cases, further education facilitated their professional advancement. So many former members of a Cortland, New York literary society were at Amherst College in the 1890s that they formed an alumni club. Similar "colonies" of Illinois State Normal graduates appeared at Harvard, Swarthmore, and the University of Michigan. During his nine years in teaching, Farmington's Ezra Elliot attended Bryant and Stratton Business College, the Waterville Classical Institute, and Colby College, where he earned a bachelor of art's degree. In 1889, 11 Farmington men held bachelor's degrees, 7 had done some college work, 11 more were attending college, and 24 others had undertaken academic "work not classified." While working as school administrators, several Oshkosh graduates enrolled in summer university courses: Frederick Oldenburg attended seven summer sessions at the University of Wisconsin; George Price enrolled at the universities of Chicago and Wisconsin and took manual-training courses at Stout Institute, Bradley Institute, and Armour Institute; and Walter Hagman attended the University of Chicago for three consecutive summers. Westfield's Marcus White took time out from teaching in order to earn a bachelor's degree at Wesleyan University in 1888; his class included another Westfield graduate, Henry P. Griffin '83. And, between two principalships, Oneonta's Wilbur Lynch entered Harvard in 1895 and completed a bachelor's degree there, while Frank Sincerbeaux graduated second in his class at Yale.[73]

Advanced education led some normal-school men to successful careers as scholars and writers. After completing the classical and scientific course at the University of Kansas, L. L. Dyche (Emporia '77) joined the faculty there; in 1889, he was "professor of anatomy and physiology, and taxidermist." Emory D. Kirby (Emporia '81) taught briefly before attending Battle Creek College and the University of Michigan; in 1889, he was "professor of Latin and Greek in Battle Creek College." Walter B. Ford (Oneonta '93) joined Lynch at Harvard after two years of study at Amherst College; he went on to teach at the University of Michigan, study at the University of Paris and complete a Ph.D. at Harvard, and become a well-known mathematician. Another Oneonta man, Emory B. Pottle '95 graduated from Amherst in 1899, after which he published stories in *Harper's Bazaar*, *Atlantic*, and other magazines, edited *The Criterion*, and taught English at Columbia University. Geneseo's Herbert Millholen, who had taught for just one year, earned a bachelor's in philosophy degree at Cornell University in 1889, and then worked for 35 years as a staff writer and editor for the *New York Evening Post*, the *New York Commercial Advertiser*, and the Associated Press. William E. Ritter (Oshkosh '84) studied chemistry and zoology at the University of California and Harvard, and then advanced from instructor to professor of zoology in addition to directing the Scripps Institution for Biological Research at the University of California. Among the other Oshkosh graduates who became academics were Ezra Thayer Towne '94, professor of economics and political science at Carleton College and author of *The Organic Theory of Society*, and Otto Louis Kowalke '97, professor of chemical engineering at the University of Wisconsin and author of various publications in his field. In addition, the first Rhodes Scholar from Oklahoma was Walter S. Campbell, a 1908 Weatherford graduate. After earning an advanced degree from the University of Oxford, he wrote respected histories of the American West under the pen name Stanley Vestal and taught writing at the University of Oklahoma.[74]

The legal profession was a very popular area of study and work among men who had attended state normal schools. Over 30 percent of male San Jose graduates between the 1870s and the 1900s who undertook advanced studies focused on law, and 12 percent of the men who graduated from Oshkosh between 1871 and 1900 were working as attorneys in 1906. One of the oldest, John F. Burke '75 practiced in the Pabst Building in Milwaukee. One of the youngest, George J. Danforth '00 held an L. L. B. from the University of Wisconsin and was embarking upon a successful career in Sioux Falls, South Dakota. Bernard R. Goggins '84 practiced in Grand Rapids, Wisconsin, where he had been district attorney in the mid-1890s and mayor in the early 1900s. After Yale, Oneonta's Sincerbeaux completed law school at Columbia University and practiced in New York, where he cofounded the law firm of Sincerbeaux and Shrewsbury. John B. Donovan (Farmington '76) earned a law degree at Boston University in 1880; he served in the Maine legislature in 1883 and was practicing law in 1889. At that time, another Farmington graduate, Arthur C. Rounds '80 was midway through the three-year course at Harvard Law School, where he edited the *Harvard Law Review*. Also in 1889, five male Emporia graduates were working as attorneys, including M. Luther Rees, who was using his legal expertise to manage an investment company in Dallas. Florence, Alabama counted a slew of lawyers among its alumni. They included Robert T. Simpson, Jr. '82, who, in addition to practicing law in Florence, served two terms as solicitor for a judicial circuit, president of the Florence Chamber of Commerce, and an elder in the Florence Presbyterian church, and J. M. Pratt '05, who served in the Alabama legislature while attending law school and then practiced law in Carrollton while also serving as the town's mayor and later editing the local newspaper. Even the African American normal school at Pine Bluff, Arkansas produced a few lawyers: members of the class of 1884 were practicing law in Pine Bluff and Chicago by 1900, and O. Benjamin Jefferson '99 earned an L. L. B. and practiced in Oklahoma after working as a teacher.[75]

Nearly as popular as law, medicine attracted many male normal-school graduates. More than 20 percent of the men who graduated from San Jose between the 1870s and the 1900s and undertook advanced studies focused on medicine, and over 10 percent of the male Oshkosh graduates between 1871 and 1900 were working as physicians in 1906. They included John Tarleton Scollard '81, M. D., who practiced in Milwaukee after teaching Obstetrics at Wisconsin College of Physicians and Surgeons and Diseases of the Chest at Milwaukee Medical College. Wenzel M. Wochos '99 was a "Physician and Surgeon" in Kewaunee, Wisconsin; he held an M. D. from the University of Illinois and would undertake post-graduate studies at Johns Hopkins and other hospitals. Samuel Middleton and Samuel Courtney (Westfield '71 and '82, respectively) earned M. D. degrees at Harvard, and Emanuel Kauffman (Emporia '85) worked as a physician after taking "a course of medical lectures at Jefferson Medical College, Philadelphia." Of the 11 men profiled for the 50-year reunion of Geneseo's class of 1884, 2 became medical doctors. Frank Sidney Fielder "practiced in New York and was Medical Inspector in Health Department from 1894 till his death" in 1917. Meanwhile, Herbert W. Hoyt "practiced in Rochester 1891–1925; six years in general practice and after study in New York and abroad became specialist in Ear, Nose and Throat." Hoyt married a medical school classmate, and "after his retirement, he and Mrs. Hoyt took a years [*sic*] round the world trip." Three members of the class of 1896 and three members of the class of 1906 were among the Florence alumni who worked as physicians, and several others practiced dentistry. Eight Pine Bluff graduates of the 1880s and 1890s became M.D.s, while John W. Corbin '88 and Harrison Duke '99 went into dentistry.[76]

In addition to medicine, law, and academia, male normal-school graduates pursued a wide range of other occupations that fit within urban middle-class professional culture. A few went into the ministry, others became engineers, and many engaged in various business or financial pursuits. Almost as many San Jose men worked in business as in law. For example, Ralph B. Matthews '01 earned a bachelor's degree at Stanford after teaching for few years, and then worked for a San Francisco insurance company, while Alexander Cuthbertson '02 also graduated from Stanford and attended Yale before gaining "considerable business experience, in real estate and as a chemist," and then returning to teaching in the mid-1920s. The 1889 listing of the current occupations of 117 men who graduated from Emporia between its opening and 1885 included 24, or just over 20 percent, who worked in insurance, real estate, or banking. A former school superintendent, John F. Kirker '75 "engaged in hardware business . . . for some time," and then "engaged in the banking business." While also superintending the local schools, Oshkosh's John Calnan ran a "Hardware, Furniture and Undertaking Business" in Berthold, North Dakota, where he also was chief of the fire department and a member of the executive board of the North Dakota Funeral Directors' Association. A handful of the men profiled for the fiftieth anniversary celebration of Geneseo's class of 1884 had careers in business; among them were John A. Milroy, who was "actively and successfully engaged in real estate in Houston, Texas," and Walter S. Palmer, who for 25 years "was prominent and successful in his work with Standard Life Insurance Company." Florence graduates also made their mark in financial pursuits; in 1898, S. H. Morris '80 was "merchandizing in Lacon, where by his enterprise he has contributed largely in building up his town," and Benjamin Sherrod '88 was "employed in banks," while Sterling P. McDonald '95 would become a bank president within ten years of his graduation.[77]

In all professional fields, there were a few normal-school men who gained statewide or even national prominence, leaving no doubt that they had achieved middle-class, or higher, status. Attorney George B. Cortelyou (Westfield '82) was secretary to President William McKinley, a member of Theodore Roosevelt's cabinet, postmaster general, secretary of the United States Treasury, and, finally, president of Consolidated Gas Company. Among the many normal-school men elected to state office, attorney Edward B. Almon (Florence '82) was speaker of the Alabama House of Representatives in 1911 and then represented the Eighth Congressional District of Alabama in the U.S. House for more than fifteen years. Physician, medical inspector, and Geneseo graduate Frank Fielder was a "recognized authority on smallpox vaccine and rabies" who published "a long list" of articles and held a high post at the Bureau of Laboratories. After graduating with honors from Harvard and working as private secretary to Lee Higgenson, the head of Boston's largest stock brokerage firm, Seth T. Gano (Oneonta '00) established himself as a leading businessman in Massachusetts. Gano's classmate at Oneonta and at Harvard, physicist Harvey Hayes won the civilian Distinguished Service Award from the U.S. Navy in 1945. In each of these cases, attendance at normal school was early in a succession of experiences that culminated in high achievements. It is no surprise that

J. S. Nasmith, who attended Platteville, Wisconsin in the early 1870s, reflected after 60 years that normal school "woke me up; made me desire to be somebody, do something worthwhile." For him, this meant preaching as a Baptist minister on the frontier and helping his wife raise four children, who were, he reported proudly, "all college graduates and all but one married to college graduates." Nasmith, like many other normal-school men, had clearly moved his family into the middle class.[78]

Women

Women graduated from state normal school with two obligations: their alma mater required them to teach for a couple or a few years as repayment for an almost-free education, and society expected them to marry within a number of years. Most women fulfilled one or both obligations, in a manner that confirmed their middle-class sensibilities and their notion of individual autonomy. While teaching did not guarantee a middle-class income, it did square with middle-class gender ideology while also allowing women a degree of independence. Normal-school principals throughout the country testified to the high rates at which their graduates, the majority of whom were women, taught. San Jose's principal Charles A. Allen reported in 1884 that "about sixty percent of all the graduates of the school since its beginning in 1862, taught in the State last year," and a teacher of ancient languages at Geneseo in 1891 found that 98 percent of the school's graduates had taught, in nearly every county of New York as well as in other states. The Commissioners of Higher Education in Kansas reported in 1913 that a "full 95 per cent" of graduates of the state's normal schools "have become teachers and have devoted an average of five and one-half years to this profession," and "more than one-half of the whole number are now actually engaged in the profession of teaching." Indeed, the vast majority of female normal-school graduates followed through on their promise to teach. Approximately 90 percent of the women who graduated from San Jose between the 1870s and the 1900s taught for some period of time, and 94 percent of the women who graduated from Normal, Illinois between the 1860s and the 1890s taught for three or more years. Female graduates of both Bridgewater and Oshkosh during the last three decades of the nineteenth century taught for an average of well over eight years. The women in Oshkosh's class of 1878 averaged as many as 15.7 years in the schools, while the class of 1890 averaged only 6.4 years; 44 percent of the women who graduated in 1896, 50 percent of those who graduated in 1898, and 57 percent of those who graduated in 1900, were still working in education in 1906.[79]

The majority of normal-school women also married, and marriage was a primary determinant of women's social standing during this period. As historian Geraldine Jonçich Clifford has pointed out, "Unlike working-class women generally, teachers were better positioned to 'marry up' as a consequence of their greater visibility, geographic mobility, and genteel image"; indeed, normal-school alumnae tended to marry members of the middle class. An 1889 list of the "occupations of the husbands of the graduates" of the Western Maine Normal School in Farmington indicated that the largest group of "husbands," 41 out of 172, was farmers. Still, almost 90 Farmington women, or over 50 percent, were married to men who engaged in occupations that were more clearly middle class, as the list also included: 27 merchants, 11 clergymen, 11 manufactures, 10 lawyers, 2 superintendents of schools, and 25 or so others who were insurance agents, real estate brokers, bankers, engineers, clerks, and the like. Similarly, biographical information on graduates of Mankato, Minnesota published in 1891 indicated that more than a dozen female graduates of the 1870s were married to farmers, while two were married to lawyers and among the other husbands were a druggist, a "real estate dealer," a banker, a merchant, a dentist, a physician, a teacher, a school superintendent, a judge, and a state senator. Several women's profiles for the 50-year reunion of Geneseo's class of 1884 included indications that their husbands' occupations or social positions were middle class or higher. Jennie E. Webster married "a leading merchant," Georgiana Wilkie married "a business man," and the husband of Reta A. Butler, William J. Hoyt was "a prominent business man and an upright and influential citizen," who had also attended Geneseo. Lina K. Scott (Mrs. Ednor A. Marsh) had "a perfectly good husband . . . a lawyer and a prominent citizen of Rochester." Margaret E. Kavanagh's profile did not state her husband's occupation, but did mention "a very pleasant home on North Ashland Avenue, Chicago" and their "delightful year of European travel and much travel in this country."[80]

While female normal-school graduates generally wed within the middle class, their decisions regarding whether and when to do so reflected a strong sense of autonomy. At a time when approximately 90 percent of American women married at some point in their life course, the marriage rates for normal-school women ranged between 40 and 80 percent. As the *Crucible* at Greeley, Colorado reported in 1904, "Of the 586 female students graduated from the Normal . . . 415 still remain unmarried. . . . Let those who lament we are losing our graduates to marriage take courage." Seventy-nine percent of the women who graduated from Florence between 1877 and 1899 and 62 percent of those who graduated between 1890 and 1904 were married by 1914. Outside the South, normal-school women wed in smaller numbers: 63 percent of the women who graduated from Castleton, Vermont before 1910 and between 58 and 66 percent of the women who graduated from Normal, Illinois during the 1870s, 1880s, and 1890s, married. Rates were even lower for Oshkosh women: only 44 percent of the graduates of the mid-1870s through the mid-1890s married by 1906. Thus, normal-school women—like female college graduates—married at rates significantly below the national norm.[81]

In addition, at a time when American women wed, on average, by their early 20s, female normal-school graduates—again like college women—tended to delay their nuptials until they were more mature. Tempe, Arizona graduate Effie Richardson reflected in later years that she had viewed normal school as "a way to wait until I was older to marry. With a Normal education

I could get a job easily, and I didn't have to marry just anyone." Another Tempe graduate, Irene Bishop remarked, "I suppose I always thought I would get married. But I wanted to be on my own for a while. I wanted to make my own money and be independent." Married Oshkosh alumnae of the 1870s to the mid-1900s averaged 5.6 years of teaching experience, which suggests that most were in their mid-20s or older before they wed. Of the 70 percent of the women in the class of 1908 at California, Pennsylvania who reported at the time of their 20-year reunion that they had married, half had waited 10 or more years after leaving the normal to do so. There is no doubt that Callie Curtis, an 1876 graduate of West Liberty, West Virginia, was a mature bride, as she waited four decades until she finally married. Being single was an important prerequisite for female autonomy in the late nineteenth and early twentieth centuries, and, according to historian Barbara Miller Solomon, "College education made it possible for a woman to approach single-hood in a new way." Normal-school women remained single at higher rates than most, and some would continue to indulge their expansive spirit after marriage. In the words of historians Madelyn Holmes and Beverly J. Weiss, they "seized the opportunities for 'autonomy and advancement' and the responsibilities that were open to them."[82]

Marrying at an older age was a gentle, and probably unintended, challenge to societal expectations regarding marriage. The many alumnae who delayed their nuptials often pursued further education and had fairly significant careers during the period between normal school and marriage, indirectly further testing the boundaries of middle-class gender norms. On the West Liberty faculty from 1887 until she resigned in 1916 to get married, Callie Curtis taught several subjects and served as both assistant and acting principal. Mary B. Rawson, an 1878 Westfield graduate, taught school and studied Greek and Latin until she married in 1882. Her classmate Hattie T. Chapin spent the ten years between graduation and her marriage teaching in Massachusetts and then at the "state school for feeble-minded children" in Illinois. After graduating from Geneseo in 1884, Alice C. DeVoe "taught for four years at Flushing, Long Island, two years New Paltz Normal, then State Normal at Florence, Alabama, where she taught four years" before marrying in 1895. During the 20 years that DeVoe's classmate Jennie McLaughlin waited before marrying, she taught in Michigan and three different regions of New York State, as well as "traveled to Europe and studied in Paris." While still single, Flora Emogene Chapman (Oshkosh '88) taught at a high school, then attended Cornell University for three years before teaching at Milwaukee Normal and Teachers College of Columbia University, and Annie Lou McIntyre (Florence '90) taught for five years in the public schools of Birmingham and Montgomery, also attending summer school at Cook County Normal in Illinois. Georgina Groleau of California's class of 1908 taught in the public schools for six years while attending summer sessions at the University of West Virginia and Grove City College; in 1914, she graduated from the latter institution and married a local school superintendent the following day.[83]

A small number of these women took the more radical step of teaching or doing other paid work after they married. Descriptions of the activities of Farmington alumni published in 1889 indicated that a half-dozen or so of the women taught while married. Clara F. Elliot '73 married classmate Enos Floyd in California in 1877, "and taught with him nearly 200 weeks." After marrying in 1882, Emily J. Richards '78 was "Assistant Principal two years, and Principal for two years of the Evening High School" in Fall River, Massachusetts, where she had also taught before her marriage. Alice M. Norton '79 and Alice I. Foster '80, who married in 1880 and 1883 respectively, were both teaching in Rangeley, Maine in 1889; Norton was also a mother. Margaret Lyons, who graduated from Whitewater, Wisconsin in 1874 and married L. P. Wilcox in 1877, continued her education and her career with her husband at her side. According to a brief biographical account published in 1893, they both taught at Oshkosh Normal and then in Milwaukee though 1880. "The next year was spent at the Wisconsin University. They then entered the University of Michigan, where both husband and wife graduated from the Literary Dep't in '85 and from the Law Dep't in '87. Since '87 Mrs. Wilcox has taught Latin, Greek, and mathematics in the Lake High School" in Chicago. Two dozen or so graduates of San Jose between the 1870s and the 1900s taught after they married, including Katie Mitchess '76, who "taught in Lassen, and Contra Costa Counties. Went to the Hawaiian Islands in December 1888, on account of her husband's health, and taught there a short time." Her classmate, Mary Packham married soon after graduation and had one child, yet had a long career as a teacher at San Jose High School. More than one-third of the married graduates of Normal, Illinois in the 1870s, 1880s, and 1890s as well as of California, Pennsylvania in 1908 taught or engaged in other work for pay after marriage. Georgina Groleau, who had since become the mother of two sons, in 1923 returned to teaching as an instructor of summer sessions at a local college, while her classmate Victoria Wilson, a wife and mother of four, taught from 1923 to 1925 and then worked as a register assessor in her township.[84]

Most of the Pine Bluff alumnae who taught before marriage continued to teach afterward, as it was fairly common for married black women at the time to work. For example, Josie Pierce '93 taught in various Arkansas towns before and after she became Mrs. Woolfork. Another southern black woman, Olivia Davidson, who had journeyed north and graduated from Framingham, Massachusetts in 1881, taught at Tuskegee Institute in Alabama before and after her marriage to Booker T. Washington. Even some southern white women taught after marrying; their ranks included 31 Florence alumnae whose listings in a 1914 "Register of the Alumni" indicated that they were both wives and teachers. Theresa Kachelhofer '79 "continued to teach before and after her marriage" in Kentucky. Two other Florence alumnae, Ella Lowery '83 and Mattie Wesson '88, taught before marrying and had careers as writers afterward.[85]

The high percentage of normal-school graduates who did not marry usually had to support themselves, and teaching was a good means for doing so.

Many logged a remarkable number of years and a wide variety of experiences in the classroom. Christabel Campbell, who was a member of Geneseo's class of 1884, taught for four years in western New York State, and then for 32 more years in Colorado. The fiftieth-anniversary profile stated that her classmate Sara M. Harrington had a "remarkable career," as she gave to Victor, New York "a lifetime of service both in the school and in the village . . . She has been preceptress and later vice-principal of the school, teaching German and then history," while logging 17 years as president of the Women's Club and 10 years as village historian. During her 50 years of teaching and community activities, Harrington also found time for further study, taking on "summer work at Chautauqua, library work in Albany and work at the University of Rochester." Another member of the same class, Margaret B. Mann taught in small towns in New York, Yankton, South Dakota, Great Falls, Montana, and then spent 25 years at Schenectady High School before finally retiring in 1930. Along the way, Mann earned degrees from the New York State College for Teachers in Albany and Columbia University, and traveled to Yellowstone National Park and throughout the Great Lakes; in 1931, she toured Europe. After graduating from Oneonta in 1901, Anne Scott taught in public schools on Long Island and then at the state normal in New Paltz before returning in 1913 to Oneonta, where she was a critic teacher and, beginning in 1917, head of the geography department. Before her retirement in 1940, Scott also completed B.A. and M.A. degrees at New York University, studied at four other universities, and traveled considerably to improve her knowledge of geography. Meanwhile, Emma Bunn Matteson (Oneonta '02) served on the faculty at the state normal in Winona, Minnesota, at Simmons College, and at George Peabody Teachers College, and authored *A Laboratory Manual of Foods and Cookery*. In addition to the Oswego graduates who staffed normal schools throughout the country, Harriet Stevens, Caroline V. Sinnamon, and Caroline L. G. Scales devoted a total of more than 130 years to teaching at their alma mater.[86]

Of course, alumnae of normal schools outside New York had extensive teaching careers as well. An 1891 graduate of Pine Bluff, Luella Allen taught in Hot Springs, Arkansas, from the early 1890s into the late 1920s, while Sarah Murphy '96 taught in Arkansas and Oklahoma for upward of thirty years. A few Florence graduates of the 1890s and 1900s logged decades in teaching, taking time away only to earn bachelor's degrees at the University of Alabama, and almost ninety women who graduated from Normal, Illinois between the 1870s and the 1890s went on to teach for twenty-five or more years. After more than thirty years of teaching in "Dakota," Wisconsin and Minnesota, Libbie Reid (Mankato '72) was awarded a gold watch with the inscription, "Presented by the St. Paul Dispatch to Miss Libbie G. Reid, as the most popular teacher in Minneapolis, July 11, 1890." She wore the watch proudly to her new job as a school principal in Shakopee, Minnesota. From the time of her graduation from Oshkosh in 1896, Ella Elizabeth Harrington taught history and English at a high school in Manitowoc, Wisconsin. She spent one summer at the University of Chicago and, in 1910, became the head of the history department at another high school in Manitowoc. San Jose's Julia Hauck '75 had a long and varied teaching career that spanned more than half a century and two continents; a 1927 description focused on the highlights:

> Miss Hauck's history includes nine years of study [, teaching] and travel in Europe. While teaching in Oakland she took a course at the State University. For eleven years she has been connected with the San Jose Evening Schools where her proficiency with modern languages and her background of experience among the peoples of Europe have made her an invaluable instructor. . . . her influence in adult education has been a marked piece of work in the San Jose school department.[87]

Other normal-school alumnae rose to great heights in the profession as administrators. Like Mankato's Libbie Reid, many served as principals of elementary and even high schools. Eight women in Oneonta's class of 1891 stepped immediately into positions as principals or assistant principals following their graduation. After teaching for a number of years, Mary Jones (Florence '77) assumed the principalship of the Primary Department of the Normal and Industrial College at Milledgeville, Georgia; she went on to earn a diploma at Teachers College, serve as principal of the Primary Department at Peabody Normal College, and, finally, teach at Florence. Oswego instructor A. P. Hollis wrote in 1898, "Lady graduates from Oswego have frequently been pioneers in securing recognition for their sex as school officers. One lady graduate of the first class served five years as county superintendent on the Washington Territorial Board of Education. Another lady graduate had been superintendent of public schools in Iowa City, Iowa," and others "have served as county superintendents in New York State." Lucile Rogers and Willie Saathoff of the class of 1908 at San Marcos, Texas also served as county superintendents. More than thirty-five women who graduated from San Jose between the 1870s and the 1900s filled superintendent positions, while Mary Madden '78 became an "Educational Executive" in San Francisco and Maude Murchie '07 was the State Supervisor of Teacher Training. Hollis also boasted that "another [Oswego woman] was State Institute coordinator of Minnesota . . . while a recent lady graduate was a member of the State Council of Nebraska." Isabelle Eckles, in 1895 the first graduate of the New Mexico Normal School in Silver City, taught in the public schools for 13 years; then, during the next two and a half decades, she held positions including elected county superintendent of schools, president of the New Mexico Educational Association, registrar and acting president of her alma mater, elected state superintendent of public instruction, and a post in the National Education Association.[88]

Women like Eckles perceived, and seized, a world of possibilities. Attending normal school set many alumnae who remained single, as well as a few who married, on a path to all sorts of new places. Further education was one means of exploration employed by graduates like Mary Jones, Julia Hauck, and Ella Elizabeth Harrington. They made their way to institutions throughout the United States and, occasionally, in other countries. Geneseo graduate Minnie Mason Beebe's further education took her all the way to the University of Zurich, where she earned a Ph.D. in 1900. As of 1889, two Farmington women held bachelor's degrees, five had done some college work, three more were attending college, and seventy others had undertaken academic "work not classified." Female graduates of Normal, Illinois between the 1860s and the 1890s earned a total of 41 bachelor's degrees, 13 master's degrees, and five M.D. degrees.[89]

Many alumnae welcomed geographical relocations; in 1877, the *Albany Journal* noted, "Normal graduates flee with joy to the West." While higher salaries offered incentive for graduates in the East to head to the West, they also accepted jobs in other regions out of a sense of independence and adventure. An Oshkosh alumna wrote to the school's president in 1906, "I am anxious to get a position in the West in the near future. I've wanted some such Western experience for some time and wish to get a taste of life on the coast if possible." She and countless others accepted teaching jobs far from home and occasionally in very different communities than where they had lived previously. As of the late 1880s, more than half of Oswego's graduates had taught or were teaching outside New York State. Elizabeth Hyde (Framingham '75) taught African and Native American students at Hampton Institute in Virginia for over thirty years, and Minnie Hubbard (Farmington '82) found herself in the late 1880s in Meridian, Mississippi, teaching in a "colored school" with the sponsorship of the American Missionary Association. Facing primitive teaching conditions in a small Illinois town during the 1900s, Oshkosh's Daisy Chapin described her school's poor lighting and ventilation, the lack of teaching materials, and her difficulties in disciplining two older boys, yet concluded, "In short considering the difficulties I labor under, I am enjoying my work exceedingly." Throughout this period, normal-school alumnae went as teachers and missionaries to Puerto Rico, the Dominican Republic, Mexico, Brazil, Chile, Turkey, Japan, India, and other countries. Mary S. Morrill (Farmington '84) left in the late 1880s for a ten-year stint "as a missionary to North China under an appointment by the American Board of Foreign Missions," and religious missionary work took a Pine Bluff alumna to Liberia. When the education minister of Argentina in the mid-1870s invited normal-school graduates to inaugurate teacher education in his country, alumnae of Oswego and Winona, Minnesota rose to the challenge; some spent decades in the South American country, establishing normal schools and improving education in general, in part by breaking down the traditional notion that women were not to leave their homes for schooling.[90]

While alumnae explored new places from the American West to Argentina and beyond with a pioneering spirit, a few female normal-school graduates became pioneers in middle-class, male-dominated career fields. Some went into publishing or business. Westfield graduates Cassie Upson '71 and Ada Van Valkenburg '77 wrote for and edited national publications, while

Alicie A. Manter (Farmington '79) became a "general agent" for a Chicago publishing house. After six years of teaching, New Britain, Connecticut graduate Adelaide Pender worked as an editor and a writer for a school journal, a publishing company, and then a daily paper. Emma Schaeffer (Oshkosh '94) worked as an advertising manager and superintendent of a mail-order department, and Maud S. McIntyre (San Jose '01) operated "a delightful resort, Lometa Vista, in Los Gatos," California. After teaching for nine years, Ethel Lockridge of the California, Pennsylvania class of 1908 became the head bookkeeper for a bank and helped to organize the Quota Club, a civic organization for businesswomen. Other alumnae pursued careers in government or law. Three of Lockridge's classmates worked for the federal government in Washington, D.C. during World War I, and one continued her government work until she married in 1923. Four women who graduated from San Jose between the 1870s and the 1900s earned law degrees and worked as attorneys, as did Marion Harris, Pine Bluff '93, and Lurie Pounders, Florence '98. Clara Hapgood (Framingham '72) was the first woman lawyer in New England, and Lucille Pugh, who graduated from Greensboro, North Carolina in 1902 and then earned a law degree at New York University, was the first female attorney to represent a defendant accused of murder in New York. In 1898, one of the first women to be elected to the Colorado legislature was Oshkosh alumna Mary Barry.[91]

Female normal-school graduates were also pioneers in medicine and other areas. Among the almost two dozen San Jose alumnae of the 1870s to the 1900s who worked as physicians were Dr. Mary Bird Bowers '74, who practiced medicine with her husband, and Dr. Mariana Bertola '89, who held a fellowship in the American Medical Association and served as physician at Mills College. Harriet Clark (Oneonta '02) served as the resident physician at a school in Cambridge, Massachusetts, and Margaret Castex (Greensboro '05) was a professor at the Woman's Medical College of Pennsylvania, where she had earned her M.D. degree, and chief of gynecology and obstetrics at two Philadelphia hospitals. Whitewater graduate Alice A. Ewing '76 taught for four years, and then studied medicine in Chicago and New York. In 1893, she had "for five years been practicing medicine in Chicago," where she was also "clinical lecturer on aural diseases in the Chicago post-graduate School of Medicine" as "a specialist in diseases of the ear, nose, and throat." Two years behind Ewing at Whitewater, Alice L. Sherman taught for several years, and then studied dentistry at the University of Michigan. In 1893, she was married, a mother of one, and practicing dentistry in Lake Geneva, Wisconsin. Westfield graduate Nettie Stevens '83, Ph.D., discovered a sex-linked chromosome and became world famous as a geneticist. The ranks of normal-school alumnae even included an early female aviator: Bernetta Adams Miller, who attended Geneseo in the early 1900s, was the third woman to fly solo in the United States. Miller and other female graduates of state normal schools ascended smoothly, but probably not stridently, into nontraditional fields. "Not necessarily bloomer women," female normal-school graduates exercised great autonomy in their decisions regarding marriage and careers.[92]

The public sphere at state normal schools in the late nineteenth and early twentieth centuries initiated students into middle-class society while also fostering an expansive vision of women's roles in that society. Female, as well as male, students' experiences with gender in the public sphere helped to shape their relaxed view of middle-class gender expectations and underlying belief in women's autonomy. When historians Holmes and Weiss wrote that normal-school alumnae and other women teachers "did not fret about constraints," but "were consumed by what they could and did accomplish,"[93] they could have been discussing male as well as female graduates and the constraints of social class as well as gender, for the life paths of many normal-school graduates traversed a world of experiences and professional accomplishments. After the 1900s, the state normal schools would begin a march toward college, and eventually university, status. Institutional changes that accompanied this rise in status would erase many of the aspects of the normals' public sphere that challenged gender and social-class boundaries.

Epilogue: "Lots of Pep! Lots of Steam!"

At the beginning of the twentieth century, state normal schools thrived throughout the United States. A century later, most of these institutions had higher stature as state colleges or regional universities, and yet had lost much of their distinctive identity. In November 2000, *The Chronicle of Higher Education* reported that, as mere "lesser versions of their states' flagship universities," many state colleges were again "looking for a makeover." In such recent discussions of program expansion and name changes at these institutions, reporters and scholars have invoked the term "mission creep."[1] The history of the American state normal school suggests that "mission creep" more accurately characterizes much of the early decades of many regional public colleges and universities, and actually helped to create their distinctive identity. Education reformers of the early nineteenth century and state legislators, education leaders, and normal-school principals of the middle and late nineteenth century intended to establish and maintain single-purpose institutions focused on teacher preparation. After a few decades in which they did little more than instill the notion of teaching as a calling, state normal schools did create a strong professional spirit through teacher-education coursework, observation and practice teaching, and student activities. While teacher preparation remained their official purpose, however, the normal schools were never really single-purpose institutions. From the beginning, their students' weak academic background and expansive interests, along with the desires of the communities in which the normals were located, forced these institutions to do more than simply train teachers.

By the 1870s, mission creep was full-blown at the state normal schools. Because some of their students were members of minority racial and ethnic groups, most were women, and the majority were struggling financially, mature in age, and lacking in sophistication, welcoming "nontraditional" students with flexible admissions standards, low fees, and a supportive atmosphere was an important part of the normal schools' informal mission. Mission creep also added a considerable dose of general academic, or liberal arts education to the teacher-training function of the normals. Required studies included English, history, mathematics, and the sciences, and students were able to pursue more advanced, elective studies in these areas as well as modern and ancient languages. In addition, student activities, especially the very popular literary societies augmented the liberal-arts curriculum with investigations of literature, music, art, history, and current events. As the acquisition of high-status cultural knowledge helped acquaint students with the mores of the middle class, their involvement in the public sphere of the towns in which the normal schools were located and in campus life further socialized them for participation in middle-class society, and fostered a sense of female autonomy. Thus, the informal mission of the state normal schools also incorporated enabling class mobility and flexibility regarding gender roles.

The state normal school first appeared during a time in the United States when the curriculum and functions of academies, high schools, and colleges overlapped considerably, and there were no research universities; in short, higher education lacked stratification and hierarchy. In the early twentieth century, the situation was very different, as universities dominated and in many ways directed other institutions of higher education, which lined up in hierarchical order; what David Riesman would call "the academic procession" had taken shape.[2] The formation of the academic procession had important consequences for state normal schools. The low status that had plagued them since their beginnings was now of greater consequence. At the same time, the increasing popularity of public high schools as the first wrung on the higher-education ladder produced growing numbers of graduates and created teaching positions. Thus, as the normal schools jockeyed for position in the procession, they sought to gain status and students by situating themselves as colleges—specifically, "teachers colleges" (and, later, state colleges and regional universities). While a handful of normal schools became teachers colleges before 1920, the majority, including those in Pennsylvania, Alabama, Kansas, Wisconsin, and California, changed their names during the 1920s. Many normal schools, such as those in Massachusetts and Maryland, waited until the 1930s, and the institutions in Maine, Vermont, and most of state normals in New York did not become teachers colleges until the 1940s or later.[3]

Regardless of when they would officially make the transition, virtually all state normal schools began by the 1910s to consciously emulate collegiate institutions. They began, or prepared, to grant bachelor's degrees, and required, or discussed requiring, a high school diploma for admission. The curriculum dropped lower-level work and began to differentiate into separate courses of study, and the extracurriculum emulated student activities on college campuses. As the institutions' presidents—no longer principals—celebrated higher standards and increased state funding, the student yearbook for 1922 at the newly minted Chico State Teachers College in California chimed in, "How we have grown! We have cast aside the swaddling clothes of infancy and donned the vestments of manhood. . . . The change has come about quickly. Yesterday we were a Normal—today we are a College."[4] Instead of mere mission *creep*, this change constituted a conscious *leap* into a new realm. And, unlike mission creep of the nineteenth century, mission leap in the early twentieth century did little to expand the intellectual and social worlds of economically disadvantaged and female students; in fact, mission leap counteracted many elements of earlier mission creep. As the normal schools strove full-steam toward collegiate stature, they initiated the loss of what had become their informal institutional identity.

Mission leap lessened the prominence of the liberal arts and lead to a differentiated and stratified curriculum at the normal schools/teachers colleges. In a few extreme cases, normal-school leaders saw the changes in the higher-education climate as an opportunity to realize the original vision of single-purpose teacher-education institutions, albeit in college form. In Massachusetts and New York, administrators asserted that the requirement of a high-school degree for admission removed any need to offer liberal education, and began by the late 1900s to strictly limit their curricula to professional studies in education. As a result, New York's normals entered curricular "dark ages" according to Oneonta's historian, and Westfield, Massachusetts suffered from "the emasculation of its academic curriculum," in the words of historian Robert T. Brown. Westfield's curriculum in 1914 consisted of little but "content and methods" courses; even instruction in "Community civics" was limited to "those forms of civil life which fall within the personal experience of everyday life, with special reference and application to public school pupils." In Alabama by the early 1910s, according to Troy State Normal president Edward M. Shackelford, "the other higher institutions were beginning to feel the effects of the competition of the normals for students; and board members . . . were anxious to reduce the competition as much as possible. To do so, they tried to make the course of study of the normals as unattractive as possible to non-professional students." In 1914, Troy reduced offerings in higher mathematics and eliminated foreign languages. While most other normal schools did not gut the academic curriculum do the extent that Oneonta, Westfield, and Troy did and these institutions would reintroduce liberal-arts courses in a couple of decades, it was very common for state normals by the 1910s to reduce their liberal-arts offerings in the interest of focusing more exclusively on teacher education; in the process, they also lessened the intellectual vitality that had inspired earlier students.[5]

Preparation for teaching by the 1910s featured manual training, domestic science, agriculture, and commercial education much more prominently than in earlier decades. These subjects were becoming widespread in the public schools, and, as a consequence, in the normal schools; their expansion drew attention and resources away from what remained of the liberal arts. Commissioner of Education in Massachusetts from 1910 to 1916, vocational-education advocate David Snedden saw to it that the state's normal schools prioritized instruction in these areas. The courses he mandated in household, manual, and commercial arts, as well as gardening, basket-making, and chicken-raising corresponded to Snedden's belief that normal-school students had "distinct social limitations" and were "of only average capacity for work and abstract thinking." San Jose, California began in 1911 to offer advanced courses in industrial and household arts, and by the 1920s, "many beautiful examples of mechanical skill" had "been produced by the women as well as the men students" in industrial arts, and household arts had expanded from cooking and sewing to also include "the entire field of home life from budget making to the care and rearing of children." The Southwest Texas State Normal School in San Marcos offered 13 courses in home economics and 11 in industrial arts by 1917–1918, and 31 in home economics and 12 in industrial arts by 1922–1923. The increase in courses at San Marcos and elsewhere resulted in part from the passage in 1917 of the Smith-Hughes Act, which provided federal assistance for vocational education in industrial arts, home economics, and agriculture. As Smith-Hughes emphasized training in trade skills, the industrial-arts courses at Oswego, New York, and undoubtedly other normal schools, focused increasingly on training teachers of specific trades rather than general skills. Such an emphasis, as well as the overall displacement of the liberal arts by vocationally oriented instruction, allowed little opportunity for normalites to reach beyond their "distinct social limitations."[6]

This shift in the curriculum also had profound gender ramifications. Although women were able to study industrial arts at San Jose and elsewhere, the expectation was that they and the male students would adhere to gender conventions. At Oswego, women had to receive special permission to enroll in manual training or industrial arts; after the shift to trade skills, they were not welcome at all. At the African American state normal in Pine Bluff, Arkansas by the 1910s, the vocationally oriented curriculum mandated the separation of the male and female students: for three years, the men spent four hours per week in the shops, while the women spent eight hours working in the sewing room. Another southern institution, Florence, Alabama by the mid-1910s also required manual training of males and domestic science of females. In domestic science, each senior woman made her own dress for graduation, spending no more than five dollars for materials. "Away Up North in Florence," by Lloyd Garrison of the class of 1915, probably best represents the prominence of vocational courses and the resulting gender distinctions. Sung to the tune of "Dixie," the song had several verses, but only the following one referred to the curriculum:

> In S. N. S. our course is good,
> The boys are taught to work in wood;
> Come along! Come along!
> Come along to Florence,
> The girls are taught to sew and cook,
> And the cakes they bake, how good they look!
> Come along! Come along!
> Get a wife from Florence.[7]

The rise of manual training and domestic science was part of a larger trend by the 1910s: the separation of expanding subject offerings into growing numbers of highly differentiated courses of study. Students in different courses had fewer and fewer academic classes in common, which contributed to the decline of the general intellectual climate of the normals. Illinois State Normal University in Normal offered 58 elective classes in 1905, 90 in 1910,

126 in 1920, and as many as 230 electives by 1930. While students had some freedom to choose according to their interests, "recommended electives" as early as 1905 strongly encouraged them to make selections according to their future teaching plans. By 1908, "elective groups" of three or four classes were requirements in major areas of study. Other normal schools began "grouping" subjects into parallel tracks according to what students were to teach. In 1911, Oshkosh initiated primary, intermediate, grammar, and high-school "groups," as well as a course to prepare teachers for country schools. A few years later, this Wisconsin normal also offered courses for graded-school principals, teachers of industrial subjects, and even two-year programs in commerce, journalism, engineering, pre-medicine, and other subjects designed for students preparing to matriculate as juniors at colleges and universities. Beginning in 1912, San Marcos had five "groups": Agriculture, which prepared teachers of agriculture and rural-school teachers; Industrial Arts, which was subdivided according to whether students were to teach manual training or home economics; Language, which prepared teachers of English, Latin, and German; Primary, Elementary, and Art, which prepared elementary-school, music, and art teachers; and Science–Mathematics, which prepared teachers of these subjects. The History–English group debuted the following year, and by 1920, San Marcos students selected majors and minors from an even longer list of groups.[8]

Other normal schools followed the path of Oshkosh and San Marcos, and by the late 1910s, most normalites were grouped in courses of study that corresponded to different future plans, much like they would have been at colleges or universities. Students in different courses had increasingly different academic experiences; by the early 1920s, the administration at Oshkosh was able to report: "In the various curricula there is a sharp differentiation in the type of subject matter." Not only did students across campus no longer share a common academic foundation or intellectual culture, but, reflective of the larger society, social-class background and gender had a hand in determining the course of study they pursued. Because there was little variation in social origins among normal-school students, class probably helped—by influencing attitudes like David Snedden's—to limit overall choices rather than assign individual students to particular curricular tracks. Gender segregation, however, was quite prevalent. At Oshkosh in 1918, for example, a total of 123 women, but no men, graduated from the courses for primary, grammar-grade, and rural-school teachers, while 15 men and a single woman graduated from the course for teachers of manual arts. Other courses of study were more evenly balanced by gender, but graduated men far out of proportion to their representation on campus; these courses were for: graded-school principals (9 women and 10 men); high-school teachers (16 women and 7 men), and the "college course" (1 woman and 3 men).[9]

Also crucial to the reduction in intellectual vitality as well as the overall professional spirit at the normal schools were substantial changes in student activities. An educational researcher reported in 1929 that students and faculty at the newly christened teachers colleges had "not attempted to develop a new type of student activity. They have instead taken over, by imitation of the nonprofessional college and university, the good and the bad in their extra-curricular practices. These activities . . . have created a hodge-podge of activity which is time consuming and questionable in its ultimate effect upon the individual and institutional life." A significant contribution to the "hodge-podge" was the literary societies' leap to become collegiate fraternities and sororities. San Jose was in the vanguard of this trend. Its Allenian Rhetorical Society had begun to change as early as 1903, when *The Normal Pennant* reported: "It was at first of a debating and oratorical nature, but has since developed a social side which makes her semi-annual functions and affairs toward which many look forward." Six years later, the *Pennant* stated: "At the present time our societies exist for a closer association of the students, for recreation and for pleasure. With the exception of the literary and musical programs which we give occasionally the educational phase of our societies is nil." The Browning Society changed its pin from the letter "B" to a Greek letter, and the Allenian officially dropped "Rhetorical" from its name. By the mid-1910s, fraternities replaced San Jose's Young Men's Normal Debate Society, and yearbook descriptions of the Allenian, Browning, Sappho, Erosophian, and Dailean societies consisted of accounts of initiations, rushes, cotillions, carnivals, teas, and parties.[10]

The literary societies of other state normal schools soon fell into step behind San Jose's, and intellectual topics moved to the periphery of their activities. When "ball games and picnics" interrupted a few meetings of San Marcos' Every Day Society in 1911–1912, members reported, "it was school patriotism that distracted our attention and we lost nothing by it." Similarly, at a meeting in 1917, Oshkosh's all-female Alethean society "made and carried" a motion "that we dispense with the program in order that we may attend a basketball game." Meanwhile, the Agonian Society at Geneseo, New York studied "modern authors," but focused on "sisterhood, which," members explained, "after all, is our chief sorority aim." Instead of literary programs or debates on current events or educational issues, societies staged increasing numbers of entertainment shows and dances. A summary of society activities at Oneonta in 1920–1921 included dances, teas, and social events, but only one literary program. At Millersville, Pennsylvania, societies often did not meet at all because members had failed to prepare programs, and in 1921 the faculty gave up on the annual intersociety debate, for which it had awarded a silk banner to the winners. At Kirksville, Missouri, fraternities arose independently of literary societies; their appearance in the 1915 yearbook signaled their acceptance by the school administration. At Normal, Illinois, where the administration refused to permit students to establish fraternities and sororities, the old societies limped along until they finally faded away in the 1920s.[11]

Regardless of how administrators felt about it, the complete transformation of the literary societies into sororities and fraternities at most normal schools took less than a decade. Oshkosh's long-standing, coeducational Lyceum became an all-male fraternal organization and Phoenix became its all-female sister in 1917. Phoenix's activities for 1919–1920 consisted of a dance for the faculty and new students, "dancing, eating, chatting" at a member's home, a "movie party," an informal dance with Lyceum, a party on Valentine's Day, a formal tea, and a group excursion to a basketball game. A San Marcos instructor later remembered literary societies of the early 1920s to be "social clubs . . . the parent organizations of our present fraternities and sororities." At Florence, where the old societies had ceased to exist in the late 1910s, Junior Clubs formed in the mid-1920s to "give the students an excellent opportunity to more quickly form acquaintances and more easily get into the social life of the institution." Dances and formal events invariably required proper attire. A professor at Cedar Falls, Iowa observed that as "Secret societies, fraternities, sororities" and other groups "sprang up like mushrooms" on campus, "Spike-tail coats, cut-a-way vests and more elaborate gowns began to be in evidence in all social functions. The cosmetic trade at the stores materially increased." The typical normal-school student had very little money for frills, and must have found these developments challenging, and possibly very uncomfortable. It is hardly surprising that several students in the early 1930s at Oneonta, where sororities were moving into their own houses, protested the sororities on the grounds that they were exclusive.[12]

While the transformation of literary societies into fraternities and sororities helped to insulate normalites from the intellectual world beyond campus, the appearance and expansion of campus residential facilities helped to insulate students from the social world beyond campus. Although they continued to enjoy public events in the cities and towns in which their schools were located, students' lives became less intertwined in the public sphere of local middle-class society as increasing numbers of them lived on campus. From Farmington, Maine to Troy, Alabama and from Frostburg, Maryland to Cheney, Washington, campuses that had never had residential space built their first dormitories in the 1910s. The majority of these new facilities were for women only, which subjected them to a higher standard of behavioral scrutiny than the men. In 1913, Oshkosh converted a private residence into a women's dorm, and Florence began construction of its first one. The administration at the latter institution soon issued the instruction, "All nonresident young ladies will board in the dormitory unless excused by the President," and promised that, in the new dorm, the "social and religious interests of the young ladies will be carefully looked after." The new residence hall for women at Carbondale, Illinois was named, rather ironically, for Susan B. Anthony.[13]

Reinforcing the decline of intellectual life, insulation from the larger world, and restriction of female normalites, state normal schools leapt with both feet to embrace the cult of football that reigned at colleges and universities. Women students continued during the 1910s to engage in all sorts of athletic endeavors, but female athletes faced increasing restrictions and found themselves in the ever-darker shadow of men's sports. As normal schools established interscholastic athletic leagues for men, they banned inter-school contests among women. Both men's and women's teams at Ypsilanti, Michigan engaged in intercollegiate sports until 1910, when Director of Women's Athletics Fannie Burton announced the end of such competition for women for fear that they would overexert themselves. Wisconsin's state normal schools inaugurated in internormal basketball tournament for men in 1912, and in the very same year prohibited internormal play among the women. Meanwhile, the faculty at Ellensburg, Washington ruled in 1914 that female students were not to play basketball against teams from other schools but encouraged interclass competition. Women continued to participate in basketball tournaments and a variety of other intermural sports, but without the spectators they once had—the crowds, including many female students, were too busy watching men's sports. A new way for individual women to seize the spotlight soon appeared, however: in 1926, freshman Hazel Cline was crowned the fist homecoming queen at Southwestern State Teachers College in Weatherford, Oklahoma. Team captain Swede Umbach crowned her with a standard football helmet.[14]

Like the women, male normalites had been active in athletics for decades, but it was not until the early twentieth century that the normal schools imitated the spirit of "pep" that surrounded men's collegiate sports. Wisconsin's normals used a 1912 state law requiring them to provide training for physical-education teachers as an excuse to begin hiring full-time coaches. The Wisconsin Normal Athletic Conference for men's sports formed in 1913, the same year that the Whitewater institution received an appropriation to upgrade its athletic field and build a large concrete grandstand. A new gymnasium at Kirksville in the late 1910s had plenty of room for spectators, and the institution's first men's varsity basketball team received much attention. Men's basketball, baseball, and, especially, football began to receive the lion's share of the coverage in student magazines and yearbooks, as well as the adulation of growing crowds. Oshkosh's *The Normal Advance* reported in 1911 on a "mass meeting" in the auditorium the evening before a football game: "School songs were sung, yells given, and pictures of the football team and members of the faculty were thrown upon the screen. The different members of the team [as well as a few professors] gave short speeches, . . . Much enthusiasm was aroused at this meeting." Similar rallies took place at other normals, and San Jose students in 1922 formed a " 'Pep' Society," which performed "rally and bleacher stunts"; their yearbook praised the male class president for doing "much to place athletics on a collegiate basis." Press coverage of the new football culture included women students only to the extent that they were enthusiastic fans. After a victory, *The Normal Advance* praised "the members of the gentler sex" who "turned out in large numbers" for the game, and stated, "it is not improbable that it was their presence on the side lines that was one of the reasons why our team won. . . . Although the latter [women] can not play the game themselves, they come as near to it as they can, that is,—they play the game in spirit." The rise of school spirit and "pep" elevated physically strong male students above other men and female students, contributing to a new hierarchy in campus life.[15]

Finally, state normal schools also transformed their graduation celebrations as they became teachers colleges. Instead of student orations and essays, commencement exercises began to feature speeches by distinguished guests. Graduates wore caps and gowns proudly, not only at commencement, but also for their yearbook photographs and other campus events. A long-serving professor at Cedar Falls remembered the senior's "glorious privilege of wearing a cap and gown at chapel . . . and inestimable privilege of receiving his diploma on Commencement Day and immediately thereafter removing the tassel from the right side of his cap to the left."[16] The achievement of earning a bachelor's degree, especially by women and men from nontraditional backgrounds, was indeed worthy of celebration. But mission leap also brought the glorification of football, the rise of sororities and fraternities, and the shift away from liberal studies and toward a stratified curriculum—developments that were several steps removed from the normals' de facto role, shaped by mission creep, of serving the needs of economically disadvantaged and female students. In transforming themselves into teachers colleges—and later into state colleges and regional universities—normal schools erased qualities that had made them distinctive, all in the interest of institutional status in the academic procession. Whereas earlier normalites had celebrated the expansion of their intellectual and social worlds in clever and spirited debates, speeches, poems, and even cheers, all that one class at Ellensburg in the 1910s could manage was:

Lots of pep!
Lots of steam!
Senior class '16![17]

Appendix: State Normal Schools in the United States

State	Date open	Location	Names and dates of name changes, with current title in bold	Institution: Women; African American; Native American
Ala.	1907	Daphne	State Normal School (Closed)	
	1873	Florence	State Normal School; State Normal College (1889); State Normal School (1912); State Teachers College (1929); State College (194?); **University of North Alabama** (1967)	
	1875	Normal	State Normal School for Colored Students; State Normal and Industrial School (1885); State Agricultural and Mechanical College for Negroes (1896); State Agricultural and Mechanical Institute for Negroes (1919); Alabama Agricultural and Mechanical College (1948); **Alabama A & M University** (1969)	African American
	1883	Jacksonville	State Normal School; State Teachers College (1929); State College (1947); **Jacksonville State University** (1966)	
	1883	Livingston	State Normal College (for Girls); Alabama Normal College (1900; coeducational beginning in 1900); State Normal School (1907); State Teachers College (1929); State College (1957); **Livingston University** (1969)	Women
	1873	Montgomery	Lincoln Normal University (located in Marion, 1873–1887); State Colored Normal School (1886); State Teachers College (1929); State College (1946); **Alabama State University** (1969)	African American

Continued

Continued

State	Date open	Location	Names and dates of name changes, with current title in bold	Institution: Women; African American; Native American
	1910	Moundville	State Normal College; State Normal School (1912) (Closed)	
	1887	Troy	State Normal School; State Normal College (1893); State Normal School (1911); State Teachers College (1927); State College (1957); **Troy State University** (1967)	
Alaska	Alaska did not establish state normal schools.			
Ariz.	1899	Flagstaff	Northern Arizona Normal School; Northern State Teachers College (1925); Northern State College (1945); **Northern Arizona University** (1966)	
	1886	Tempe	Territorial Normal School/Normal School of Arizona; State Teachers College (1925); Arizona State College (1945); **Arizona State University** (1958)	
Ark.	1908	Conway	State Normal School; State Teachers College (1925); State College (1967); **University of Central Arkansas** (1975)	
	1876	Pine Bluff	Branch Normal College (of Arkansas Industrial University); Arkansas Agricultural, Mechanical and Normal School (1921); Arkansas Agricultural, Mechanical and Normal College (1928); **University of Arkansas at Pine Bluff** (1972)	African American
Calif.	1914	Arcata	Humboldt State Normal School; Humboldt State Teachers College (1921); Humboldt StateCollege (1935); **Humboldt State University** (1972)	
	1889	Chico	State Normal School; State Teachers College (1921); State College (1935); **California State University at Chico** (1972)	

	1911	Fresno	State Normal School; State Teachers College (1921); State College (1935); **California State University at Fresno** (1972)
	1882	Los Angeles	State Normal School (Closed in 1919; site became first location of University of California at Los Angeles)
	1898	San Diego	State Normal School; State Teachers College (1921); State College (1935); **San Diego State University** (1972)
	1899	San Francisco	State Normal School; State Teachers College (1921); State College (1935); **California State University at San Francisco** (1972)
	1862	San Jose	State Normal School (located in San Francisco, 1862–1871); State Teachers College (1921); State College (1935); **San Jose State University** (1972)
	1910	Santa Barbara	State Normal School of Manual Arts and Home Economics; State Normal School (1919); State Teachers College (1921); State College (1935); College of the University of California (1944); **University of California at Santa Barbara** (1958)
Colo.	1890	Greeley	State Normal School; State Teachers College (1911); State College of Education (1935); **University of Northern Colorado** (1957)
	1901	Gunnison	State Normal School; **Western State College of Colorado** (1923)
Conn.	1904	Danbury	State Normal (Training) School; State Teachers College (1937); State College (1959); **Western Connecticut State University** (1983)
	1850	New Britain	State Normal (Training) School; Teachers College of Connecticut (1933); Central Connecticut State College (1959); **Central Connecticut State University** (1983)

Continued

Continued

State	Date open	Location	Names and dates of name changes, with current title in bold	Institution: Women; African American; Native American
	1893	New Haven	State Normal (Training) School; State Teachers College (1937); Southern State College (1959); **Southern Connecticut State University** (1983)	
	1889	Willimantic	State Normal (Training) School; State Teachers College (1937); Eastern State College (1959); **Eastern Connecticut State University** (1983)	
Del.	Delaware did not establish state normal schools.			
Fla.	1887	DeFuniak Springs	State Normal College/State Normal School for White Students (Merged with Florida State College in Tallahassee in 1905)	
	1887	Tallahassee	State Normal School for Colored Teachers; State Normal and Industrial College for Colored Students (1891); State Normal and Industrial School (1903); Florida Agricultural and Mechanical College for Negroes (1909); **Florida A & M University** (1953)	African American
Ga.	1917	Albany	Georgia Normal and Agricultural College; State College (1943); **Albany State University** (1996)	African American
	1890	Milledgeville	State Normal and Industrial College; State College for Women (1922); Women's College of Georgia (1961); Georgia College (1967); **Georgia College & State University** (1996)	Women
	1913	Valdosta	Southern State Normal College; State Women's College (1922); State College (1950); **Valdosta State University** (1993)	

Hawaii	1896	Honolulu	Territorial Normal and Training School (Merged with University of Hawaii in 1931 as University of Hawaii Teachers College)
Idaho	1894	Albion	State Normal School; Southern Idaho College of Education (1947) (Closed in 1951)
	1896	Lewiston	State Normal School; Northern Idaho College of Education (1947; Closed, 1951–1955); Lewis-Clark State Normal School (1955); **Lewis-Clark State College** (1971)
Ill.	1874	Carbondale	Southern State Normal University; **Southern Illinois University** (1947)
	1899	Charleston	Eastern State Normal School; Eastern State Teachers College (1921); Eastern State College (1947); **Eastern Illinois University** (1957)
	1899	De Kalb	Northern State Normal School; Northern State Teachers College (1921); Northern State College (1955); **Northern Illinois University** (1957)
	1902	Macomb	Western State Normal School; Western State Teachers College (1921); Western State College (1947); **Western Illinois University** (1957)
	1857	Normal	State Normal University; **Illinois State University** (1967)
Ind.	1918	Muncie	State Normal School, Eastern Division; Ball Teachers College, Eastern Division (1922); Ball State Teachers College (1929); **Ball State University** (1965)
	1870	Terre Haute	State Normal School; State Teachers College (1929); **Indiana State University** (1965)
Iowa	1876	Cedar Falls	State Normal School; State Teachers College (1909); Northern State College (1961); **University of Northern Iowa** (1967)

Continued

Continued

State	Date open	Location	Names and dates of name changes, with current title in bold	Institution: Women; African American; Native American
Kans.	1865	Emporia	State Normal School; State Teachers College (1923); State College (1974); **Emporia State University** (1977)	
	1902	Hays	Western Branch State Normal School; Fort Hays State Normal School (1913); State Teachers College (1923); State College (1931); **Fort Hays State University** (1977)	
	1903	Pittsburg	Auxiliary Manual Training Normal School; State Teachers College (1923); State College (1959); **Pittsburg State University** (1977)	
Ky.	1907	Bowling Green	Western State Normal School; Western State Normal School and Teachers College (1922); Western State Teachers College (1930); Western State College (1948); **Western Kentucky University** (1966)	
	1887	Frankfort	State Normal School for Colored Persons; Kentucky Normal and Industrial Institute for Colored Persons (1902); State Industrial College for Colored Persons (1926); State College for Negroes (1938); State College (1952); **Kentucky State University** (1972)	African American
	1923	Morehead	State Normal School; State Normal School and Teachers College (1926); State Teachers College (1930); State College (1948); **Morehead State University** (1966)	
	1923	Murray	State Normal School; State Normal School and Teachers College (1926); State Teachers College (1930); State College (1948); **Murray State University** (1966)	

	1907	Richmond	Eastern State Normal School; Eastern State Normal School and Teachers College (1922); Eastern State College (1948); **Eastern Kentucky University** (1966)
La.	1885	Natchitoches	State Normal School; State Normal College (1921); Northwestern State College (1944); **Northwestern State University** (1970)
Maine	1867	Castine	Eastern State Normal School (Closed in 1942; site became Maine Maritime Academy)
	1864	Farmington	Western State Normal School; State Normal (and Training) School (1879); State Teachers College (1945); State College of the University of Maine (1968); **University of Maine at Farmington** (1970)
	1878	Fort Kent	Madawaska Training School; State Normal School (1956); State Teachers College (1962); State College (1964); State College of the University of Maine (1968); **University of Maine at Fort Kent** (1970)
	1879	Gorham	State Normal and Training School; Western State Normal School (1886); State Teachers College (1946); State College (1964); State College of the University of Maine (1968); University of Maine-Portland/Gorham (1970); **University of Southern Maine** (1978)
	1910	Machias	Washington State Normal School; Washington State Teachers College (1952); Washington State College (1965); Washington State College of the University of Maine (1968); **University of Maine at Machias** (1970)
	1903	Presque Isle	Aroostook State Normal School; Aroostook State Teachers College (1952); Aroostook State College (1965); Aroostook State College of the University of Maine (1968); **University of Maine at Presque Isle** (1970)

Continued

Continued

State	Date open	Location	Names and dates of name changes, with current title in bold	Institution: Women; African American; Native American
Md.	1866	Baltimore	State Normal School (moved to Towson in 1916); State Teachers College (1935); Towson State College (1963); State University (1976); **Towson University** (1997)	
	1914	Bowie	Maryland Normal and Industrial School; State Teachers College (1938); State College (1963); **Bowie State University** (1988)	African American
	1902	Frostburg	State Normal School; State Teachers College (1935); State College (1963); **Frostburg State University** (1987)	
	1925	Salisbury	State Normal School; State Teachers College (1935); State College (1963); State University (1988); **Salisbury University**	
Mass.	1873	Boston	Massachusetts Normal Art School; **Massachusetts College of Art**	
	1840	Bridgewater	State Normal School; State Teachers College (1932); **Bridgewater State College** (1960)	
	1895	Fitchburg	State Normal School; State Teachers College (1933); **Fitchburg State College** (1962)	
	1839	Framingham	State Normal School (located in Lexington, 1839–1844 and West Newton, 1844–1853); State Teachers College (1932); **Framingham State College** (1960)	Women
	1897	Hyannis	State Normal School (Closed)	
	1897	Lowell	State Normal School; State College (1932); University of Lowell (1973); **University of Massachusetts at Lowell** (1991)	

	1897	North Adams	State Normal School; State Teachers College (1932); State College (1968); **Massachusetts College of Liberal Arts** (1997)	
	1854	Salem	State Normal School (Coeducational beginning in 1899); State Teachers College (1932); **Salem State College** (1960)	Women
	1839	Westfield	State Normal School (located in Barre, 1839–1841; Closed, 1841–1844); State Teachers College (1932); **Westfield State College** (1960)	
	1874	Worcester	State Normal School; State Teachers College (1932); **Worcester State College** (1960)	
Mich.	1904	Kalamazoo	Western State Normal School; Western State Teachers College (1927); Western Michigan College of Education (1941); Western Michigan College (1955); **Western Michigan University** (1957)	
	1899	Marquette	Northern State Normal School; Northern State Teachers College (1927); Northern Michigan College of Education (1941); Northern Michigan College (1955); **Northern Michigan University** (1963)	
	1895	Mt. Pleasant	Central State Normal School; Central State Teachers College (1927); Central Michigan College of Education (1940); Central Michigan College (1955); **Central Michigan University** (1959)	
	1853	Ypsilanti	State Normal School; State Normal College (1899); Eastern Michigan College (1956); **Eastern Michigan University** (1959)	
Minn.	1919	Bemidji	State Normal School; State Teachers College (1921); State College (1957); **Bemidji State University** (1976)	
	1902	Duluth	State Normal School; State Teachers College (1921); **University of Minnesota, Duluth** (1947)	

Continued

Continued

State	Date open	Location	Names and dates of name changes, with current title in bold	Institution: Women; African American; Native American
	1868	Mankato	State Normal School; State Teachers College (1921); State College (1957); State University (1975); **Minnesota State University, Mankato**	
	1888	Moorhead	State Normal School; State Teachers College (1921); State College (1957); State University (1975); **Minnesota State University, Moorhead**	
	1869	St. Cloud	State Normal School; State Teachers College (1921); State College (1957); **St. Cloud State University** (1975)	
	1860	Winona	State Normal School (Closed, 1862–1864); State Teachers College (1921); State College (1957); **Winona State University** (1975)	
Miss.	1912	Hattiesburg	Mississippi Normal School; Mississippi Normal College (1918); State Teachers College (1924); Mississippi Southern College (1940); **University of Southern Mississippi** (1962)	
	1871	Holly Springs	State (Colored) Normal School (Closed)	African American
Mo.	1873	Cape Girardeau	Southeast State Normal School; State Normal School (third district) (1879); Southeast State Teachers College (1919); Southeast State College (1946); **Southeast Missouri State University** (1972)	
	1879	Jefferson City	Lincoln Normal Institute; **Lincoln University** (1921)	African American
	1871	Kirksville	North State Normal School; State Normal School (first district) (1880); Northeast State Teachers College (1919); Northeast State College (1968); Northeast State University (1972); **Truman State University** (1996)	

	1906	Maryville	State Normal School (fifth district); State Teachers College (1919); State College (1949); **Northwest Missouri State University** (1970)
	1906	Springfield	State Normal School (fourth district); Southwest State Teachers College (1919); Southwest State College (1945); **Southwest Missouri State University** (1972)
	1871	Warrensburg	South State Normal School; State Normal School (second district) (1878); Central State Teachers College (1919); Central State College (1946); **Central Missouri State University** (1972)
Mont.	1927	Billings	Eastern State Normal School; Eastern Montana College of Education (1949); Eastern Montana College (1966); **Montana State University-Billings** (1994)
	1897	Dillon	State Normal School; State Normal College (1903); Western Montana College of Education (1949); Western Montana College (1965); **Western Montana College of the University of Montana** (1988)
Nebr.	1911	Chadron	State Normal School; State Teachers College (1921); **Chadron State College** (1963)
	1905	Kearney	State Normal School; State Teachers College (1921); State College (1963); **University of Nebraska at Kearney** (1991)
	1867	Peru	State Normal School; State Teachers College (1921); **Peru State College** (1963)
	1910	Wayne	State Normal School; State Normal School and Teachers College (1921); State Teachers College (1949); **Wayne State College** (1963)
Nev.	Nevada did not establish a separate normal school; prospective teachers were educated at the state university.		

Continued

Continued

State	Date open	Location	Names and dates of name changes, with current title in bold	Institution: Women; African American; Native American
N.H.	1909	Keene	State Normal School; State Teachers College (1939); **Keene State College** (1963)	
	1871	Plymouth	State Normal School; State Teachers College; State College; **Plymouth State University** (2003)	
N.J.	1923	Glassboro	Glassboro Normal School; State Teachers College (1937); State College (1958); Rowan College of New Jersey (1992); **Rowan University** (1996)	
	1927	Jersey City	State Normal School; State Teachers College (1935); State College (1958); **New Jersey City University** (1998)	
	1908	Montclair	State Normal School; State Teachers College (1929); State College (1958); **Montclair State University** (1994)	
	1913	Newark	State Normal School; State Teachers College (1937); State College (1958; moved to Union, 1958); Kean College of New Jersey (1974); **Kean University of New Jersey** (1997)	
	1923	Paterson	State Normal School; State Teachers College (1937); State College (1958); William Paterson College of New Jersey (1971); **William Paterson University of New Jersey** (1997)	
	1855	Trenton	State Normal (and Model) School; State Normal School and Teachers College (1929); State Teachers College (1952); State College (1958); **The College of New Jersey** (1996)	
N.Mex.	1913	El Rita	Spanish–American Normal School (Closed)	
	1898	Las Vegas	New Mexico Normal School; New Mexico Normal University (1899); **New Mexico Highlands University** (1941)	

	1894	Silver City	New Mexico Normal School; State Teachers College (1921); **Western New Mexico University** (1963)
N.Y.	1844	Albany	State Normal School; State Normal College (1890); State College for Teachers (1914); State University College of Education (1959); State University College (1961); **State University of New York at Albany** (1962)
	1867	Brockport	State Normal School (and Training) School; State Teachers College (1942); **State University of New York College at Brockport**
	1871	Buffalo	State Normal (and Training) School; State Teachers College (1928); State College for Teachers (1946); State University College for Teachers (1951); State University College of Education (1959); **State University of New York College at Buffalo** (1962)
	1869	Cortland	State Normal (and Training) School; State Teachers College (1941); State College of Education (1959); **State University of New York College at Cortland** (1961)
	1868	Fredonia	State Normal (and Training) School; State Teachers College (1948); **State University of New York College at Fredonia** (1961)
	1871	Geneseo	State Normal (and Training) School; State Teachers College (1942); **State University of New York College at Geneseo** (1961)
	1897	Jamaica	State Normal (and Training) School (Control transferred to New York City in 1905; became city training school for teachers)
	1886	New Paltz	State Normal (and Training) School; State Teachers College (1942); State College of Education (1959); **State University of New York College at New Platz** (1961)

Continued

Continued

State	Date open	Location	Names and dates of name changes, with current title in bold	Institution: Women; African American Native American
	1889	Oneonta	State Normal (and Training) School; State Teachers College (1938); State University Teachers College (1948) State University College of Education (1951); **State University of New York College at Oneonta** (1961)	
	1866	Oswego	State Normal (and Training) School; State Teachers College (1948); **State University of New York College at Oswego** (1962)	
	1890	Plattsburgh	State Normal (and Training) School; State Teachers College (1942); State University Teachers College (1948); State University College of Education (1959); **State University of New York College at Plattsburgh** (1961)	
	1869	Potsdam	State Normal (and Training) School; State Teachers College (1942); State University Teachers College (1948); State University College of Education (1959); **State University of New York College at Potsdam** (1961)	
N.C.	1903	Boone	Appalachian State Normal School; Appalachian State Teachers College (1929); **Appalachian State University** (1967)	
	1905	Cullowhee	Normal and Industrial School; State Normal School (1925); Western Carolina Teachers College (1929); Western Carolina College (1953); **Western Carolina University** (1967)	
	1891	Elizabeth City	State Colored Normal School; State Teachers College (1939); State College (1963); **Elizabeth City State University** (1969)	African American
	1877	Fayetteville	State (Colored) Normal School; State Teachers College (1944); State College (1963); **Fayetteville State University** (1969)	African American
	1887	Goldsboro	State (Colored) Normal School (Closed)	African American

	1892	Greensboro	State Normal and Industrial School/College for Girls; North Carolina College for Women (1919); Women's College of the University of North Carolina (1932); **University of North Carolina at Greensboro** (1963)	Women
	1909	Greenville	East Carolina Teachers Training School (1909); East Carolina State Teachers College (1921); East Carolina College (1951); **East Carolina University** (1967)	
	1913	Pembroke	Cherokee Indian State Normal School; State College for Indians (1941); State College (1949); **Pembroke State University** (1969)	Native American
	1881	Plymouth	State (Colored) Normal School/College (Closed in 1905)	African American
	1881	Salisbury	State (Colored) Normal School (Closed in 1905)	African American
	1895	Winston-Salem	Slater Industrial and State Normal School; State Teachers College (1925); State College (1963); **Winston-Salem State University** (1969)	African American
N.Dak.	1918	Bottineau	State Normal School (Closed)	
	1918	Dickenson	Normal School; State Teachers College (1931); State College (1963); **Dickenson State University** (1987)	
	1907	Ellendale	State Normal and Industrial School (Closed)	
	1890	Mayville	State Normal School; **Mayville State University**	
	1913	Minot	State Normal School/College; State Teachers College (1924); State College (1964); **Minot State University** (1987)	
	1890	Valley City	State Normal School; State Teachers College (1921); State College (1964); **Valley City State University** (1987)	
Ohio	1914	Bowling Green	State Normal College; State College (1929); **Bowling Green State University** (1935)	
	1913	Kent	State Normal School; State Normal College (1915); State College (1929); **Kent State University** (1935)	

Continued

Continued

State	Date open	Location	Names and dates of name changes, with current title in bold	Institution: Women; African American; Native American
Okla.	1909	Ada	East Central State Normal School; East Central State Teachers College (1919); East Central State College (1939); **East Central University** (1974)	
	1897	Alva	Northwestern Territorial Normal School; Northwestern State Normal School (1904); Northwestern State Teachers College (1919); Northwestern State College (1939); **Northwestern Oklahoma State University** (1974)	
	1909	Durant	Southeastern State Normal School; Southeastern State Teachers College (1921); Southeastern State College (1939); **Southeastern Oklahoma State University** (1974)	
	1891	Edmond	Territorial Normal School; Central State Normal School (1904); Central State Teachers College (1919); Central State College (1939); Central State University (1974); **University of Central Oklahoma** (1990)	
	1897	Langston	Colored Agricultural and Normal University; **Langston University** (1941)	African American
	1909	Tahlequah	Northeastern State Normal School; Northeastern State Teachers College (1919); Northeastern State College (1939); **Northeastern State University** (1974)	Native American
	1903	Weatherford	Southwestern State Normal School; Southwestern State Teachers College (1920); Southwestern Institute of Technology (1941); Southwestern State College (1949); **Southwestern Oklahoma State University** (1974)	
Oreg.	1882	Ashland	State Normal School (Closed, 1890–1895); Southern State Normal School (1895; Closed, 1909–1926); Southern	

State	Year	Location	Names	Notes
			Oregon College of Education (1939); Southern Oregon College (1956); Southern State College (1975); **Southern Oregon University** (1996)	
	1885	Drain	State Normal School; Central State Normal School (1899) (Closed in 1909)	
	1883	Monmouth	State Normal School (Closed, 1909–1911); Oregon College of Education (1939); Western Oregon College (1981); **Western Oregon University**	
	1885	Weston	Eastern State Normal School (Closed in 1909)	
Pa.	1869	Bloomsburg	State Normal School (sixth district); State Teachers College (1927); State College (1960); **Bloomsburg University of Pennsylvania** (1983)	
	1874	California	Southwestern Normal College; Southwestern State Normal School (1878); State Normal School (1914); State Teachers College (1928); State College (1959); **California University of Pennsylvania** (1983)	
	1921	Cheyney	State Normal School; State Teachers College (1951); State College (1959); **Cheyney University of Pennsylvania** (1983)	African American
	1887	Clarion	State Normal School (thirteenth district); State Teachers College (1927); State College (1959); **Clarion University of Pennsylvania** (1983)	
	1893	East Stroudsburg	State Normal School; State Teachers College (1927); State College (1960); **East Stroudsburg University of Pennsylvania** (1983)	
	1861	Edinboro	(Northwestern) State Normal School (twelfth district); State Teachers College (1926); State College (1960); **Edinboro University of Pennsylvania** (1983)	
	1875	Indiana	State Normal School; State Teachers College (1927); State College (1960); **Indiana University of Pennsylvania** (1965)	

Continued

Continued

State	Date open	Location	Names and dates of name changes, with current title in bold	Institution: Women; African American; Native American
	1866	Kutztown	State Normal School; Keystone State Normal School (1871); State Teachers College (1928); State College (1960); **Kutztown University of Pennsylvania** (1983)	
	1877	Lock Haven	Central State Normal School; State Teachers College (1927); State College (1962); **Lock Haven University of Pennsylvania** (1983)	
	1862	Mansfield	State Normal School (fifth district); State Teachers College (1927); State College (1960); **Mansfield University of Pennsylvania** (1983)	
	1859	Millersville	(First) State Normal School (second district); State Teachers College (1927); State College (1959); **Millersville University of Pennsylvania** (1983)	
	1873	Shippensburg	Cumberland Valley State Normal School; State Teachers College (1927); State College (1960); **Shippensburg University of Pennsylvania** (1983)	
	1889	Slippery Rock	State Normal School (eleventh district); State Teachers College (1926); State College (1960); **Slippery Rock University of Pennsylvania** (1983)	
	1871	West Chester	State Normal School; State Teachers College (1927); State College (1960); **West Chester University of Pennsylvania** (1983)	
R.I.	1871	Providence	State Normal School; State College of Education (1920); **Rhode Island College** (1959)	
S.C.	1896	Orangeburg	State Colored Normal, Industrial, Agricultural, and Mechanical College; State College (1954); **South Carolina State University** (1992)	African American

	1891	Rock Hill	Winthrop Normal (and Industrial) College (of South Carolina); Winthrop College for Women (1920); Winthrop College (1974); **Winthrop University** (1988)	Women
S.Dak.	1901	Aberdeen	Northern Normal and Industrial School; Northern State Teachers College (1939); Northern State College (1964); **Northern State University** (1989)	
	1884	Madison	State Normal School; General Beadle State College; Dakota State College (1969); **Dakota State University** (1989)	
	1884	Spearfish	State Normal School; Black Hills State Teachers College (1941); Black Hills State College (1964); **Black Hills State University** (1989)	
	1897	Springfield	State Normal School (Closed)	
Tenn.	1911	Johnson City	East State Normal School; East State Teachers College (1925); East State College (1943); **East Tennessee State University** (1963)	
	1912	Memphis	West State Normal School; West State Teachers College (1929); State College (1941); State University (1957); **University of Memphis** (1994)	
	1911	Murfreesboro	Middle State Normal School; Middle State Teachers College (1925); Middle State College (1943); **Middle Tennessee State University** (1965)	
	1912	Nashville	Agricultural and Industrial State Normal School for Negroes; Agricultural and Industrial State Normal College (1924); Agricultural and Industrial State College (1927); Agricultural and Industrial State University (1951); **Tennessee State University** (1969)	African American
Tex.	1920	Alpine	State Normal School for Teachers; Sul Ross State Teachers College (1923); Sul Ross State College (1949); **Sul Ross State University** (1969)	

Continued

Continued

State	Date open	Location	Names and dates of name changes, with current title in bold	Institution: Women; African American; Native American
	1910	Canyon	West State Normal College; West State Teachers College (1923); West State College (1949); West State University (1963); **West Texas A & M University** (1992)	
	1917	Commerce	East State Normal College; East State Teachers College (1923); East State College (1957); East Texas State University (1965); **Texas A & M University—Commerce** (1995)	
	1901	Denton	North State Normal College; North State Teachers College (1923); North State College (1949); North State University (1965); **University of North Texas** (1988)	
	1878	Prairie View	Normal Institute; State Normal and Industrial College (1889); Agricultural and Mechanical College (1947); **Prairie View A & M University** (1970)	African American
	1879	Huntsville	Sam Houston State Normal Institute; Sam Houston State Teachers College (1923); Sam Houston State College (1965); **Sam Houston State University** (1969)	
	1903	San Marcos	Southwest State Normal School; Southwest State Teachers College (1923); Southwest State College (1959); Southwest State University (1969); **Texas State University-San Marcos** (2003)	
Utah	1903	Cedar City	Southern Branch Normal School; Branch Agricultural College (1913); College of South Utah (1953); Southern State College (1969); **Southern Utah University** (1992)	
Vt.	1867	Castleton	State Normal School; State Teachers College (1947); **Castleton State College** (1962)	

	1867	Johnson	State Normal School; State Teachers College (1947); **Johnson State College** (1962)	
	1911	Lyndon	Lyndon Institute; State Teachers College (1949); **Lyndon State College** (1962)	
	1867	Randolph	State Normal School (Closed)	
Va.	1884	Farmville	State (Female) Normal School; State Teachers College (1924); Longwood College (1949); **Longwood University**	Women
	1911	Fredericks-burg	State Normal (and Industrial) School for Women; State Teachers College (1924); Mary Washington College (of the University of Virginia) (1938); **University of Mary Washington** (2004)	Women
	1909	Harrisonburg	State Normal (and Industrial) School for Women; State Teachers College (1924); Madison College (1938); **James Madison University** (1977)	Women
	1883	Petersburg	Virginia Normal and Collegiate Institute; Virginia Normal and Industrial Institute (1902); State College for Negroes (1930); State College (1946); **Virginia State University** (1979)	African American
	1913	Radford	State Normal and Industrial School for Women; State Teachers College (1924); Radford College (1944); **Radford University** (1979)	Women
Wash.	1899	Bellingham	New Whatcom State Normal School; Western Washington College of Education (1937); Western State College (1961); **Western Washington University** (1977)	
	1890	Cheney	State Normal School (and Training) School; Eastern Washington College of Education (1937); Eastern State College (1961); **Eastern Washington University** (1977)	
	1891	Ellensburg	State Normal (and Training) School; Central Washington College of Education (1937); Central State College (1961); **Central Washington University** (1977)	
W.Va.	1875	Athens	Concord State Normal School; Concord State Teachers College	

Continued

Continued

State	Date open	Location	Names and dates of name changes, with current title in bold	Institution: Women; African American; Native American
			(1931); Concord College (1943); **Concord University** (2004)	
	1895	Bluefield	Colored Institute; State Teachers College (1931); **Bluefield State College** (1943)	African American
	1867	Fairmont	State Normal School; State Teachers College (1931); State College (1943); **Fairmont State University**	
	1873	Glenville	State Normal School; **Glenville State College** (1930)	
	1867	Huntington	Marshall College State Normal School; **Marshall University** (1961)	
	1891	Institute	West Virginia Colored Institute; West Virginia Collegiate Institute (1912); State College (1929); **West Virginia State University** (2004)	African American
	1872	Shepherds-town	Shepherd College, State Normal School; Shepherd College, Teachers College (1930); Shepherd College (1950); **Shepherd University** (2004)	
	1870	West Liberty	State Normal School; State Teachers College (1931); **West Liberty State College** (1943)	
Wisc.	1917	Eau Claire	State Normal School; State Teachers College (1927); State College (1951); **University of Wisconsin-Eau Claire** (1972)	
	1909	La Crosse	State Normal School; State Teachers College (1927); State College (1951); **University of Wisconsin-La Crosse** (1972)	
	1885	Milwaukee	State Normal School; State Teachers College (1927); State College (1951); **University of Wisconsin-Milwaukee** (1972)	
	1871	Oshkosh	State Normal School; State Teachers College (1927);	

State	Date	Town	Names
			State College (1951); **University of Wisconsin-Oshkosh** (1972)
	1866	Platteville	State Normal School; State Teachers College (1927); State College (1951); **University of Wisconsin-Platteville** (1972)
	1875	River Falls	State Normal School; State Teachers College (1927); State College (1951); **University of Wisconsin-River Falls** (1972)
	1894	Stevens Point	State Normal School; State Teachers College (1927); State College (1951); **University of Wisconsin-Stevens Point** (1972)
	1896	Superior	State Normal School; State Teachers College (1927); State College (1951); **University of Wisconsin-Superior** (1972)
	1868	Whitewater	State Normal School; State Teachers College (1927); State College (1951); **University of Wisconsin-Whitewater** (1972)
Wyo.	Wyoming did not establish a separate normal school; prospective teachers were educated at the state university.		

Note: If a state normal school grew out of a private academy or seminary, the date it became a state normal school is reported as the date of opening. The names of states and towns in which the institutions were located are left out of the reported names, while the present name of each institution is written in full.

Sources: United States Commissioners of Education, *Reports and Bulletins, 1875–1928–1930* (Washington: U.S. Government Printing Office, 1876–1932); Harry Christopher Humphreys, *The Factors Operating in the Location of State Normal Schools* (New York: Teachers College, 1923), appendix, 144–147; Esek Ray Mosher, "The Rise and Organization of State Teachers Colleges" (Ph.D. diss., Harvard University, 1923); Julian B. Roebuck and Komanduri S. Murty, *Historically Black Colleges and Universities: Their Place in American Higher Education* (Westport, CT: Praeger, 1993), chapter 3; college and university catalogs, histories, and websites.

Abbreviations of Archival Collections

CASUNYG	College Archives, Milne Library, State University of New York College at Geneseo, Geneseo, New York
CSCA	Castleton State College Archives, Vermont Room, Coolidge Library, Castleton, Vermont
LCHS	Livingston County Historical Society, Geneseo, New York
SJSUA	San Jose State University Archives, Special Collections, SJSU Library, San Jose, California
SWTA	Southwest Texas State University Archives, Special Collections, Alkek Library, San Marcos, Texas
UCUNA	University Collection, Collier Library Archives, University of North Alabama, Florence, Alabama
UWOA	University of Wisconsin-Oshkosh Archives, Area Research Center, Polk Library, Oshkosh, Wisconsin
WRUA	Wisconsin Room-University Archives, Karrmann Library, University of Wisconsin-Platteville, Platteville, Wisconsin

Notes

Introduction: "It Wasn't Much of a College"

1. David Riesman, *Constraint and Variety in American Higher Education* (Garden City, NY: Doubleday [1956] 1958), 21, 43, 61–62.
2. See the appendix for a state-by-state list of all state normal schools.
3. Paul Woodring, "The Development of Teacher Education," in *Teacher Education: The Seventy-fourth Yearbook of the National Society for the Study of Education*, ed. Kevin Ryan (Chicago: The University of Chicago Press, 1975), 5.
4. William Marshall French and Florence Smith French, *College of the Empire State: A Centennial History of The New York State College for Teachers at Albany* ([Albany?], 1944), no page numbers.
5. Catalogs of Wisconsin State Normal School/Teachers College at Oshkosh testify to such actions. Maps in early catalogs showed Normal Avenue running along campus; in later catalogs the same street was named College Avenue. One early catalog included a photo of a doorway with "Normal" carved into the stonework; by the mid-1990s, raised black letters in the same location spelled "Dempsy," presumably the name of a past leader or benefactor.
6. John I. Goodlad, *Teachers for Our Nation's Schools* (San Francisco: Jossey-Bass, 1990), 73. See also John I. Goodlad, "Connecting the Present to the Past," in *Places Where Teachers Are Taught*, ed. John I. Goodlad, Roger Soder, and Kenneth Sirotnik (San Francisco: Jossey-Bass, 1990).
7. In the early 1980s, Jurgen Herbst and Colin Burke urged historians to reconsider state normal schools in light of the educational advantages they brought to the sons and daughters of midwestern farmers. Geraldine Jonçich Clifford recently suggested that educational historians have created an artificial division between "higher" and "lower" education, and limited their understanding of normal schools by falsely placing them on the "lower" side of the barrier; normal schools, as well as academies and early high schools, often offered "college-like" curricula. Clifford and Sally Schwager have urged historians of education for women to include normal schools in their analysis, because the story of women and education is incomplete without an understanding of the institutions that provided thousands of women with a taste of higher education. See Jurgen Herbst, "Nineteenth-Century Normal Schools in the United States: A Fresh Look," *History of Education* 9 (September 1980): 219–227; Colin B. Burke, *American Collegiate Populations: A Test of the Traditional View* (New York: New York University Press, 1982); Geraldine Jonçich Clifford, " 'Shaking Dangerous Questions from the Crease': Gender and American Higher Education," *Feminist Issues* 3 (Fall 1983): 3–62, "No Shade in the Golden State: School and University in Nineteenth-Century California," *History of Higher Education Annual* 12 (1992): 35–68, and "*Equally in View*": *The University of California, Its Women, and the Schools* (Berkeley: Center for Studies in Higher Education and Institute for

Governmental Studies, University of California, Berkeley, 1995), 2–6; Sally Schwager, "Educating Women in America," *Signs: Journal of Women in Culture and Society* 12 (Winter 1987): 333–372. See also E. Alden Dunham, *Colleges of the Forgotten Americans: A Profile of State Colleges and Regional Universities* (New York: McGraw-Hill, 1969).

8. Robert A. Caro, *The Years of Lyndon Johnson: The Path to Power* (New York: Alfred A. Knopf, 1982), 141.
9. John R. Thelin, "Rudolph Rediscovered," prefatory essay to Frederick Rudolph, *The American College and University: A History* (Athens: The University of Georgia Press, 1990), xviii.
10. Edward Everett, "An Address by Edward Everett, Governor of Massachusetts, at the Opening of the Normal School at Barre, September 5, 1839," *American Journal of Education* 13 (December 1863): 769.

1 "To Awaken the Conscience": Establishing Teacher Education and State Normal Schools

1. Arthur C. Boyden, *The History of Bridgewater State Normal School* (Bridgewater, MA: Bridgewater Normal Alumni Association, 1933), 30–31; *Seventy-Fifth Anniversary of the State Normal School, Bridgewater, Massachusetts, June 19, 1915* (Bridgewater, MA: Arthur H. Willis, 1915), 71.
2. Lawrence A. Cremin, *American Education: The National Experience, 1783–1876* (New York: Harper & Row, 1980), chapter 4, Mann quoted on 140–141; Carl F. Kaestle, *Pillars of the Republic: Common Schools and American Society, 1780–1860* (New York: Hill and Wang, 1983), chapters 2–5; Jurgen Herbst, *And Sadly Teach: Teacher Education and Professionalization in American Culture* (Madison: The University of Wisconsin Press, 1989), chapter 1.
3. Kaestle, *Pillars*, chapters 5, 6, and 8; Herbst, *And Sadly Teach*, chapter 1.
4. John L. Rury, "Who Became Teachers? The Social Characteristics of Teachers in American History," in *American Teachers: Histories of a Profession at Work*, ed. Donald Warren (New York: MacMillan Publishing Company), 11–14; Willard S. Elsbree, *The American Teacher: Evolution of a Profession in a Democracy* (Westport, CT: Greenwood Press, [1939] 1970), parts 1 and 2, quotation on 29; Michael W. Sedlak, " 'Let Us Go and Buy a School Master,' " in *American Teachers*, 259–262; Herbst, *And Sadly Teach*, 22–24; Christopher J. Lucas, *Teacher Education in America: Reform Agendas for the Twenty-First Century* (New York: St. Martin's Press, 1997), 3–10.
5. Myra H. Strober and David Tyack, "Why Do Women Teach and Men Manage? A Report on Research on Schools," *Signs: Journal of Women in Culture and Society* 5 (1980): 495–497; Rury, "Who Became Teachers?," 15–16; Kathleen Weiler, *Country Schoolwomen: Teaching in Rural California, 1850–1950* (Stanford, CA: Stanford University Press, 1998), 9–13; Catharine Beecher, "Remedy for Wrongs to Women," in *Woman's "True" Profession: Voices from the History of Teaching*, ed. Nancy Hoffman (New York: The Feminist Press, 1981), 40.
6. Weiler, *Country Schoolwomen*, 9–16, Mann quoted on 12; Willard quoted in David B. Tyack and Elisabeth Hansot, *Learning Together: A History of Coeducation in American Public Schools* (New Haven, CT: Yale University Press, 1991), 42; Beecher, "Remedy for Wrongs to Women," 36–56, quotation on 45; Geraldine Jonçich Clifford, "Man/Woman/Teacher: Gender, Family, and Career in American Educational History," in *American Teachers*, 299–305;

Herbst, *And Sadly Teach*, 27–29; Rury, "Who Became Teachers?," 15–16; Strober and Tyack, "Why Do Women Teach," 495–497; Kaestle, *Pillars*, 123–125. See also Lucas, *Teacher Education*, 12–15; Donald H. Parkerson and Jo Ann Parkerson, *Transitions in American Education: A Social History of Teaching* (New York: RoutlegeFalmer, 2001), chapter 4; Kathryn Kish Sklar, *Catharine Beecher: A Study in American Domesticity* (New Haven, CT: Yale University Press, 1973).

7. Herbst, *And Sadly Teach*, 24–27; Polly Welts Kaufman, *Women Teachers on the Frontier* (New Haven, CT: Yale University Press, 1984), xviii; Clifford, "Man/Woman/Teacher," 293–295; Patricia A. Schmuck, "Women School Employees in the United States," in *Women Educators: Employees of Schools in Western Countries*, ed. Patricia A. Schmuck (Albany: State University of New York Press, 1987), 76; Kaestle, *Pillars*, 125; Rury, "Who Became Teachers?," 17–20; Strober and Tyack, "Why Do Women Teach," 498. See also Keith E. Melder, "Woman's High Calling: The Teaching Profession in America, 1830–1860," *American Studies* 13 (Fall 1972): 19–32; Myra H. Strober and A. G. Langford, "The Feminization of Public School Teaching: Cross-Sectional Analysis," *Signs: Women and Culture in Society* 11 (1980): 212–235. For further discussion of the statistical analysis of the feminization of teaching, see Joel Perlman and Robert A. Margo, *Women's Work? American Schoolteachers, 1650–1920* (Chicago: The University of Chicago Press, 2001), and John L. Rury, review of Perlman and Margo, *Women's Work*, in *History of Education Quarterly* 42 (Summer 2002): 292–295.
8. Herbst, *And Sadly Teach*. See also Kaestle, *Pillars*, 124–125.
9. Herbert Kliebard states, "Market forces and a gender-related conception of teaching notwithstanding, the single greatest influence on the feminization of teaching was probably professionalization." Herbert M. Kliebard, "The Feminization of Teaching on the American Frontier: Keeping School in Otsego, Wisconsin, 1867–1880," *Journal of Curriculum Studies* 27 (1995): 558.
10. Franklin quoted in Merle L. Borrowman, *The Liberal and Technical in Teacher Education: A Historical Survey of American Thought* (New York: Teachers College Bureau of Publications, 1956), 35; the article in *Massachusetts Magazine* was published anonymously, but the author was probably Elisha Ticknor, quoted in J. P. Gordy, *Rise and Growth of the Normal-School Idea in the United States* (U.S. Bureau of Education Circular of Information No. 8, Washington: Government Printing Office, 1891), 9; Olmstead quoted in Gordy, *Rise and Growth*, 10; Borrowman, *The Liberal and Technical*, 35; Paul H. Mattingly, *The Classless Profession: American Schoolmen in the Nineteenth Century* (New York: New York University Press, 1975), 28–32; Herbst, *And Sadly Teach*, 22; Charles A. Harper, *A Century of Public Teacher Education: The Story of the State Teachers Colleges as They Evolved from the Normal Schools* (Washington, DC: American Association of Teachers Colleges, 1939), 13–14; Samuel R. Hall, *Lectures on School-Keeping* (Boston: Richardson, Lord and Holbrook, 1829), iv.
11. Hall, *Lectures*, iv; Herbst, *And Sadly Teach*, 30–31; Gordy, *Rise and Growth*, 13–14; Harper, *A Century*, 14–16, Gallaudet quoted on 15. See also Mattingly, *The Classless Profession*, 23–27.
12. Willis Rudy, "America's First Normal School: The Formative Years," *Journal of Teacher Education* 5 (December 1954): 263–264; Edward Everett, "An Address by Edward Everett, Governor of Massachusetts, at the Opening of the Normal School at Barre, September 5, 1839," *American Journal of Education* 14

(December 1863): 758–759; Gordy, *Rise and Growth*, 17–18; Herbst, *And Sadly Teach*, 32–34, Dwight quoted on 32.

13. Gordy, *Rise and Growth*, 19–20; Herbst, *And Sadly Teach*, 21–22, 35–50, quotation on 48; Harper, *A Century*, 17–20.
14. Harper, *A Century*, 16–17, 22–24, Adams quoted on 23–24; Walter H. Ryle, *Centennial History of the Northeast Missouri State Teachers College* (Kirksville, MO: Northeast Missouri State Teachers College, 1972), 19–32.
15. Dwight quoted in Herbst, *And Sadly Teach*, 60; Harper, *A Century*, 21–25, 51, Mann quoted on 21–22; Everett, "An Address," 769, 760.
16. George Frederick Miller, *The Academy System of New York State* (Albany, NY: J. B. Lyon Company, 1922), 131; Dixon quoted in Bjarne R. Ullsvik, *A History of the Platteville Academy, 1839–1866* (Platteville, WI, 1994), 14; Barbara Miller Solomon, *In the Company of Educated Women: A History of Women and Higher Education in America* (New Haven, CT: Yale University Press, 1985), 18–21. For more on academies, see Kim Tolley, "Mapping the Landscape of Higher Schooling, 1727–1850," in *Chartered Schools: Two Hundred Years of Independent Academies in the United States, 1727–1925*, ed. Nancy Beadie and Kim Tolley (New York: RoutledgeFalmer, 2002), 19–43.
17. Frederick Rudolph, *The American College and University: A History* (Athens: The University of Georgia Press, [1962] 1990), 217; Brockport catalog quoted in Thomas E. Finegan, *Teacher Training Agencies: A Historical Review of the Various Agencies of the State of New York Employed in Training and Preparing Teachers for the Public Schools of the State* (Albany, NY: The University of the State of New York, 1917), 206; Geneseo Academy–Temple Hill Academy (a.k.a. Livingston County High School), binder 974.785, LCHS; *Catalogue of the Officers and Students of the Platteville Academy*, 1849, Platteville Academy, Annual Catalogs, PL Series 16, Box 1, WRUA; Ullsvik, *A History*, 20; Richard D. Gamble, *From Academy to University, 1866–1966: A History of Wisconsin State University, Platteville, Wisconsin* (Platteville, WI: Wisconsin State University, 1966), 71–72, 76; Castleton catalog quoted in *Proceedings of the Laying of the Cornerstone and Dedication of the New Administration Building* (Castleton Normal School, 1926), CSCA, 17–18; Mary Clough Cain, *The Historical Development of State Normal Schools for White Teachers in Maryland* (New York: Teachers College Bureau of Publications, 1941), 37–47. For more on teacher education in academies, see Christine A. Ogren, "Betrothed to the State?: Nineteenth-Century Academies Confront the Rise of the State Normal Schools," in *Chartered Schools*, 284–303.
18. Marvin G. Maiden, "History of the Professional Training of Teachers in Virginia" (Ph.D. diss., University of Virginia, 1927), 92–93; *Catalogue of the Officers and Students of the Platteville Academy*, 1856, 1857, 1859, 1860, 1861, 1862; *Catalogue of the Officers and Students of Geneseo Academy*, 1855, 1856, 1857, 1860, 1861, LCHS; Willis F. Dunbar, *The Michigan Record in Higher Education* (Detroit, MI: Wayne State University Press, 1963), 6; Christine A. Ogren, "Normal Departments," in *Historical Dictionary of Women's Education in the United States*, ed. Linda Eisenmann (Westport, CT: Greenwood Press, 1998), 305–307. On teacher education in liberal arts colleges, see Charles Burgess, "Abiding by the 'Rule of Birds': Teaching Teachers in Small Liberal Arts Colleges," in *Places Where Teachers Are Taught*, ed. John I. Goodlad, Roger Soder, and Kenneth A. Sirotnik (San Francisco: Jossey-Bass, 1990), 87–135.
19. Miller, *The Academy System*, 19, 137–144, 152–161; William Marshall French, "How We Began to Train Teachers in New York," *Proceedings of the New York*

State Historical Association 34 (1936): 183–186, Regents' minutes quoted on 183; Finegan, *Teacher Training Agencies*, 21, 31–37, 41, 47; *Catalogue of the Officers and Students of Geneseo Academy*, 1855, 1856, 1857, 1860, 1861. See also Ogren, "Betrothed," 286–687.

20. William S. Taylor, *The Development of the Professional Education of Teachers in Pennsylvania* (Philadelphia, PA: J. B. Lippincott Company, 1924), 77–80; Virginia legislative act quoted in Maiden, "History," 78–79; Albert Salisbury, *Historical Sketch of Normal Instruction in Wisconsin* (n.p., 1893), 13–14; William Kittle, *A Brief History of the Board of Regents of Normal Schools of Wisconsin, 1857–1925* (n.p., 1924), 3, 7; George C. Purington, *History of the State Normal School, Farmington, Maine* (Farmington, ME: Knowlton, McLeary & Co., 1889), 9–10; George Frank Sammis, "A History of the Maine Normal Schools" (Ph.D. diss., University of Connecticut, 1970), 43; James R. Dotson, "The Historical Development of the State Normal School for White Teachers in Alabama" (Ed.D. diss., University of Alabama, 1961), 164–165.
21. Folder: "Platteville Academy—Commencement, 1854, 55, 63," Platteville Academy, Annual Catalogs and Misc., PL Series 16, Box 2, WRUA; commencement programs in Geneseo Academy–Temple Hill Academy, binder 974.785; Maiden, "History," 92; Taylor, *The Development*, 159; David F. Allmendinger, Jr., "Mount Holyoke Students Encounter the Need for Life Planning, 1837–1850," *History of Education Quarterly* 19 (Spring 1979): 40. See also Anne Firor Scott, "The Ever-Widening Circle: The Diffusion of Feminist Values from the Troy Female Seminary, 1822–1872," *History of Education Quarterly* 19 (Spring 1979): 3–25.
22. Harper, *A Century*, 51; Elsbree, *The American Teacher*, 155; French, "How We Began," 190; Kaestle, *Pillars*, 129. See also Henry Barnard, *Normal Schools and Other Institutions, Agencies, and Means Designed for the Professional Education of Teachers* (Hartford, CT: Case, Tiffany and Company, 1851).
23. Elsbree, *The American Teacher*, 156; Ullsvik, *A History*, 16; Irving G. Hendrick, *California Education: A Brief History* (San Francisco: Boyd & Fraser, 1980), 42; Deward Homan Reed, *The History of Teachers Colleges in New Mexico* (Nashville, TN: George Peabody College for Teachers, 1948), 7–8; Rhode Island law quoted in Elsbree, *The American Teacher*, 156; Clarence H. Dempsey, "The Castleton Normal School," in *Proceedings*, 29; John W. Payne, "Poor-Man's Pedagogy: Teachers' Institutes in Arkansas," *The Arkansas Historical Quarterly* 14 (1955): 195–206; Gamble, *From Academy*, 90.
24. Elsbree, *The American Teacher*, 157–159, discussion questions quoted on 159; Payne, "Poor-Man's Pedagogy," 196; Barnard quoted in Roy C. Woods, "Private Normal Schools in West Virginia," *West Virginia History* 15 (October 1953): 108.
25. Illinois superintendent and Henry Barnard quoted in Kaestle, *Pillars*, 129; West Virginia superintendent quoted in Woods, "Private," 114; attendance figures reported in Kaestle, *Pillars*, 129; Mattingly, *The Classless Profession*, 71. For a more detailed discussion of teachers' institutes, see Mattingly, *The Classless Profession*, chapter 4.
26. James Robert Overman, *The History of Bowling Green State University* (Bowling Green, OH: Bowling Green University Press, 1967), 5; Travis Edwin Smith, *The Rise of Teacher Training in Kentucky* (Nashville, TN: George Peabody College for Teachers, 1932), 149–153; R. H. Eckelberry, "The McNeely Normal School and Hopedale Normal College" *Ohio Archeological and Historical Publications* 40 (1931): 86–136; Ryle, *Centennial History*, 56–70; Elsbree, *The American Teacher*, 312–313.

27. Elsbree, *The American Teacher*, 152–153; Finegan, *Teacher Training Agencies*, 314–315; Donald R. Raiche, *From a Normal Beginning: The Origins of Kean College of New Jersey* (Rutherford, NJ: Fairleigh Dickinson University Press, 1980), 57, 154; Herbst, *And Sadly Teach*, 92–94.
28. Taylor, *The Development*, 65, 85–88, 107–108; Lee Graver, *A History of the First Pennsylvania State Normal School, Now the State Teachers College at Millersville* (Millersville, PA: State Teachers College, 1955), 20–30; Elsbree, *The American Teacher*, 150; Herbst, *And Sadly Teach*, 100–101, 144–145.
29. Taylor, *The Development*, 89, 94; Ryle, *Centennial History*, 19–32; Salisbury, *Historical Sketch*, 9–13, Ladd quoted on 11; Finegan, *Teacher Training Agencies*, 193.
30. J. L. Meader, *Normal School Education in Connecticut* (New York: Teachers College Bureau of Publications, 1928), 11–12; Egbert R. Isbell, *A History of Eastern Michigan University, 1849–1965* (Ypsilanti, MI: Eastern Michigan University Press, 1971), 8–9, 16, legislation quoted on 8.
31. Glen E. Hickman, "A History of Teacher Education in Nebraska" (Ph.D. diss., The University of Oregon, 1947), 21–23; I. F. Boughter, ed., *Fairmont State Normal School: A History* (Fairmont, WV: Fairmont State Normal School, 1929), 30–32; Cain, *The Historical Development*, 55–56; Maiden, "History," 141.
32. Everett, "An Address," 760; Mark K. Fritz, "The State Normal Schools: Teaching Teachers and Others," *Pennsylvania Heritage* 11 (Fall 1985): 4; Taylor, *The Development*, 103–106, 110; Lee Graver, "Kutztown State College—1864–1964," *Historical Review of Berks County* 29 (Autumn 1964): 107; Dempsey, "The Castleton Normal School," 29–30; "Castleton Normal School: Its History and Outlook," *The Normal Student* 2 (Nov. 21, 1902): 1, CSCA; Harper, *A Century*, 66, 75, 88; Isbell, *A History*, xii–xiii. For more on the semiprivate nature of normal schools in Pennsylvania, see Linda Eisenmann, "The Influence of Bureaucracy and Markets: Teacher Education in Pennsylvania," in *Places Where Teachers Are Taught*, 287–329.
33. Charles A. Harper, *Development of the Teachers College in the United States, with Special Reference to the Illinois State Normal University* (Bloomington, IL: McKnight & McKnight, 1935), 23.
34. David B. Potts, " 'College Enthusiasm!' As Public Response, 1800–1860," *Harvard Educational Review* 47 (February 1977): 41.
35. "The Normal School," *Grant County Witness*, Dec. 7, 1865, WRUA, 2.
36. Herbert E. Fowler, *A Century of Teacher Education in Connecticut: The Story of the New Britain State Normal School and the Teachers College of Connecticut, 1849–1949* (New Britain, CT: The Teachers College of Connecticut at New Britain, 1949), 26–27, legislation quoted on 26; Isbell, *A History*, 11; Philip R. Leavenworth, "The School in Retrospect," in *Proceedings*, 18; Conrad E. Patzer, *Public Education in Wisconsin* (Madison, WI: John Callahan, State Superintendent, 1924), 145–146; Janette Bohi, "Whitewater, A Century of Progress (1868–1968)," in *History of the Wisconsin State Universities*, ed. Walker D. Wyman (River Falls, WI: River Falls State University Press, 1968), 57–58, 62, newspaper quoted on 58.
37. Bohi, "Whitewater," 59–60; Fritz, "The State Normal Schools," 5; Taylor, *The Development*, 108; Graver, "Kutztown State College," 105–107; Sammis, "A History," 46–47, 83–84; Ernest Longfellow, *The Normal on the Hill: One Hundred Years of Peru State College* (Grand Island, NE: The Augustine Company, 1967), 6, 11–13; Boughter, *Fairmont State Normal*, 20; Victoria Ann Smith,

"A Social History of Marshall University During the Period at the State Normal School, 1867–1900," *West Virginia History* 25 (October 1963): 32; Gamble, *From Academy*, 93; Finegan, *Teacher Training Agencies*, 136, 206–208. See also Ogren, "Betrothed," 292–297.

38. Gordy, *Rise and Growth*, 51–54; Harper, *A Century*, 35; Herbst, *And Sadly Teach*, 64; Vernon Lamar Mangun, *The American Normal School: Its Rise and Development in Massachusetts* (Baltimore, MD: Warwick & York, Inc., 1928), chapters 8–10; Fowler, *A Century*, 59; J. M. McKenzie, *History of the Peru State Normal* (Auburn, NE: The Nemaha County Republican, 1911), 24; Edwin H. Cates, *A Centennial History of St. Cloud State College* (Minneapolis, MN: Dillon Press, 1968), 27–34; Finegan, *Teacher Training Agencies*, legislative resolution quoted on 118, 137. In New York, some academies became openly critical of the normal schools, which they accused of stealing students by offering too much general academic work in addition to teacher training. Finegan, *Teacher Training Agencies*, 118.
39. Kirksville newspaper quoted in Ryle, *Centennial History*, 98; Jean Talbot, *First State Normal School 1860, Winona State College, 1960* (Winona, MN: *Quarterly Bulletin of Winona State College* 55 [August 1959]—56 [August 1960]), 3–4, 6–7; John Swett, *Public Education in California: Its Origin and Development, With Personal Reminiscences of Half a Century* (New York: American Book Company, 1911), 264.
40. Everett, "An Address," 758, 764; Cyrus Peirce, Letter to Henry Barnard, Jan. 1, 1841, in *The First State Normal School in America: The Journals of Cyrus Peirce and Mary Swift*, ed. Arthur O. Norton (Cambridge: Harvard University Press, 1926), l.
41. Everett, "An Address," 765–768; Harper, *Development*, 88. The main distinction that historians of teacher education have emphasized in the normal schools' approaches to teacher education in the mid-nineteenth century, is between followers of the Bridgewater and Westfield traditions. Followers of Nicolas Tillinghast at Bridgewater, according to this analysis, were more likely to focus on the preparation of elementary-school teachers, while followers of the Westfield approach thought more expansively, eventually focusing on the preparation of high-school teachers and administrators. See Borrowman, *The Liberal and Technical*, chapter 2; Harper, *Development*, chapter 5; Herbst, *And Sadly Teach*, chapter 4.
42. William F. Phelps, *David P. Page: His Life and Teachings* (New York: E. L. Kellogg & Co., 1892), 7–9, Mann quoted on 9; Norton, ed., *The First State Normal*, xxv; John W. Cook and James V. McHugh, *A History of the Illinois State Normal University* (Normal, IL: Illinois State Normal University, 1882), 28; David Nelson Camp, *David Nelson Camp: Recollections of a Long and Active Life* (New Britain, CT: Privately printed, 1917), chapter 1; Arthur Clarke Boyden, *Albert Gardner Boyden and the Bridgewater State Normal School: A Memorial Volume* (Bridgewater, MA: Arthur H. Willis, 1919), 18; Edward Austin Sheldon, *Autobiography of Edward Austin Sheldon*, ed. Mary Sheldon Barnes (New York: Ives-Butler Company, 1911), chapters 1–7; Helen E. Marshall, *Grandest of Enterprises: Illinois State Normal University, 1857–1957* (Normal, IL: Illinois State Normal University, 1956), 82–83.
43. Phelps, *David P. Page*, 9–10; Cook and McHugh, *A History*, 28–30; Louie G. Ramsdell, "First Hundred Years of the First State Normal School in America: The State Teachers College at Framingham, Massachusetts—1839–1939," in *First State Normal School in America: The State Teachers College*

at Framingham, Massachusetts (Framingham, MA: The Alumnae Association of the State Teachers College at Framingham, Massachusetts, 1959), 1–54; Marshall, *Grandest*, 85; Isbell, *A History*, 109, 113; Norton, *The First State Normal*, xxvi; Robert T. Brown, *The Rise and Fall of the People's Colleges: The Westfield Normal School, 1839–1914* (Westfield, MA: Institute for Massachusetts Studies, Westfield State College, 1988), 31; W. Wayne Dedman, *Cherishing This Heritage: The Centennial History of the State University College at Brockport, New York* (New York: Appleton-Century-Crofts, 1969), 94–95; McCormick quoted in Marshall, *Grandest*, 101. On the early normal-school principals, see also Mattingly, *The Classless Profession*, chapter 7.

44. Harper, *A Century*, 28; Marshall, *Grandest*, 84–85; Estelle Greathead, *The Story of an Inspiring Past: Historical Sketch of the San Jose State Teachers College from 1862 to 1928 with an Alphabetical List of Matriculants and Record of Graduates by Classes* (San Jose, CA: San Jose State Teachers College, 1928), 129; Purington, *History*, 20–28; Talbot, *First State Normal*, 4; *A History of the State Normal School of Kansas for the First Twenty-Five Years* (Emporia Kansas, 1889), 14, 52; Salisbury, *Historical Sketch*, 39; Cain, *The Historical Development*, 74–75.
45. Dorothy Rogers, *Oswego: Fountainhead of Teacher Education: A Century in the Sheldon Tradition* (New York: Appleton-Century-Crofts, Inc., 1961), 23, 46; Harper, *The Development*, 72; Cain, *The Historical Development*, 73; Phelps, *David P. Page*, 22; Daniel Putnam, *A History of the Michigan State Normal School at Ypsilanti, Michigan, 1849–1899* (Ypsilanti, MI: Michigan State Normal College, 1899), 211.
46. Isbell, *A History*, 15; Graver, *A History*, 69; Talbot, *First State Normal*, 7; Edgar B. Wesley, *NEA: The First Hundred Years: The Building of the Teaching Profession* (New York: Harper & Brothers, 1957), 391; Richard Edwards, "Normal Schools in the United States," in *Teacher Education in America: A Documentary History*, ed. Merle L. Borrowman (New York: Teachers College Press, 1965), 79.
47. Everett, "An Address," 765–766; Edwards, "Normal Schools," 80; 1844 Westfield catalog quoted in Brown, *The Rise and Fall*, 35; 1854–1855 Westfield catalog quoted in James Ralph Fiorello, "General Education in the Preparation of Teachers at Westfield State College, 1839–1960" (Ph.D. diss., University of Connecticut, 1969), 296; Boyden, *Albert Gardner Boyden*, 138.
48. Peirce, Letter to Henry Barnard, li, lii; "The Journal of Cyrus Peirce," in *The First State Normal School in America: The Journals of Cyrus Peirce and Mary Swift*, 35, 43, 53, 54.
49. "The Journal of Mary Swift," in *The First State Normal School in America: The Journals of Cyrus Peirce and Mary Swift*, 157, 198, 147, 163, 193–194.
50. David P. Page, *Theory and Practice of Teaching*, Seventh Edition (Syracuse, NY: Hall & Dickenson, 1847), chapters 6–8, 10, 14, quotations on 84, 86, 107, 112, 139, 218, 292, 297, 307, 311.
51. Fowler, *A Century*, Barnard quoted on 36–37, 1850 curriculum listed on 37, Philbrick quoted on 49.
52. Taylor, *The Development*, 122–123; Sammis, "A History," 87; Harper, *The Development*, 132; Bridgewater catalog quoted in Boyden, *The History*, 38; *Report of the Principal of the Kansas State Normal School to the Board of Directors for the Year 1865* (Emporia, KS, 1865), 8; Westfield catalog quoted in Fiorello, "General Education," 298.
53. Elsbree, *The American Teacher*, 229–234; Harper, *A Century*, 17–18; Rogers, *Oswego*, 19. See also Herbst, *And Sadly Teach*, chapter 2.

54. "The Journal of Mary Swift," 146–147; Page, *Theory and Practice*, 86–88, 98; Dickinson quoted in Work Projects Administration in the State of Massachusetts, *The State Teachers College at Westfield* (Boston: State Department of Education, 1941), 35–36.
55. Rogers, *Oswego*, 5–6, 16–20, Sheldon's principles listed on 20; Gordy, *Rise and Growth*, 61–75, committee report quoted on 67; Harper, *A Century*, 122–124; Borrowman, *The Liberal and Technical*, 115–117; Sheldon, *Autobiography*.
56. Welch quoted in Isbell, *A History*, 17; Putnam, *A History*, Board of Education quoted on 51, curriculum listed on 52; C. O. Ruggles, *Historical Sketch and Notes: Winona State Normal School, 1860–1910* (Winona, MN: Jones & Kroeger Co., 1910), 123; *Report of the Principal*, 8–9; 1866–1867 Westfield catalog quoted in Fiorello, "General Education," 299.
57. Edwards quoted in Harper, *The Development*, 130; Columbus story in Rogers, *Oswego*, 21; Baltimore catalog quoted in Cain, *The Historical Development*, 149.
58. Everett, "An Address," 766–768.
59. "The Journal of Mary Swift," 100–102, 106–107, 117–118, 132–138, quotations on 100, 134; "The Journal of Cyrus Peirce," 73.
60. Page, *Theory and Practice*, chapters 2, 3, 9, quotations on 39, 170, 189; 1854–1855 Westfield catalog quoted in Fiorello, "General Education," 296; Graver, *A History*, 70; McKenzie, *History*, 28.
61. Ruggles, *Historical Sketch*, 123; Sammis, "A History," 87; Ypsilanti catalog for 1868–1869 quoted in Putnam, *A History*, 58; Isbell, *A History*, 53.
62. Everett, "An Address," 768; Edwards, "Normal Schools," 81; Connecticut legislation quoted in Fowler, *A Century*, 23; Missouri legislation quoted in Ryle, *Centennial History*, 75.
63. Maude B. Gerritson, "Training School, 1839–1939, Especially 1914–1939," in *First State Normal*, 55–56; Isbell, *A History*, 33–34; Harper, *A Century*, 64; Cain, *The Historical Development*, 149–150.
64. "The Journal of Cyrus Peirce," 12, 19; Peirce, Letter to Henry Barnard, liii–lv.
65. Philbrick quoted in Fowler, *A Century*, 49; Baltimore catalog quoted in Cain, *The Historical Development*, 149; Graver, *A History*, 191.
66. Ramsdell, "The First Hundred Years," 7–8; Brown, *The Rise and Fall*, 79, 42–43, 55–56.
67. Longfellow, *The Normal*, 36; Fiorello, "General Education," 116; Conant quoted in Boyden, *The History*, 32; "The Journal of Mary Swift," 90; Edwards, "Normal Schools," 81–82.
68. Marshall, *Grandest*, 93; Salisbury, *Historical Sketch*, 63.
69. Page, *Theory and Practice*, 10, 3; Peirce, Letter to Henry Barnard, lv.
70. Page, *Theory and Practice*, 336–346, quotations on 356; "The Journal of Mary Swift," 82–83; Hanna quoted in Marshall, *Grandest*, 101.
71. *Semi-Centennial History of the Illinois State Normal University, 1857–1907* (Normal, IL: Illinois State Normal University, 1907), 197.
72. "The Journal of Cyrus Peirce," 6, 7, 17, 52, 68; Edwards quoted in Sandra D. Harmon, " 'The Voice, Pen and Influence of Our Women Are Abroad in the Land': Women and the Illinois State Normal University, 1857–1899," in *Nineteenth-Century Women Learn to Write*, ed. Catherine Hobbs (Charlottesville: University of Virginia Press, 1995), 89; Phelps quoted in Borrowman, *The Liberal and Technical*, 44–45.
73. Mattingly, *The Classless Profession*, 139, 141; Maine legislation quoted in Sammis, "A History," 83.

74. Nancy Beadie, "Internal Improvement: The Structure and Culture of Academy Expansion in New York State in the Antebellum Era, 1820–1860," in *Chartered Schools*, 89–115; Tyack and Hansot, *Learning Together*, chapter 5; William J. Reese, *The Origins of the American High School* (New Haven, CT: Yale University Press, 1995); David F. Allmendinger, *Paupers and Scholars: The Transformation of Student Life in Nineteenth-Century New England* (New York: St. Martin's Press, 1975); Solomon, *In the Company*, chapter 2; Margaret A. Nash, " 'A Triumph of Reason': Female Education in Academies in the New Republic," in *Chartered Schools*, 64–86.
75. George W. Neel, "A History of the State Teachers College at Edinboro, Pennsylvania" (Ed.D. diss., Rutgers University, 1950), 29–30; Camp, *David Nelson Camp*, 33; Brown, *The Rise and Fall*, 60–61, 127, 46; superintendent quoted in Ruggles, *Historical Sketch*, 171–172; Purington, *History*, 179; Minns quoted in Maxine Ollie Merlino, "A History of the California State Normal Schools—Their Origin, Growth, and Transformation into Teachers Colleges" (Ed.D. diss., University of Southern California, 1962), 36; William A. Stone, *The Tale of a Plain Man* (Self-published, 1917), 95–96; William Marshall French and Florence Smith French, *College of the Empire State: A Centennial History of The New York State College for Teachers at Albany* ([Albany?], 1944), 95, 257; Ramsdell, "First Hundred Years," 5; George H. Martin, "The Bridgewater Spirit," in *Seventy-Fifth Anniversary*, 14.
76. "The Journal of Cyrus Peirce," 39, 40; student quoted in W. Charles Lahey, *The Potsdam Tradition: A History and a Challenge* (New York: Appleton-Century-Crofts, 1966), 77; Stone, *The Tale*, 96; catalog quoted in Brown, *The Rise and Fall*, 53.
77. Westfield catalogs summarized in Fiorello, "General Education," appendix J; New Britain curriculum printed in Fowler, *A Century*, 37; Ypsilanti circular quoted at length in Putnam, *A History*, 52; *Report of the Principal*, 7–8; Report of Principal Phelps quoted at length in Ruggles, *Historical Sketch*, 123; 1860 Normal catalog summarized in Harper, *Development*, 57; "Castleton Seminary" / "State Normal School," Flier, 1868, CSCA.
78. Michigan's normal-school act stated that the Ypsilanti institution would "give instructions in the mechanic arts, and in the arts of husbandry and agricultural chemistry," and 1857 legislation in Illinois "provided that 'natural science, including agricultural chemistry and animal and vegetable physiology,' be taught" at Illinois State Normal University. Michigan legislation quoted in Isbell, *A History*, 8; Harper, *Development*, 361. Because such statements distinguished normal-school legislation in the Midwest from that in the East, historians including Harper and Herbst have claimed that midwestern normals functioned more as general-purpose institutions while eastern normals retained a focus on teacher education. However, the background and interests of students in all parts of the country resulted in extensive academic—but not agricultural and mechanical—offerings alongside teacher education; there were few regional differences in curriculum. See Harper, *Development*; Jurgen Herbst, "Nineteenth-Century Normal Schools in the United States: A Fresh Look," *History of Education* 9 (1980): 219–227.
79. 1854–1855 Westfield catalog quoted in Fiorello, "General Education," 296; *Report of the Principal*, 8; Taylor, *The Development*, 123; Harper, *Development*, 363, 371; Isbell, *A History*, 240–241.
80. Norton quoted in Harmon, " 'The Voice, Pen and Influence' " 94; Boyden, *Albert Gardner Boyden*, 116–117, 138–139; Putnam, *A History*, 215–217; Marshall,

Grandest, 42, 46–47; Albert Reynolds Taylor, *Autobiography of Albert Reynolds Taylor* (Decatur, IL: Review Printing and Stationery Co., 1929), 35.

81. Brown, *The Rise and Fall*, 30; Rogers, *Oswego*, 62; "The Journal of Mary Swift," 98, 89; Speech by A. J. Hutton in "The Golden Jubilee, 1866–1916," ed. J. A. Wilgus, *Bulletin: State Normal School, Platteville, Wisconsin* 14 (June 1917), 20–21.
82. Debate topics quoted in Cook and McHugh, *A History*, 142–143; debate topics quoted in Putnam, *A History*, 219; Marshall, *Grandest*, 47; Work Projects Administration, *The State Teachers College*, 37; *Historical Sketch of the State Normal School at San Jose, California, with a Catalogue of Its Graduates and a Record of their Work for Twenty-Seven Years* (Sacramento: J.D. Young, 1889), 98–99.
83. "The Journal of Cyrus Peirce," 49; Lyceum records quoted in Putnam, *A History*, 217, 220; Graver, *A History*, 222; Marshall, *Grandest*, 70; debate topics quoted in Cook and McHugh, *A History*, 105; *A History of the State Normal School of Kansas*, 62–63, 17.
84. "The Journal of Mary Swift," 97; Rogers, *Oswego*, 69; Cook and McHugh, *A History*, 105; *A History of the State Normal School of Kansas*, 63; Isbell, *A History*, 83–84; Marshall, *Grandest*, 135; Graver, *A History*, 225.
85. Norton, *The First State Normal*, xxxi; Stone, *The Tale*, 105–106, 110, 116, 122–184.
86. Herbst, *And Sadly Teach*, 41.

2 "The Masses and Not the Classes": A Tradition of Welcoming Nontraditional Students

1. *Semi-Centennial History of the Illinois State Normal University, 1857–1907* (Normal, IL: Illinois State Normal University, 1907), 202.
2. Carl F. Kaestle, *Pillars of the Republic: Common Schools and American Society, 1780–1860* (New York: Hill and Wang, 1983), epilogue; David Tyack, *The One Best System: A History of American Urban Education* (Cambridge: Harvard University Press, 1974), part II; James D. Anderson, *The Education of Blacks in the South, 1860–1935* (Chapel Hill: The University of North Carolina Press, 1988), chapters 1 and 5; Robert H. Wiebe, *The Search for Order, 1877–1920* (New York: Hill and Wang, 1967); Burton J. Bledstein, *The Culture of Professionalism: The Middle Class and the Development of Higher Education in America* (New York: W. W. Norton & Company, Inc., 1976).
3. *New York State Teachers College At Buffalo: A History, 1871–1946* (Buffalo: New York State Teachers College at Buffalo, 1946), 17; Irving G. Hendrick, "Teacher Education and Leadership in Major Universities," in *Places Where Teachers Are Taught*, ed. John I. Goodlad, Roger Soder, and Kenneth A. Sirotnik (San Francisco: Jossey-Bass, 1990), 236–284; Geraldine Jonçich Clifford and James W. Guthrie, *Ed School: A Brief for Professional Education* (Chicago: The University of Chicago Press, 1988), chapter 2.
4. Kent D. Beeler and Philip C. Chamberlain, " 'Give a Buck To Save a College': The Demise of Central Normal College," *Indiana Magazine of History* 57:2 (June 1971): 117–128; Anthony O. Edmonds and E. Bruce Geelhoed, *Ball State University: An Interpretive History* (Bloomington, IN: Indiana University Press, 2001), 52; Travis Edwin Smith, *The Rise of Teacher Training in Kentucky*

(Nashville, TN: George Peabody College for Teachers, 1932), 149–153, 161–162; Anderson, *The Education of Blacks*, 134; Titus Brown, "A New England Missionary and African-American Education in Macon: Raymond G. Von Tobel at the Ballard Normal School, 1908–1935," *The Georgia Historical Quarterly* 82:2 (Summer 1998): 283–304; Joe M. Richardson, "Allen Normal School: Training 'Leaders of Righteousness,' 1885–1933," *The Journal of Southwest Georgia History* 12 (Fall 1997): 1–26; Willard S. Elsbree, *The American Teacher: Evolution of a Profession in a Democracy* (Westport, CT: Greenwood Press, [1939] 1970), 329–330, 365; Jurgen Herbst, *And Sadly Teach: Teacher Education and Professionalization in American Culture* (Madison: The University of Wisconsin Press, 1989), 100–102, chapter 6; Pamela Claire Hronek, "Women and Normal Schools: Tempe Normal, a Case Study, 1885–1925" (Ph.D. diss., Arizona State University, 1985), 74–75; Melvin Frank Fiegel, "A History of Southwestern State College, 1903–1953" (Ed.D. diss., Oklahoma State University, 1968), 2; Joe C. Jackson, "Summer Normals in Indian Territory After 1898," *The Chronicles of Oklahoma* 37 (Autumn 1959): 307–329; R. McLaran Sawyer, "No Teacher for the School: The Nebraska Junior Normal School Movement," *Nebraska History* 52:2 (Summer 1971): 191–203; Andrea Radke, " 'I Am Very Aspiring': Muirl Dorrough & the Alliance Junior Normal School," *Nebraska History* 81 (Spring 2002): 2–11.

5. For more details, see appendix. Only four states never established normal schools: Alaska, Delaware, Nevada, and Wyoming. The two states that did not establish normal schools until the 1910s were Ohio and Tennessee. The large number of private normal schools in Ohio probably helps to explain why that state did not establish public state normal schools until the 1910s. In 1875, Tennessee established the State Normal College, which in 1889 became Peabody Normal College, at the University of Nashville. Despite its name, this institution was privately run with almost no financial support or oversight from the state. Alfred L. Crabb, *The Genealogy of George Peabody College for Teachers* (Nashville, 1935), 31–34; Allison Norman Horton, "Origin and Development of the State College Movement in Tennessee" (Ed.D. diss., George Peabody College for Teachers, 1953), 16, 22–26; Sherman Dorn, *A Brief History of Peabody College* (Nashville, TN: Peabody College of Vanderbilt University, 1996).
6. Charles A. Harper, *Development of the Teachers College in the United States, with Special Reference to the Illinois State Normal University* (Bloomington, IL: McKnight & McKnight, 1935), 90–91.
7. Irving H. Hart, *The First 75 Years* (Cedar Falls, IA: Iowa State Teachers College, 1951), 6–7, State Superintendent Abernathy quoted on 6.
8. George Frank Sammis, "A History of the Maine Normal Schools" (Ph.D. diss., University of Connecticut, 1970), 149, 201; Thomas E. Finegan, *Teacher Training Agencies: A Historical Review of the Various Agencies of the State of New York Employed in Training and Preparing Teachers for the Public Schools of the State* (Albany: The University of the State of New York, 1917), 148; Albert Salisbury, *Historical Sketch of Normal Instruction in Wisconsin* (n.p., 1893), 42–43; May A. Greene, "Danbury State Teachers College," *Teacher Education Quarterly* (Fall 1949): 31; C. Francis Willey, "Willimantic State Teachers College," *Teacher Education Quarterly* (Fall 1949): 15–16; Terry A. Barnhart, "Educating the Masses: The Normal-School Movement and the Origins of Eastern Illinois University, 1895–1899," *Journal of Illinois History* 4 (Autumn 2001): 195–202, *Mattoon Gazette* quoted on 199; John Lankford, " 'Culture

and Business': The Founding of the Fourth State Normal School at River Falls," *Wisconsin Magazine of History* 47 (1963): 29–31; Willis F. Dunbar, *The Michigan Record in Higher Education* (Detroit, MI: Wayne State University Press, 1963), 200–203.

9. Walter H. Ryle, *Centennial History of the Northeast Missouri State Teachers College* (Kirksville, MO: Northeast Missouri State Teachers College, 1972), 71–77; Susan Vaughn, "The History of State Teachers College, Florence, Alabama," *Bulletin of the State Teachers College, Florence, Alabama* 18 (Supplemental, 193?): 1–13; Robin O. Harris, " 'To Illustrate the Genius of Southern Womanhood': Julia Flisch and Her Campaign for the Higher Education of Georgia Women," *The Georgia Historical Quarterly* 80 (Fall 1996): 515; Elisabeth Ann Bowles, *A Good Beginning: The First Four Decades of the University of North Carolina at Greensboro* (Chapel Hill: The University of North Carolina Press, 1967), 7; Fiegel, "A History," 4–13; James H. Thomas and Jeffrey A. Hurt, "Southwestern Normal School: The Founding of an Institution," *The Chronicles of Oklahoma* 54 (Winter 1976–1977): 462–463; Jerry G. Nye, *Southwestern Oklahoma State University: The First 100 Years* (Weatherford, OK: Southwestern Oklahoma State University, 2001), 9–13.

10. Rudolph Jones, "The Development of Negro State Colleges and Normal Schools in North Carolina," *The Quarterly Review of Higher Education Among Negroes* (Charlotte, NC: Johnson C. Smith University) 6 (April 1938): 132–133, 136; Mark Andrew Huddle, "To Educate a Race: The Making of the First State Colored Normal School, Fayetteville, North Carolina, 1865–1877," *The North Carolina Historical Review* 74:2 (April 1997): 140, 150–153; E. Louise Murphy, "Origin and Development of Fayetteville State Teachers College, 1867–1959—A Chapter in the History of the Education of Negroes in North Carolina" (Ph.D. diss., New York University, 1960), 70–90, 97; Arkansas legislation quoted in Thomas Rothrock, "Joseph Carter Corbin and Negro Education in the University of Arkansas," *The Arkansas Historical Quarterly* 30 (Winter 1971): 282; Willis L. Brown and Janie M. McNeal-Brown, "Oklahoma's First Comprehensive University: Langston University, The Early Years," *The Chronicles of Oklahoma* 74:1 (1996): 35–43; Second Morrill Act quoted in Christopher J. Lucas, *American Higher Education: A History* (New York: St. Martin's Griffin, 1994), 164. See also Frank Bowles and Frank A. DeCosta, *Between Two Worlds: A Profile of Negro Higher Education* (New York: McGraw Hill Book Company, 1971), 31–32.

11. Roy Wilson McNeal, *Southern Oregon College Cavalcade* (Ashland, OR: Southern Oregon College Foundation, 1972), 7–11; William Pierce Tucker, "Ashland Normal School, 1869–1930," *The Oregon Historical Quarterly* 32 (March and June 1931): 46–47; Ellis A. Stebbins, *The OCE Story* (Monmouth, OR: Oregon College of Education, 1973), 33–37; John C. Almack, "History of Oregon Normal Schools," *The Quarterly of the Oregon Historical Society* 21 (June 1920): 112; Maxine Ollie Merlino, "A History of the California State Normal Schools—Their Origin, Growth, and Transformation into Teachers Colleges" (Ed.D. diss., University of Southern California, 1962), 89–90; Cecil Dryden, *Light for an Empire: The Story of Eastern Washington State College* (Cheney, WA: Eastern Washington State College, 1965), 8–16, 30–33.

12. Dudley S. Brainard, *History of St. Cloud State Teachers College* (St. Cloud, MN, 1953), 17–18; New York legislative report quoted in Finegan, *Teacher Training*, 125; I. F. Boughter, ed., *Fairmont State Normal School: A History* (Fairmont,

WV: Fairmont State Normal School, 1929), 36; Ryle, *Centennial History*, 115–117; Hronek, "Women," 97; Almack, "History," 121–134, charges listed on 134.

13. Carey W. Brush, *In Honor and Good Faith: A History of the State University College at Oneonta, New York* (Oneonta, NY: The Faculty-Student Association of State University Teachers College at Oneonta, Inc., 1965), 2; Bessie L. Park, *Cortland—Our Alma Mater: A History of Cortland Normal School and the State University of New York Teachers College at Cortland, 1869–1959* (Cortland, NY, 1960), 46; Ryle, *Centennial History*, 157–158; Hart, *The First 75 Years*, 13; J. Orin Oliphant, *History of the State Normal School at Cheney, Washington* (Spokane: Inland-American Printing Company, 1924), chapters 3 and 4.
14. Dryden, *Light*, 46–50, 79; Oliphant, *History*, 33, 63; Deward Homan Reed, *The History of Teachers Colleges in New Mexico* (Nashville, TN: George Peabody College for Teachers, 1948), 73–74; Almack, "History," 122–131; Tucker, "Ashland," 168–174; Stebbins, *The OCE Story*, 45–51.
15. Marcus M. Wilkerson, *Thomas Duckett Boyd: The Story of a Southern Educator* (Baton Rouge: Louisiana State University Press, 1935), 95; Ernest Longfellow, *The Normal on the Hill: One Hundred Years of Peru State College* (Grand Island, NE: The Augustine Company, 1967), 17; Ryle, *Centennial History*, 103; Herbert E. Fowler, *A Century of Teacher Education in Connecticut: The Story of the New Britain State Normal School and the Teachers College of Connecticut, 1849–1949* (New Britain, CT: The Teachers College of Connecticut at New Britain, 1949), 65; John Edward Merryman, "Indiana University of Pennsylvania: From Private Normal School to Public University" (Ph.D. diss., University of Pittsburgh, 1972), 104–105; Robert McGraw, "A Century of Service," in *The First 100 Years: Worcester State College*, ed. Herb Taylor (Worcester, MA: Worcester State College, Office of Community Services, 1974), no page numbers; Merlino, "A History," 75, 92, 160; Brush, *In Honor*, 1; Reed, *The History*, 74; Ernest J. Hopkins and Alfred Thomas, Jr., *The Arizona State University Story* (Phoenix, AZ: Southwest Publishing Co., Inc., 1960), 115; "Old Main," *ASU Vision: The Magazine of the ASU Alumni Association* 1 (Summer 1997): 17.
16. David Sands Wright, *Fifty Years at the Teachers College: Historical and Personal Reminiscences* (Cedar Falls, IA: Iowa State Teachers College, 1926), 82, 137; George Kimball Plochmann, *The Ordeal of Southern Illinois University* (Carbondale, IL: Southern Illinois University Press, 1957), 15; Sandra D. Harmon, " 'The Voice, Pen and Influence of Our Women Are Abroad in the Land': Women and the Illinois State Normal University, 1857–1899," in *Nineteenth-Century Women Learn to Write*, ed. Catherine Hobbs (Charlottesville: University of Virginia Press, 1995), 203; William Frederick Hartman, "The History of Colorado State College of Education: The Normal School Period, 1890–1911" (Ph.D. diss., Colorado State College of Education, 1951), 83, 89, newspaper quoted on 90; Frank A. Cooper, *The Plattsburgh Idea in Education, 1889–1964* (Plattsburgh, NY: Plattsburgh College Benevolent and Educational Association, Inc., 1964), 25–26; S. E. Rothery, "Some Educational Institutions: Pilgrimages About San Jose," *The Overland Monthly* 30 (July 1897): 74.
17. On nontraditional students, see John P. Bean and Barbara S. Metzner, "A Conceptual Model of Nontraditional Undergraduate Student Attrition," *Review of Educational Research* 55 (1985): 485–540; Arthur Levine and Associates, *Shaping Higher Education's Future: Demographic Realities and*

Opportunities, 1990–2000 (San Francisco: Jossey-Bass, 1989); Barbara S. Metzner and John P. Bean, "The Estimation of a Conceptual Model of Nontraditional Undergraduate Student Attrition," *Research in Higher Education* 27 (1987): 15–38; Ernest T. Pascarella and Patrick T. Terenzini, "Studying College Students in the 21st Century: Meeting New Challenges," *The Review of Higher Education* 21 (1998): 151–165. For a more detailed discussion of literature on nontraditional students in the context of state normal schools, see Christine A. Ogren, "Rethinking the 'Nontraditional' Student from a Historical Perspective: State Normal Schools in the Late Nineteenth and Early Twentieth Centuries," *The Journal of Higher Education* 74 (Nov./Dec. 2003): 640–664.

18. Barbara Miller Solomon, *In the Company of Educated Women: A History of Women and Higher Education in America* (New Haven, CT: Yale University Press, 1985), chapters 4–7; Lynn D. Gordon, *Gender and Higher Education in the Progressive Era* (New Haven, CT: Yale University Press, 1990), chapter 1; Helen Lefkowitz Horowitz, *Campus Life: Undergraduate Cultures from the End of the Eighteenth Century to the Present* (Chicago: The University of Chicago Press, 1987), chapter 9. See also Thomas Woody, *A History of Women's Education in the United States*, 2 vols. (New York: Science Press, 1929); Mabel Newcomer, *A Century of Higher Education for American Women* (New York: Harper & Brothers, 1959); Patricia Albjerg Graham, "Expansion and Exclusion: A History of Women and American Higher Education," *Signs: Journal of Women in Culture and Society* 3 (Summer 1978): 759–773; Patricia A. Palmieri, "From Republican Motherhood to Race Suicide: Arguments on the Higher Education of Women in the United States, 1820–1920," in *Educating Men and Women Together: Coeducation in a Changing World*, ed. Carol Lasser (Urbana: The University of Illinois Press, 1987), 49–64; Elizabeth Seymour Eschbach, *The Higher Education of Women in England and America, 1865–1920* (New York: Garland Publishing, Inc., 1993); Charlotte Williams Conable, *Women at Cornell: The Myth of Equal Education* (Ithaca: Cornell University Press, 1977); Dorothy Gies McGuigan, *A Dangerous Experiment: 100 Years of Women at the University of Michigan* (Ann Arbor, MI: Center for Continuing Education of Women, 1970).

19. Ryle, *Centennial History*, 112; Wright, *Fifty Years*, 130; Hartman, "The History," 153; George E. Bates, Jr., "Winona Normal School Student Profile, 1860–1900," *Journal of the Midwest History of Education Society* 7 (1979): 11; United States Commissioner/Bureau of Education, *Reports and Bulletins* (Washington, D.C.: U.S. Government Printing Office, 1868–1945).

20. Anderson, *The Education of Blacks*, chapters 2 and 7. See also Bowles and DeCosta, *Between Two Worlds*; Julian B. Roebuck and Komanduri S. Murty, *Historically Black Colleges and Universities: Their Place in American Higher Education* (Westport, CT: Praeger, 1993). Two states also established normal schools for Native-American students: Oklahoma opened Northeastern State Normal School in 1909, and North Carolina opened in the Cherokee Indian State Normal School in 1913. These institutions opened too late to be included in this analysis.

21. Lucas, *American Higher Education*, 207–209; Victoria-Maria MacDonald and Teresa Garcia, "Historical Perspectives on Latino Access to Higher Education, 1848–1990," in *The Majority in the Minority: Expanding the Representation of Latina/o Faculty, Administrators, and Students in Higher Education*, ed. Jeanett Castellanos and Lee Jones (Sterling, VA: Stylus Publishing, 2003); Horowitz, *Campus Life*.

22. Percentage of Catholic students at Providence calculated from data in Len West, "Teacher Education for Americanizing Immigrants in the Public Schools, 1871–1920: The Rhode Island Normal School Programs" (Ph.D. diss., University of Connecticut, 1995), 159; Percentage of immigrant students calculated from data in *Biographical Directory and Condensed History of the State Normal School, Mankato, Minn, 1870–1890* (Mankato, MN: Mankato State Normal School Alumni Association, 1891); McGraw, "A Century," section 4, no page numbers; William Marshall French and Florence Smith French, *College of the Empire State: A Centennial History of The New York State College for Teachers at Albany* ([Albany?], 1944), 95, 258; Oswego students quoted in Dorothy Rogers, *Oswego: Fountainhead of Teacher Education: A Century in the Sheldon Tradition* (New York: Appleton-Century-Crofts, Inc., 1961); Douglas R. Skopp, *Bright With Promise: From the Normal and Training School to SUNY Plattsburgh, 1889–1989* (Norfolk, VA: The Donning Company, 1989), 45; Robert T. Brown, *The Rise and Fall of the People's Colleges: The Westfield Normal School, 1839–1914* (Westfield, MA: Institute for Massachusetts Studies, Westfield State College, 1988), 61–62; McGraw, "A Century," section 4, no page numbers; Louie G. Ramsdell, "First Hundred Years of the First State Normal School in America: The State Teachers College at Framingham, Massachusetts—1839–1939," in *First State Normal School in America: The State Teachers College at Framingham, Massachusetts* (Framingham, MA: The Alumnae Association of the State Teachers College at Framingham, Massachusetts, 1959), 11; Helen E. Marshall, *Grandest of Enterprises: Illinois State Normal University, 1857–1957* (Normal, IL: Illinois State Normal University, 1956), 132–133, 239; Elizabeth Tyler Bugaighis, "Blackboard Diplomacy: The Role of American Normal Schools in Exporting Education to Latin America, 1891–1924" (paper presented at the annual meeting of the American Educational Research Association, New Orleans, April 2000); Evans quoted in Tom W. Nichols, *Rugged Summit* (San Marcos, TX: The University Press, South West Texas State University, 1970), 94.
23. Bledstein, *The Culture of Professionalism*; Oscar Handlin and Mary F. Handlin, *The American College and American Culture: Socialization as a Function of Higher Education* (New York: McGraw-Hill Book Company, 1970); Arthur Levine and Jana Nidiffer, *Beating the Odds: How the Poor Get to College* (San Francisco: Jossey-Bass Publishers, 1996), 37–48; Horowitz, *Campus Life*, 51; Solomon, *In the Company*, 68–71.
24. Blair quoted in William P. Turner, *A Centennial History of Fairmont State College* (Fairmont, WV: Fairmont State College, 1970), 21–22; Class Day Program, 1888, UWOA, 13; Percentages of parents engaged in certain occupations at Providence calculated from data in West, "Teacher Education," 158; David A. Gould, "Policy and Pedagogues: School Reform and Teacher Professionalization in Massachusetts, 1840–1920" (Ph.D. diss., Brandeis University, 1977), 87; Brown, *The Rise and Fall*, 79; statistics on parents of students at San Marcos calculated from data in *Announcement of the Southwest Texas State Normal School for the Annual Session 1904–1905, 1905–1906, 1906–1907, 1907–1908, 1908–1909, 1909–1910, 1910–1911* (Austin, 1903–1910), SWTA.
25. Rosalind R. Fisher, *". . . the stone strength of the past . . .": Centennial History of State University College of Arts and Science at Geneseo, New York* (Geneseo, NY, 1971), 146; Hart, *The First 75 Years*, 116; *A History of the State Normal School of Kansas for the First Twenty-Five Years* (Emporia, KS, 1889), 45; Pamela Dean, "Covert Curriculum: Class, Gender, and Student Culture at a New South

Woman's College, 1892–1910" (Ph.D. diss., University of North Carolina at Chapel Hill, 1994), 108–109, 114; Elizabeth L. Wheeler, "Isaac Fisher: The Frustrations of a Negro Educator at Branch Normal College, 1902–1911," *The Arkansas Historical Quarterly* 41 (Spring 1982): 5.

26. Robert A. Caro, *The Years of Lyndon Johnson: The Path to Power* (New York: Alfred A. Knopf, 1982), 142; Morey quoted in C. O. Ruggles, *Historical Sketch and Notes: Winona State Normal School, 1860–1910* (Winona, MN: Jones & Kroeger Co., 1910), 210; J. S. Nasmith, "An Open Letter From J. Nasmith," *Platteville Witness* LXIII (April 13, 1932), 2; newspaper quoted in Irene Goldgraben, "And the Glory of the Latter House Shall Be Greater Than That of the Former," in *And the Glory of the Latter House Shall Be Greater Than That of the Former: An Informal History of Castleton State College*, ed. Holman D. Jordan (Castleton, VT: Castleton State College, 1968), 19; Brown, *The Rise and Fall*, 89; Rogers, *Oswego*, 58; cheer quoted in *The Arkansasyer* (Pine Bluff, AR: Faculty of Arkansas Agricultural, Mechanical and Normal College) 1 (1928): 1, in "Keepers of the Spirit: The L. A. Davis, Sr. Historical Collection," Exhibit, Isaac S. Hathaway-John M. Howard Fine Arts Center, University of Arkansas at Pine Bluff, Pine Bluff, AR, May 1995.

27. West, "Teacher Education," 157; Dignam quoted in Work Projects Administration in the State of Massachusetts, *The State Teachers College at Westfield* (Boston: State Department of Education, 1941), 53; Hopkins and Thomas, *The Arizona State University Story*, 85–86; Class Day Program (Oshkosh), 1888, 18; Hartman, "The History," 154; Brush, *In Honor*, 45, 46, 114; students' ages at San Marcos calculated from data in Student Registers, 1903–1910, box 70, series 3, SWTA; *Annual Catalogue of the State Normal School at Oshkosh, Wis., for the School Year 1894–95, 1899–1900*, UWOA; *Oshkosh State Normal School Bulletin* 2 (June 1905), 7 (June 1910), UWOA; *Announcement of the Southwest Texas State Normal School for the Annual Session 1905–1906, 1910–1911*, SWTA; *The Normal School Bulletin* 4 (July 1, 1915), 8 (July 1919), SWTA; Skopp, *Bright*, 62–63.

28. Rogers, *Oswego*, 59; Farmington statistics based on information in George C. Purington, *History of the State Normal School, Farmington, Maine* (Farmington, ME: Knowlton, McLeary & Co., 1889), 141–174; Brown, *The Rise and Fall*, 71; Mankato statistics calculated from data in *Biographical Directory*; Oshkosh statistics calculated from *Annual Catalogue of the State Normal School at Oshkosh, Wis., for the School Year 1879–80, 1884–85, 1889–90, 1891–92, 1893–94, 1895–96, 1897–98, 1899–1900, 1901–1902*, and *Oshkosh State Normal School Bulletin* 1 (June 1904); 3 (June 1906); 5 (June 1908); quotations from Class Day Programs (Oshkosh), 1895, 18; 1896, 16.

29. Hopkins and Thomas, *The Arizona State University Story*, 85–86; Bowles, *A Good Beginning*, 15–16; Florence data calculated using statistics in State Superintendent of Education (AL), *Reports, 1899 and 1900*, 31–32; *1901–1902*, 117. San Marcos statistics calculated from information in Student Registers (San Marcos), 1903–1910. Wilson quoted in Lowell H. Harrison, "Gordon Wilson's Normal Education: Western Kentucky State Normal School, 1908–1913," *Register of the Kentucky Historical Society* 86 (Winter 1988): 27–28.

30. Hewett quoted in Harmon, " 'The Voice' "; Egbert R. Isbell, *A History of Eastern Michigan University, 1849–1965* (Ypsilanti, MI: Eastern Michigan University Press, 1971), 138; State Superintendent of Education (Alabama), *Report, 1901–1902* (Montgomery, AL, n.d.), 117; Stebbins, *The OCE Story*, 46;

Brush, *In Honor*, 114; Rogers, *Oswego*, 122; Nasmith, "An Open Letter," 2; Morey quoted in Ruggles, *Historical Sketch*, 210.

31. George H. Martin, "The Bridgewater Spirit," in *Seventy-Fifth Anniversary of the State Normal School, Bridgewater, Massachusetts, June 19, 1915* (Bridgewater, MA: Arthur H. Willis, 1915), 14; Dixon quoted in Arthur Clarke Boyden, *Albert Gardner Boyden and the Bridgewater State Normal School: A Memorial Volume* (Bridgewater, MA: Arthur H. Willis, 1919), 143; *Annual Catalogue of the State Normal School, Florence, Alabama, 1867–1877, 1883 and 1884*, UCUNA; *Annual Catalogue of the State Normal College, Florence, Alabama, 1889–1890, 1894–95, 1899–1900, 1904–1905, 1910–1911*, UCUNA; State Superintendent of Education (AL), *Report, 1912*; Dean, "Covert Curriculum," 102; *Annual Catalogue of the State Normal School at Oshkosh, Wis., for the School Year 1879–80, 1884–85, 1889–90, 1894–95, 1899–1900; Oshkosh State Normal School Bulletin* 2 (June 1905); 7 (June 1910); Merlino, "A History," 94–95. Kathleen Underwood paints a different picture of the female students at Greeley, Colorado in the 1890s, emphasizing their middle-class "urban" (greater than 2,500 population) backgrounds. See Kathleen Underwood, "The Pace of Their Own Lives: Teaching Training and the Life Course of Western Women," *Pacific Historical Review* 55 (November 1986): 513–530.
32. Johnson quoted in Charles H. Coleman, "Eastern Illinois State College: Fifty Years of Public Service," *Eastern Illinois State College Bulletin* 189 (January 1, 1950): 65.
33. Caro, *The Years of Lyndon Johnson*, 142.
34. McKenzie, *History*, 22–23, 96; Principal's annual report, 1910–1911, quoted in Arthur Charles Forst, Jr., "From Normal School to State College: The Growth and Development of Eastern Connecticut State College From 1889 to 1959" (Ph.D. diss., University of Connecticut, 1980), 101; President Matthews at Tempe quoted in Hronek, "Women," 165; Mead quoted in Ruggles, *Historical Sketch*, 212.
35. Class Day Program (Oshkosh), 1888, 17–18.
36. Ryle, *Centennial History*, 151–152; Edwin H. Cates, *A Centennial History of St. Cloud State College* (Minneapolis, MN: Dillon Press, 1968), 37; Thomas W. Bicknell, *History of the Rhode Island Normal School* (Providence, RI, 1911), 226; Samuel R. Mohler, *The First Seventy-Five Years: A History of Central Washington State College, 1891–1966* (Ellensburg, WA: Central Washington State College, 1967), 48–49.
37. Purington, *History*, 22; M. Janette Bohi, *A History of Wisconsin State University Whitewater, 1868–1968* (Whitewater, WI: Whitewater State University Foundation, 1968), 69; Cates, *A Centennial History*, 37; Wright, *Fifty Years*, 42–43; David Sands Wright, "Iowa State Normal School," *The Palimpsest* 13 (January 1932): 6–9; Ryle, *Centennial History*, 153; W. Wayne Dedman, *Cherishing This Heritage: The Centennial History of the State University College at Brockport, New York* (New York: Appleton-Century-Crofts, 1969), 110, 165–166.
38. Ramsdell, "First Hundred Years," 8; *Biographical Directory*, 14; Mary Clough Cain, *The Historical Development of State Normal Schools for White Teachers in Maryland* (New York: Teachers College Bureau of Publications, 1941), 83; Murphy, "Origin and Development," 123–128.
39. Edwards quoted in Harper, *Development*, 107; Boyd quoted in *Thomas Duckett Boyd*, 122; Bowles, *A Good Beginning*, 32–33; McIver quoted in Pamela Dean, "Learning to Be New Women: Campus Culture at the North Carolina Normal

and Industrial College," *The North Carolina Historical Review* 68:3 (July 1991): 289; Judith Hillman Paterson, "To Teach the Negro," *Alabama Heritage* 40 (Spring 1996): 9–11.

40. Boyden, *Albert Gardner Boyden*, 44; Westfield principal quoted in Brown, *The Rise and Fall*, 73; James Ralph Fiorello, "General Education in the Preparation of Teachers at Westfield State College, 1839–1960" (Ph.D. diss., University of Connecticut, 1969), 70–71, 58; Jean Talbot, *First State Normal School 1860, Winona State College, 1960* (Winona, MN: *Quarterly Bulletin of Winona State College* 55 [August 1959]—56 [August 1960]), 12; Lyman Dwight Wooster, *Fort Hays State College: An Historical Study* (Hays, KS: Fort Hays Kansas State College, 1961), 40; Mohler, *The First Seventy-Five Years*, 65–66; Irving G. Hendrick, *California Education: A Brief History* (San Francisco: Boyd & Fraser, 1980), 24; Estelle Greathead, *The Story of an Inspiring Past: Historical Sketch of San Jose State Teachers College From 1862 to 1928* (San Jose, CA: San Jose State Teachers College, 1928), 15; Charles William Dabney, *Universal Education in the South* (New York: Arno Press & The New York Times, [1936] 1969), vol. 2, 425–429, 382–386; Melvin E. Mattox, "The Curriculum," in *Three Decades of Progress: Eastern Kentucky State Teachers College*, ed. Jonathan T. Dorris (Richmond, KY: Eastern Kentucky State Teachers College, 1936), 71; Murphy, "Origin and Development," 178; Vaughn, "The History," 25; John G. Flowers, "History and Development," in *Emphasis Upon Excellence* (San Marcos, TX: Southwest Texas State College Press, 1964), 2.

41. Adelaide R. Pender, "At the New Britain Normal School, 1886–1888," Pender Collection, University Archives, Elihu Burritt Library, Central Connecticut State University, New Britain, Connecticut, 1; Westfield catalog quoted in Fiorello, "General Education," 59; Mohler, *The First Seventy-Five Years*, 65–66. Questions for students desiring to enter the junior class at San Jose included: "How many cords of wood in a pile 18 feet long, 2 feet wide, and 4 feet high?"; "Give three ways in which a noun may be in the objective case, with an example of each"; and "Compare the length of the equatorial diameter with that of the polar diameter." Students intending to enter the middle class also had to answer: "3 is 175 per cent. of what?"; "Conjugate the verb *flow* in the indicative future, in the potential perfect, and in the subjective present"; "Give the use of the fly-wheel, the governor, and the sliding valve in a steam engine." *Catalogue and Circular of the California State Normal School, San Jose, 1879* (Sacramento, 1879), 47–51; "State Normal School, Castleton, Vermont," Flier, 1891, CSCA; Albert Salisbury, *The Normal Schools of Wisconsin: A Souvenir of the Meeting of the National Educational Association held at Milwaukee, Wis., July 6–9, 1897* (n.p., 1897), 17.

42. *Annual Catalogue of the State Normal School, Florence, Alabama, 1876–77*, 12; Fredrick Chambers, "Historical Study of Arkansas Agricultural, Mechanical and Normal College, 1873–1943" (Ed.D. diss., Ball State University, 1970), 68; *Annual Catalogue of the State Normal School, Florence, Alabama, 1883 and 1884*, 13; *Annual Catalogue of the State Normal College, Florence, Alabama, 1886–87*, 14; *Catalogue and Circular of the California State Normal School, San Jose, 1887–88*, 18; Westfield catalog quoted in Fiorello, "General Education," 256; *Annual Catalogue of the State Normal School at Oshkosh, Wis., for the School Year 1898–99*, 113.

43. Iowa board and Cedar Falls catalog quoted in Clarence Theodore Molen, "The Evolution of a State Normal School Into a Teachers College: The University of

Northern Iowa, 1876–1916" (Ph.D. diss., University of Iowa, 1974), 50–51; *Catalogue and Circular of the California State Normal School, San Jose, 1885–86*, 43; Millersville's *The Normal Journal* quoted in Elizabeth Tyler Bugaighis, "Liberating Potential: Women and the Pennsylvania State Normal Schools, 1890–1930" (Ph.D. diss., The Pennsylvania State University, 2000), 204.

44. *State Normal School, Geneseo, NY, 1901–1902*, 16; *Annual Catalogue of the State Normal School at Oshkosh, Wis., for the School Year 1898–99*, 111; *Bulletin of the State Normal College, Florence, Alabama* 1 (1912), UCUNA, 9.
45. *The First Half Century of the Oshkosh Normal School* (Oshkosh, WI: The Faculty of State Normal School, 1921), 17; Albert Reynolds Taylor, *Autobiography of Albert Reynolds Taylor* (Decatur, IL: Review Printing and Stationery Co., 1929), 57–58; *State Normal School, Geneseo, NY, 1907–1908*, 19; Isbell, *A History*, 166; Hopkins and Thomas, *The Arizona State University Story*, 171.
46. Bridgewater catalog quoted in *As We Were . . . As We Are: Bridgewater State College, 1840–1876* (Bridgewater, MA: Alumni Association, Bridgewater State College, 1876), 79; Bicknell, *History*, 32; Murphy, "Origin and Development," 101; pledge quoted in Hronek, "Women," 94.
47. Bridgewater catalog quoted in *As We Were*, 79; Bicknell, *History*, 33; Clayton C. Mau, *Brief History of the State University Teachers College, Geneseo, New York* (Geneseo, NY, 1956), 6; *A History of the State Normal School of Kansas*, 29; Murphy, "Origin and Development," 102; Reed, *The History*, 57.
48. Isbell, *A History*, 138; *Catalogue of the California State Normal School, San Jose, 1895* (Sacramento: State Printing Office, 1895), 9; *1900* (1900), 11; Murphy, "Origin and Development," 167; Dean, "Covert Curriculum," 118; Hronek, "Women," 111; Michael Francis Bannon, "A History of State Teachers College, Troy, Alabama" (Ed.D. diss., George Peabody College for Teachers, 1954), 47; *Announcement of the Southwest Texas State Normal School for the Annual Session 1908–1909*, 44.
49. Chambers, "Historical Study," 80; Vaughn, "The History," 38; Greathead, *The Story*, 57–58; Coleman, "Eastern Illinois State College," 92; Bannon, "A History," 79; Hartman, "The History," 155.
50. Peru student quoted in McKenzie, *History*, 98; Dedman, *Cherishing*, 111; Rogers, *Oswego*, 71; Brush, *In Honor*, 32, 102–103; Nichols, *Rugged Summit*, 92–93; Murphy, "Origin and Development," 167; Rothrock, "Joseph Carter Corbin," 289; *State Normal School, Geneseo, NY, 1911–1912*, 28; *1912–1913*, 25.
51. Albert Salisbury, *Historical Sketches of the First Quarter-Century of the State Normal School at Whitewater, Wisconsin* (Madison, WI: Tracy, Gibbs & Co., 1893), 17.

3 "Substantial Branches of Learning" and "A Higher Degree of Culture": Academic Studies and Intellectual Life

1. Wilson quoted in Samuel R. Mohler, *The First Seventy-Five Years: A History of Central Washington State College, 1891–1966* (Ellensburg, WA: Central Washington State College, 1967), 62.
2. *A History of the State Normal School of Kansas for the First Twenty-Five Years* (Emporia, KS, 1889), 31.
3. Hector Richard Carbone, "The History of the Rhode Island Institute of Instruction and the Rhode Island Normal School as Agencies and Institutions of

Teacher Education, 1845–1920" (Ph.D. diss., University of Connecticut, 1971), 278; C. O. Ruggles, *Historical Sketch and Notes: Winona State Normal School, 1860–1910* (Winona, MN: Jones & Kroeger Co., 1910), 127; *Annual Catalogue of the State Normal School, Florence, Alabama, 1876–77*, UCUNA; James R. Dotson, "The Historical Development of the State Normal School for White Teachers in Alabama" (Ph.D. diss., University of Alabama, 1961), 207–208; Susan Vaughn, "The History of State Teachers College, Florence, Alabama," *Bulletin of the State Teachers College, Florence, Alabama* 18 (Supplemental, 193?), UCUNA, 16; *Catalogue and Circular of the California State Normal School, San Jose, 1878*, SJSUA, 44–46.

4. Carbone, "The History," 284–285; Charles A. Harper, *Development of the Teachers College in the United States, with Special Reference to the Illinois State Normal University* (Bloomington, IL: McKnight & McKnight, 1935), 122, 189; Melvin Frank Fiegel, "A History of Southwestern State College, 1903–1953" (Ed.D. diss., Oklahoma State University, 1968), 19–20; E. Louise Murphy, "Origin and Development of Fayetteville State Teachers College, 1867–1959—A Chapter in the History of the Education of Negroes in North Carolina" (Ph.D. diss., New York University, 1960), 103–104.
5. Albert Salisbury, *Historical Sketch of Normal Instruction in Wisconsin* (n.p., 1893), 51–53, report quoted on 51; Oshkosh catalogs and bulletins, UWOA; quotation from *Annual Catalogue of the State Normal School at Oshkosh, Wis., for the School Year 1888–89*, UWOA, 20.
6. Ruggles, *Historical Sketch*, 128–131; *Report of the Principal of the Kansas State Normal School to the Board of Directors for the Year 1865* (Emporia, KS, 1865), 7; *Official Reports of the State Normal School for the Academic Year Ending December 31, 1870, Emporia Kansas* (Topeka, KS, 1870), 8–9; Albert Reynolds Taylor, *Autobiography of Albert Reynolds Taylor* (Decatur, IL: Review Printing and Stationery Co., 1929), 58–59; Florence catalogs and bulletins, UCUNA; George C. Purington, *History of the State Normal School, Farmington, Maine* (Farmington, ME: Knowlton, McLeary & Co., 1889), 195; Castleton catalogs and bulletins, CSCA; George E. Gay, "Massachusetts Normal Schools," *Education* 17 (May 1897): 515; J. Orin Oliphant, *History of the State Normal School at Cheney, Washington* (Spokane: Inland-American Printing Company, 1924), 128–129.
7. Daniel Putnam, *A History of the Michigan State Normal School at Ypsilanti, Michigan, 1849–1899* (Ypsilanti, MI: Michigan State Normal College, 1899), 60–61.
8. W. Wayne Dedman, *Cherishing This Heritage: The Centennial History of the State University College at Brockport, New York* (New York: Appleton-Century-Crofts, 1969), 97, 112–113; Geneseo catalogs and circulars, CASUNYG; Clayton C. Mau, *Brief History of the State University Teachers College, Geneseo, New York* (Geneseo, NY, 1956), CASUNYG, 8; Rosalind R. Fisher, *". . . the stone strength of the past . . .": Centennial History of State University College of Arts and Science at Geneseo, New York* (Geneseo, NY, 1971), 67–68; I. F. Boucher, ed., *Fairmont State Normal School: A History* (Fairmont, WV: Fairmont State Normal School, 1929), 87–91; Frank T. Router, *West Liberty State College: The First 125 Years* (West Liberty, WV: West Liberty State College, 1963), 19–21; Victoria Ann Smith, "A Social History of Marshall University During the Period as the State Normal School, 1867–1900," *West Virginia History* 25 (October 1963): 33, 35.

9. Oshkosh catalogs and bulletins, UWOA; Salisbury, *Historical Sketch*, 53–55; John Edward Merryman, "Indiana University of Pennsylvania: From Private Normal School to Public University" (Ph.D. diss., University of Pittsburgh, 1972), 187; Jared Stallones, "Struggle for the Soul of a Normal School," *Journal of the Midwest History of Education Society* 23 (1996): 104; Elisabeth Ann Bowles, *A Good Beginning: The First Four Decades of the University of North Carolina at Greensboro* (Chapel Hill: The University of North Carolina Press, 1967), 11–12; *A History of the State Normal School of Kansas*, 30–31. Some normals also offered special courses for long-serving teachers or college graduates. In Wisconsin, a one-year teachers' "professional course" was part of the 1892 curriculum revision, and Washington's legislature mandated "an advanced course of one year for graduates from colleges and universities" in 1899. Such courses were unusual and generally unpopular. Salisbury, *Historical Sketch*, 55; Washington legislation quoted in Oliphant, *History*, 131.
10. According to Sandra Harmon, at Normal, Illinois, "Men and women took the same classes and met in the same classrooms, but Old Main, the only campus building for many years, had separate entrances, cloakrooms, and stairways for the two sexes." Sandra D. Harmon, " 'The Voice, Pen and Influence of Our Women Are Abroad in the Land': Women and the Illinois State Normal University, 1857–1899," in *Nineteenth-Century Women Learn to Write*, ed. Catherine Hobbs (Charlottesville: University of Virginia Press, 1995), 88.
11. In the instructors' grade books that were saved at Oshkosh, minor variations in the grades earned by women and men probably reflect some influence of gender ideology on instructors' expectations, and perhaps also students' assessments of their own capabilities. But the gender differences in grades were slight, suggesting relatively equal expectations for, and treatment of, female and male normalites. In English, women tended to score just slightly higher than men. In Word Analysis during the second quarter of 1893–1894, the average grade for the 23 female students was 82.8, and for the 13 males was 78.4. Ten years later, in first-quarter Composition II, 19 women averaged 83.5, and nine men averaged 82.1. Women scored a bit higher than men in pedagogy and psychology, and a bit lower in government. In Theory I in the department of pedagogy during the second quarter of 1902–1903, the average grade for the 47 women was 82.1, and for the 12 men was 78.8. In Psychology II during the following year, 27 females averaged 92.1, and 12 men 89.8. In Civil Government during the fall and winter of 1880–1881, ten women earned an average score of 70.9, while their seven male classmates earned an average score of 81.3; but during the latter part of 1881–1882, the women barely outscored the men—14 women averaged 80.8, and 12 men averaged 80.3. In mathematics, men generally earned somewhat higher grades than women. In Algebra I in 1890–1891, 18 male students earned an average grade of 74.3, and 18 female students averaged 73.9. Women and men earned remarkably similar grades in physics during the 1900s. In Advanced Physics during 1902–1903, all of the students—two women and eight men—earned perfect scores of 100; five years later in Heat and Light, a woman and a man shared the lowest score, 84.3. Calculated from grades in Registrar, Teachers' Grade Record Books, 1886–1935, boxes 1–18, series 90, UWOA.
12. Alumnus quoted in Fisher, *". . . the stone strength,"* 133.
13. Harper, *Development*, 98–99. National data compiled from the *Biennial Survey of Education 1926–1928*, Bulletin No. 16 (Washington, DC: Government Printing Office, 1930), reported in Elizabeth Tyler Bugaighis, "Liberating Potential: Women and the Pennsylvania State Normal Schools, 1890–1930"

(Ph.D. diss., The Pennsylvania State University, 2000), 149. See chapter 2 for a discussion of the gender composition of the student body. Faculty data collected from school catalogs and bulletins, as well as institutional histories.

14. *The First Half Century of the Oshkosh Normal School* (Oshkosh, WI: The Faculty State Normal School, 1921), UWOA, 68; *Oshkosh State Teachers College: The First Seventy-Five Years* (Oshkosh, WI: Oshkosh State Teachers College, 1946), 36; Robert McGraw, "A Century of Service," in *The First 100 Years: Worcester State College*, ed. Herb Taylor (Worcester, MA: Worcester State College, Office of Community Services, 1974), section 3, no page numbers; Dorothy Rogers, *Oswego: Fountainhead of Teacher Education: A Century in the Sheldon Tradition* (New York: Appleton-Century-Crofts, Inc., 1961), 48; *Catalogue of the State Normal School at Castleton, Vermont, 1883–1885*; State Superintendent of Education, *Vermont School Reports* (Montpelier, 1884–1896, biennial); Estelle Greathead, *The Story of an Inspiring Past: Historical Sketch of the State Normal School at San Jose, California, with a Catalogue of Its Graduates and a Record of Their Work for Twenty-Seven Years* (Sacramento: J.D. Young, 1889), 144–179; *Annual Catalogue of the State Normal School at Oshkosh, Wis., for the School Year 1898–1899, 1899–1900, 1900–1901, 1901–1902, 1902–1903*, UWOA; *Oshkosh State Normal School Bulletin* 1 (June 1904), UWOA; Elisabeth Ann Bowles, *A Good Beginning: The First Four Decades of the University of North Carolina at Greensboro* (Chapel Hill: The University of North Carolina Press, 1967), 41–42; other school catalogs, bulletins, and institutional histories. Female faculty members at normal schools were not the only women to teach in unconventional subject areas; pioneering women at colleges and universities also did so. Yet, while female college and university instructors usually had to be overqualified in comparison to their male colleagues in order to be hired and receive minimal support, women normal-school faculty members tended to hold credentials of lower status than their male colleagues. It was most common in the 1870s and 1880s for female faculty members to have graduated from other normal schools. Beginning in the 1890s, increasing numbers were graduates of colleges, and by the 1920s, most faculty women had bachelor's degrees, some also had master's degrees, and just a couple had doctorates. The men on normal-school faculties were more likely than the women to have bachelor's degrees in the 1870s, 1880s, and 1890s, masters degrees in the early twentieth century, and doctorates in the 1910s and 1920s. School catalogs, bulletins and institutional histories; Geraldine Jonçich Clifford, ed., *Lone Voyagers: Academic Women in Coeducational Institutions, 1870–1937* (New York: The Feminist Press, 1989), 4.

15. Frederick Rudolph, *Curriculum: A History of the American Undergraduate Course of Study Since 1636* (San Francisco: Jossey-Bass Publishers, 1977), 134–135, 145. The curricula at many colleges, where less than 50 percent of students were high-school graduates, were actually not college-level in the late nineteenth century. Geraldine Jonçich Clifford, *"Equally in View": The University of California, Its Women, and the Schools* (Berkeley: Center for Studies in Higher Education and Institute for Governmental Studies, University of California, Berkeley, 1995), 3. On the college curriculum during this period, see Rudolph, *Curriculum*, chapters 4 and 5.

16. Assorted school catalogs and institutional histories; Purington, *History*, 195; Florence catalogs; Harper, *Development*, 125; *Announcement of the Southwest Texas State Normal School for the Annual Session 1903–1904—1910–1911*, SWTA; Oliphant, *History*, 128–130; Salisbury, *Historical Sketch*, 51; Putnam, *A History*, 60–61; Ruggles, *Historical Sketch*, 129; C. Francis Willey,

"Willimantic State Teachers College," *Teacher Education Quarterly* (Fall 1949): 18; John C. Almack, "History of Oregon Normal Schools," *The Quarterly of the Oregon Historical Society* 21 (June 1920): 108; Adelaide R. Pender, "At the New Britain Normal School, 1886–1888," Pender Collection, University Archives, Elihu Burritt Library, Central Connecticut State University, New Britain, Connecticut, 3.

17. *Annual Catalogue of the State Normal School at Oshkosh, Wis., for the School Year 1871–72*, 17; Pender, "At the New Britain Normal School," 3; Curtis quoted in William Marshall French and Florence Smith French, *College of the Empire State: A Centennial History of The New York State College for Teachers at Albany* ([Albany?], 1944), 19; Westfield document quoted in Robert T. Brown, *The Rise and Fall of the People's Colleges: The Westfield Normal School, 1839–1914* (Westfield, MA: Institute for Massachusetts Studies, Westfield State College, 1988), 135; *Annual Catalogue of the State Normal School at Oshkosh, Wis., for the School Year 1897–98*, 73–74; *Annual Catalogue of the State Normal College, Florence, Alabama, 1899–1900*, 33; *Announcement of the Southwest Texas State Normal School for the Annual Session 1904–1905*, 20.
18. Assorted school catalogs and institutional histories; Louis I. Kuslan, "Science in Selected Normal Schools of the 19th Century" (Ph.D. diss., Yale University, 1954); Phelps quoted in Egbert R. Isbell, *A History of Eastern Michigan University, 1849–1965* (Ypsilanti, MI: Eastern Michigan University Press, 1971), 253; Westfield document quoted in Brown, *The Rise and Fall*, 135–136.
19. State Superintendent of Education (Alabama), *Report, 1885* (Montgomery, AL, 1885), 38; *Annual Catalogue of the State Normal College, Florence, Alabama, 1889–1890*, 18; Westfield document quoted in Brown, *The Rise and Fall*, 135–136; *Annual Catalogue of the State Normal School at Oshkosh, Wis., for the School Year 1897–98*, 66, 70; Saunders quoted in Mohler, *The First Seventy-Five Years*, 75.
20. Arthur C. Boyden, *The History of Bridgewater Normal School* (Bridgewater, MA: Bridgewater Normal Alumni Association, 1933), 49–51, quotation on 50; Emporia catalog printed in *A History of the State Normal School of Kansas*, 51; *Catalogue and Circular of the Branch Normal College of the Arkansas Industrial University for the Year Ending June 7, 1895* (Little Rock, AR, 1896), 20.
21. Knapp quoted in *Historical Sketch of the State Normal School at San Jose, California, with a Catalogue of Its Graduates and a Record of Their Work for Twenty-Five Years* (Sacramento, CA: J.D. Young, 1889), 70; Brown, *The Rise and Fall*, 113.
22. Matthis quoted in *Historical Sketch of the State Normal School at San Jose*, 98; Isbell, *A History*, 254; Ernest Longfellow, *The Normal on the Hill: One Hundred Years of Peru State College* (Grand Island, NE: The Augustine Company, 1967), 19; Greathead, *The Story*, 114; San Jose catalogs, SJSUA; J. M. McKenzie, *History of the Peru State Normal* (Auburn, NE: The Nemaha County Republican, 1911), 62.
23. Boyden, *The History*, 51; Ruggles, *Historical Sketch*, 143; C. Nicholas Raphael and James R. McDonald, "Geography and Geology at Eastern Michigan University: The First Hundred Years of the 'Normal,' " *Michigan Academician* 27 (1995): 428; Work Projects Administration in the State of Massachusetts, *The State Teachers College at Westfield* (Boston: State Department of Education, 1941), 49; *Annual Catalogue of the State Normal School at Oshkosh, Wis., for the School Year 1897–98*, 70; *Announcement of the Southwest Texas State Normal*

School for the Annual Session 1904–1905, 23; *Vermont State Normal Schools, 1909–1910*, CSCA, 12; *The Normal Index* 2 (Oct. 1886), SJSUA, 33. On the growth of natural-history museums after the Civil War, see Steven Conn, *Museums and American Intellectual Life, 1876–1926* (Chicago: The University of Chicago Press, 1998), chapter 2.

24. McGraw, "A Century," section 4, no page numbers; Knapp quoted in *Historical Sketch of the State Normal School at San Jose*, 69; *The Normal Index* 1 (April 1886): 74; State Superintendent of Education (Alabama), *Report, 1885*, 38; *Annual Catalogue of the State Normal College, Florence, Alabama, 1889–1900*, 31; *Announcement of the Southwest Texas State Normal School for the Annual Session 1904–1905*, 26; McKenzie, *History*, 63; Rogers, *Oswego*, 111; Eli G. Lentz, *Seventy-Five Years in Retrospect: Southern Illinois University, 1874–1949* (Carbondale, IL: University Editorial Board, Southern Illinois University, 1955), 46; Harper, *Development*, 364; Carey W. Brush, *In Honor and Good Faith: A History of the State University College at Oneonta, New York* (Oneonta, NY: The Faculty-Student Association of State University Teachers College at Oneonta, Inc., 1965), 61; Frank A. Cooper, *The Plattsburgh Idea in Education, 1889–1964* (Plattsburgh, NY: Plattsburgh College Benevolent and Educational Association, Inc., 1964), 38.
25. Emporia catalog printed in *A History of the State Normal School of Kansas*, 50; Phelps quoted in Isbell, *A History*, 253; *Annual Catalogue of the State Normal School, Florence, Alabama, 1884–85*, 17; *Catalogue and Circular of the California State Normal School, San Jose, 1888*, SJSUA, 25; *The Pedagogue* (1906), SWTA, 22.
26. Albee quoted in *The First Half Century of the Oshkosh Normal School*, 11; Emporia catalog printed in *A History of the State Normal School of Kansas*, 51; Normal catalog (1889) quoted in Harper, *Development*, 374; *Annual Catalogue of the State Normal College, Florence, Alabama, 1899–1900*, 30; *Announcement of the Southwest Texas State Normal School for the Annual Session 1904–1905*, 27; Matthis quoted in *Historical Sketch of the State Normal School at San Jose*, 97; *The Normal Advance* 1 (Jan.–Feb. 1895), UWOA, 13.
27. *The Normal Advance* 4 (Nov. 1897), 41; Kelley, Bean, and Allen quoted in Rogers, *Oswego*, 111, 50; Clark quoted in Albert Salisbury, *Historical Sketches of the First Quarter-Century of the State Normal School at Whitewater, Wisconsin* (Madison, WI: Tracy, Gibbs & Co., 1893), 53. The extensive offerings in science at state normal schools confirm historian Kim Tolley's observation that, contrary to present-day assumptions, science and mathematics were acceptable and popular subjects for female students in higher schools during the nineteenth century. See Kim Tolley, *The Science Education of American Girls: A Historical Perspective* (New York: RoutledgeFalmer, 2003).
28. Almack, "History," 109; Purington, *History*, 195; Putnam, *A History*, 60–61; Boyden, *The History*, 54; Westfield document quoted in Brown, *The Rise and Fall*, 134.
29. *Catalogue of the California State Normal School, San Jose, 1900*, SJSUA, 27; King quoted in Isbell, *A History*, 248; *Annual Catalogue of the State Normal College, Florence, Alabama, 1890–91*, 22; *Annual Catalogue of the State Normal School at Oshkosh, Wis., for the School Year 1897–98*, 60.
30. William Frederick Hartman, "The History of Colorado State College of Education: The Normal School Period, 1890–1911" (Ph.D. diss., Colorado State College of Education, 1951), 126; *Annual Catalogue of the State Normal School at*

Oshkosh, Wis., for the School Year 1872–73, 30, *1874–75*, 34, *1879–80*, 43, *1885–86*, 40, *1895–96*, 90, *1902–1903*, 130; Pine Bluff document quoted in Fredrick Chambers, "Historical Study of Arkansas Agricultural, Mechanical and Normal College, 1873–1943" (Ed.D. diss., Ball State University, 1970), 101; *Circular of the State Normal and Training School at Geneseo, NY* (1888), CASUNYG, 7–8; Emporia catalog printed in *A History of the State Normal School of Kansas*, 39–41, quotation on 39; Michael Francis Bannon, "A History of State Teachers College, Troy, Alabama" (Ed.D. diss., George Peabody College for Teachers, 1954), 46; Ruggles, *Historical Sketch*, 140; Hart quoted in Rogers, *Oswego*, 96.

31. Brush, *In Honor*, 22–23, Schumacher quoted on 19; Emporia catalog printed in *A History of the State Normal School of Kansas*, 39; James B. Bonder, "The Growth and Development of the State Teachers Colleges of Pennsylvania" (Ed.D. diss., Temple University, 1952), 334; Hartman, "The History," 116; Westfield document quoted in Brown, *The Rise and Fall*, 134; Boyden, *A History*, 41–42; Willimantic catalog quoted in Arthur Charles Forst, Jr., "From Normal School to State College: The Growth and Development of Eastern Connecticut State College From 1889 to 1959," (Ph.D. diss., University of Connecticut, 1980), 70–71; Pender, "At the New Britain Normal School," 7.
32. *Catalogue of the California State Normal School, San Jose, 1900*, 19; *Announcement of the Southwest Texas State Normal School for the Annual Session 1904–1905*, 17–18; Sandra D. Harmon, " 'The Voice, Pen and Influence of Our Women Are Abroad in the Land': Women and the Illinois State Normal University, 1857–1899," in *Nineteenth-Century Women Learn to Write*, ed. Catherine Hobbs (Charlottesville: University of Virginia Press, 1995), 90; Pender, "At the New Britain Normal School," 20; *Catalogue of the State Normal School at Castleton, Vermont, 1872–73*, 14; *Annual Catalogue of the State Normal College, Florence, Alabama, 1890–91*, 22.
33. Emporia catalog printed in *A History of the State Normal School of Kansas*, 30; *Catalogue and Circular of the California State Normal School, San Jose, 1890*, 31; *Vermont State Normal Schools, 1909–1910*, 11; *Annual Catalogue of the State Normal College, Florence, Alabama, 1889–1890*, 19; Pender, "At the New Britain Normal School," 8.
34. Harmon, " 'The Voice,' " 94; Bessie L. Park, *Cortland—Our Alma Mater: A History of Cortland Normal School and the State University of New York Teachers College at Cortland, 1869–1959* (Cortland, NY, 1960), 102; Fisher, *". . . the stone strength,"* 86–87; *The Lamron: 50th Anniversary, Geneseo Normal School* (Geneseo, NY, 1921), CASUNYG, 17; *Annual Catalogue of the State Normal College, Florence, Alabama, 1890–1891*, 22; Dotson, "The Historical Development," 408; *The First Half Century of the Oshkosh Normal School*, 23–24; *Annual Catalog of Branch Normal College, 1919–1920* (Little Rock, AR, 1920); Chambers, "Historical Study," 79–80; Commencement Programs, 1870s–1930s, UCUNA; graduation programs in "Keepers of the Spirit: The L. A. Davis, Sr. Historical Collection," Exhibit, Isaac S. Hathaway-John M. Howard Fine Arts Center, University of Arkansas at Pine Bluff, Pine Bluff, AR, May 1995; Commencement Programs, Castleton State Normal School, 1874–1940, CSCA; Class Day Programs, UWOA; Hendrix quoted in *Historical Sketch of the State Normal School at San Jose*, 49; *The Normal Advance* 1 (Jan.–Feb. 1895)–19 (Sept. 1912).

35. Virtually all school catalogs and bulletins list drawing and vocal music as required subjects. Hartman, "The History," 116; Gay, "Massachusetts Normal Schools," 515; Rogers, *Oswego*, 29, 97–98, McGuire quoted on 98; Ruggles, *Historical Sketch*, 174; Carbone, "The History," 349; Margaret Wilson and N. E. Gaymon, eds., *A Century of Wisdom: Selected Speeches of Presidents of Florida A & M University* (Winter Park, FL: Four-G Publishers, Inc., 1990), 21–22.
36. Bonder, "The Growth and Development," 334, 340; Dotson, "The Historical Development," 208, 300, 316–317; Paul Stoler, "Castleton Normal School in the Nineteenth Century," in *And the Glory of the Latter House Shall Be Greater Than That of the Former: An Informal History of Castleton State College*, ed. Holman D. Jordan (Castleton, VT: Castleton State College, 1968), 52; Chambers, "Historical Study," 66–68, 143; Murphy, "Origin and Development," 129, quotation on 145; Deward Homan Reed, *The History of Teachers Colleges in New Mexico* (Nashville, TN: George Peabody College for Teachers, 1948), 106–107.
37. "Decennial Address of President Edwards" (1872) quoted in Harper, *Development*, 118.
38. Quoted in Lawrence W. Levine, "William Shakespeare and the American People: A Study in Cultural Transformation," in *The Unpredictable Past: Explorations in American Cultural History* (New York: Oxford University Press, 1993), 169.
39. Burton J. Bledstein, *The Culture of Professionalism: The Middle Class and the Development of Higher Education in America* (New York: W. W. Norton & Company, Inc., 1976), 30, 55; Levine, "William Shakespeare and the American People"; Richard Ohmann, *Selling Culture: Magazines, Markets, and Class at the Turn of the Century* (New York: Verso, 1996), 157–159; Paul DiMaggio, "Cultural Entrepreneurship in Nineteenth-Century Boston: The Creation of an Organizational Base for High Culture in America," *Media, Culture and Society* 4 (1982): 33–50; Paul DiMaggio, "Cultural Entrepreneurship in Nineteenth-Century Boston, Part II: The Classification and Framing of American Art," *Media, Culture and Society* 4 (1982): 303–322; Alan Trachtenberg, *The Incorporation of America: Culture & Society in the Gilded Age* (New York: Hill and Wang, 1982), chapter 5; Oscar Handlin and Mary F. Handlin, *The American College and American Culture: Socialization as a Function of Higher Education* (New York: McGraw-Hill Book Company, 1970), 49–50. See also Robert H. Wiebe, *The Search for Order, 1877–1920* (New York: Hill and Wang, 1967). Jencks and Riesman distinguish between social stratification and cultural stratification, but acknowledge that the two are highly interrelated. Christopher Jencks and David Riesman, *The Academic Revolution* (Garden City, NY: Doubleday & Company, Inc., 1968), 64–90. According to Bourdieu, cultural capital has three forms: the "embodied state" is the culture absorbed into one's being or personality; "objectified" cultural capital is physical items such as books or paintings; and the "institutionalized" form is academic qualifications. Pierre Bourdieu, "The Forms of Capital," in *Handbook of Theory and Research for the Sociology of Education*, ed. John G. Richardson (New York: Greenwood Press, 1986), 243–249. See also David Swartz, *Culture & Power: The Sociology of Pierre Bourdieu* (Chicago: The University of Chicago Press, 1997); Pierre Bourdieu and Jean-Claude Passeron, *Reproduction in Education, Society and Culture*, translated by Richard Nice (Beverly Hills: SAGE Publications, 1977); Randall Collins, *The Credential Society: An Historical Sociology of Education and Stratification* (San Diego, CA: Academic Press, Inc., 1979), 57–62;

Alejandro Portes, "Social Capital: Its Origins and Applications in Modern Sociology," *Annual Review of Sociology* 24 (1998): 1–24.

40. Kirksville, Missouri offered the degree of Bachelor of the Arts and Philosophic Didactics, and had granted 46 such degrees as early as 1876. In later years, Kirksville offered other bachelors', and even a master's degree. The few other normal schools that officially offered some sort of bachelor's degree in the nineteenth century rarely—if ever—actually conferred them, and there is little evidence that these degrees certified college-level work. Cedar Falls, Iowa offered a Bachelor of Didactics degree for completion of the scientific course. At Troy, Alabama, President Edridge granted a bachelor of philosophy degree to his own daughter in 1889, but did have permission from the legislature to grant bachelors' degrees until four years later. Ypsilanti, Michigan awarded its first Bachelor of Pedagogics degree in 1890. Silver City, New Mexico began to offer Bachelor of Pedagogy and Master of Pedagogy degrees in 1896 and 1903, respectively, but did not actually confer them until the 1910s. After the turn of the century, more normals would offer bachelor's degrees. Walter H. Ryle, *Centennial History of the Northeast Missouri State Teachers College* (Kirksville, MO: Northeast Missouri State Teachers College, 1972), 113, 165; Clarence Theodore Molen, "The Evolution of a State Normal School Into a Teachers College: The University of Northern Iowa, 1876–1916" (Ph.D. diss., University of Iowa, 1974), 80; Bannon, "A History," 31–32; Isbell, *A History*, 136; Reed, *A History*, 41.
41. Providence catalog quoted in Carbone, "The History," 279; Harper, *Development*, 125, 339–340; Wisconsin report quoted in Salisbury, *Historical Sketch*, 51; Purington, *History*, 194; Putnam, *A History*, 61; Dedman, *Cherishing*, 97, 112.
42. Percentage of Millersville graduates calculated from statistics listed in Bugaighis, "Liberating Potential," 100; Harper, *Development*, 136; New York report printed in Thomas E. Finegan, *Teacher Training Agencies: A Historical Review of the Various Agencies of the State of New York Employed in Training and Preparing Teachers for the Public Schools of the State* (Albany: The University of the State of New York, 1917), 129; Brush, *In Honor*, 89; Molen, "The Evolution," 79–80; *A History of the State Normal School of Kansas*, 31.
43. Bonder, "The Growth and Development," 335; Harper, *Development*, 125; Chambers, "Historical Study," 78–79, 144–145; *Catalogue and Circular of the Branch Normal College of the Arkansas Industrial University for the Year Ending June 7, 1895*, 16, *June 8th, 1900*, 17; Ruggles, *Historical Sketch*, 129; Oliphant, *History*, 128–129; *State Normal School, Geneseo, New York, 1905–1906*, CASUNYG, 6–7; Reed, *History*, 112.
44. Westfield document quoted in Brown, *The Rise and Fall*, 137–140; Greathead, *The Story*, 16–17.
45. Ruggles, *Historical Sketch*, 129; Oliphant, *History*, 131; Fisher, *". . . the stone strength,"* 67–68; *Announcement of the Southwest Texas State Normal School for the Annual Session 1904–1905*, 28; Ernest J. Hopkins and Alfred Thomas, Jr., *The Arizona State University Story* (Phoenix, AZ: Southwest Publishing Co., Inc., 1960), 89, 105; Stallones, "Struggle," 105; *Annual Catalogue of the State Normal College, Florence, Alabama, 1891–92*, 27, *1895–96*, 34; Westfield document quoted in Brown, *The Rise and Fall*, 138–139; Putnam, *A History*, 61; Cooper, *The Plattsburgh Idea*, 27; Boughter, *Fairmont State Normal*, 73–74; Almack, "History," 109–110; Brush, *In Honor*, 273; Dedman, *Cherishing*, 112;

Brown explains that at normal schools in Massachusetts in 1897, "all foreign language instruction ceased, not to be resurrected for sixty years. That same year saw the end of the four-year advanced course of studies in all the schools except Bridgewater, by order of the Board." Brown, *The Rise and Fall*, 101.

46. Ticket for "Twelve Nights Entertainment," in Activities,1861–1940, Files, CSCA; Geneseo catalog quoted in Fisher, *". . . the stone strength,"* 64–65; Slippery Rock catalog quoted in Eugene Coleman Ney, "Slippery Rock State Normal School, 1890–1916: The Maltby Era" (Ph.D. diss., University of Pittsburgh, 1998), 107; *The Normal Advance* 2 (Sept.–Oct. 1895): 10; Louie G. Ramsdell, "First Hundred Years of the First State Normal School in America: The State Teachers College at Framingham, Massachusetts—1839–1939," in *First State Normal School in America: The State Teachers College at Framingham, Massachusetts* (Framingham, MA: The Alumnae Association of the State Teachers College at Framingham, Massachusetts, 1959), 9; Student Walter S. Campbell quoted in Jerry G. Nye, *Southwestern Oklahoma State University: The First 100 Years* (Weatherford, OK: Southwestern Oklahoma State University, 2001), 28; Bowles, *A Good Beginning*, 18, 118; Benjamin Franklin Gilbert, *Pioneers for One Hundred Years: San Jose State College, 1857–1957* (San Jose, CA: San Jose State College, 1957), 108; M. Janette Bohi, *A History of Wisconsin State University Whitewater, 1868–1968* (Whitewater, WI: Whitewater State University Foundation, 1968), 99.
47. Bohi, *A History*, 48; *A History of the State Normal School of Kansas*, 64; Dedman, *Cherishing*, 129; Boyden, *The History*, 51; Class Day Program (Oshkosh), 1889, 74; *Students' Hand Book, 1906–1907* (Oshkosh, WI: Students' Christian Association, 1906), UWOA, 22; "Normal Index Department," *The Pacific Coast Teacher: A Monthly Magazine Devoted to the Educational Interests of the Pacific Coast*, SJSUA, 1 (Oct. 1891): 61, 1 (Dec. 1891): 105.
48. Isbell, *A History*, 334; Brush, *In Honor*, 50; *The Pedagogue* (1908), 112; *The First Half Century of the Oshkosh Normal School*, 25 (Oshkosh's German club was probably Der Deutsche Literarische Kreis, or German literary circle); I. N. Mitchell, ed., *Quarter Century of the Milwaukee State Normal School, 1886–1911* (Milwaukee, 1911), 19; Bohi, *A History*, 92; McKenzie, *History*, 81.
49. Isbell, *A History*, 334; *The Normalian* 1 (June 20, 1900), CASUNYG, 4; *The Normal Pennant* 6 (June 1903), SJSUA, 79; *The Pedagogue* (1907), 112; Fisher, *". . . the stone strength,"* 85; Bohi, *A History*, 99; Helen E. Marshall, *Grandest of Enterprises: Illinois State Normal University, 1857–1957* (Normal, IL: Illinois State Normal University, 1956), 201; *The Normal Advance* 6 (Sept. 1899): 10–11; *Oshkosh State Normal School Bulletin* 3 (June 1907): 101. See chapter 5 for further discussion of student music groups and performances.
50. Frederick Rudolph, *The American College and University: A History* (Athens: University of Georgia Press, [1962] 1990), 137–150; Rudolph, *Curriculum*, 95–98; Helen Lefkowitz Horowitz, *Campus Life: Undergraduate Cultures from the End of the Eighteenth Century to the Present* (Chicago: The University of Chicago Press, 1987), chapter 2; Thomas S. Harding, *College Literary Societies: Their Contribution to Higher Education in the United States, 1815–1876* (New York: Pageant Press, 1976).
51. Harper, *Development*, 112.
52. Arthur Clarke Boyden, *Albert Gardner Boyden and the Bridgewater State Normal School: A Memorial Volume* (Bridgewater, MA: Arthur H. Willis, 1919),

138–139; Isbell, *A History*, 332–333; Marshall, *Grandest*, 158; Fiegel, "A History," 26–27; George Frank Sammis, "A History of the Maine Normal Schools" (Ph.D. diss., University of Connecticut, 1970), 203; *Catalogue of the California State Normal School, San Jose, 1900*, 8; J. S. Nasmith, "An Open Letter From J. Nasmith," *Platteville Witness* LXIII (April 13, 1932), 2; Pamela Dean, "Covert Curriculum: Class, Gender, and Student Culture at a New South Woman's College, 1892–1910" (Ph.D. diss., University of North Carolina at Chapel Hill, 1994), 179; *The Normal Index* 3 (Nov. 1887): 32; Hartman, "The History," 159. Very rarely did normal schools, such as the one in Castleton, Vermont, not establish literary societies. Also highly unusual was the situation at Winona, Minnesota, in which students had little interest in societies. One commentator reported, "For several years previous to 1903, the school had been without any literary societies. . . . for many years a certain amount of rhetorical work by each student had been a requirement for graduation . . . This gave rhetorical training, but it lacked the values obtainable from spontaneous participation, one of the greatest sources of benefit from this kind of work." Ruggles, *Historical Sketch*, 163.

53. Vaughn, "The History," 31–32; Ronald C. Brown, *Beacon on the Hill: Southwest Texas State University, 1903–1978* (San Marcos, TX: Southwest Texas State University, 1978), 21–25; *The Pedagogue* (1904–1910), SWTA; Mau, *Brief History*, 17; Dedman, *Cherishing*, 124; Brush, *In Honor*, 49; Bannon, "A History," 43–44; Chambers, "Historical Study," 89–90; Reuter, *West Liberty*, 129; Boyden, *Albert Gardner Boyden*, 116–117; Merryman, "Indiana University of Pennsylvania," 149.
54. Putnam, *A History*, 215–261, resolution quoted on 222–223; Marshall, *Grandest*, 158, 162–163; *A History of the State Normal School of Kansas*, 63–68; Greathead, *The Story*, 66–67, 73–77; *Normal Crescent* (Nov. 7, 1885), UWOA, 144; Edward Noyes, "Oshkosh–From Normal School to State University (1871–1968)," in *History of the Wisconsin State Universities*, ed. Walker D. Wyman (River Falls, WI: River Falls State University Press, 1968), 104, 115; *Oshkosh State Teachers College*, 65–66, 68; David Sands Wright, *Fifty Years at the Teachers College: Historical and Personal Reminiscences* (Cedar Falls, IA: Iowa State Teachers College, 1926), 116; McKenzie, *History*, 29, 45, 81.
55. Wright, *Fifty Years*, 116; *A History of the State Normal School of Kansas*, 30; Hopkins and Thomas, *The Arizona State University*, 111–112; Oregon catalog quoted in Almack, "History," 143; Ryle, *Centennial History*, 441, 567–568; Fiegel, "A History," 27; Reed, *The History*, 136; Mohler, *The First Seventy-Five Years*, 97; *Biographical Directory and Condensed History of the State Normal School, Mankato, Minn., 1870–1890* (Mankato, MN: Mankato State Normal School Alumni Association, 1891), 12.
56. Lee Graver, *A History of the First Pennsylvania State Normal School, Now the State Teachers College at Millersville* (Millersville, PA: State Teachers College, 1955), 229; Bannon, "A History," 44; Marshall, *Grandest*, 158.
57. Brush, *In Honor*, 48; John W. Cook and James V. McHugh, *A History of the Illinois State Normal University* (Normal, IL: Illinois State Normal University, 1882), 149; Ryle, *Centennial History*, 444; Wright, *Fifty Years*, 202; *The Normal Advance* 5 (March 1899): 120–123.
58. *New York State Teachers College At Buffalo: A History, 1871–1946* (Buffalo: New York State Teachers College at Buffalo, 1946), 143, 129; Bannon, "A History," 41; Irving H. Hart, *The First 75 Years* (Cedar Falls, IA: Iowa State

Teachers College, 1951), 148; Park, *Cortland*, 86–87; *The Crucible* quoted in Hartman, "The History," 169; *The Normal Advance*, UWOA.

59. *The Normal Advance* 7 (March 1901): 122; Dedman, *Cherishing*, 125–126; *The Normalian* 1 (March 5, 1900), CASUNYG, 8; Ladies' Literary Society, Minutes, May 2, 1879, UWOA, 58; *The Normal Advance* 1 (Nov.–Dec. 1894): 14, 4 (Feb. 1898): 100–101, 5 (Feb. 1899): 91–96, 5 (March 1899): 120–123; Agonian, Constitution and Alpha Chapter Secretaries' Reports, Nov. 5, 1897, CASUNYG, 276; *The Normalian* 1 (March 5, 1900): 8; Fisher, *". . . the stone strength,"* 86; Agonian, Constitution and Alpha Chapter Secretaries' Reports, Nov. 15, 1895, 192.
60. Vaughn, "The History," 31; *Oshkosh State Teachers College*, 68; Wright, *Fifty Years*, 74; *The Normal Pennant* 4 (June 1901): 18; *The Pedagogue* (1905), 61; Mohler, *The First Seventy-Five Years*, 96; Beach quoted in Salisbury, *Historical Sketches of the First Quarter-Century*, 91; Dedman, *Cherishing*, 125–126.
61. Dedman, *Cherishing*, 125; *The Pedagogue* (1909), 98–99; Mohler, *The First Seventy-Five Years*, 96; Brush, *In Honor*, 52; Agonian, Constitution and Alpha Chapter Secretaries' Reports, April 30, 1892, 30; Dixie Club, Roll call and minutes of meetings, Oct. 17, 1902, in Organizations, Files, UCUNA; *The Pedagogue* (1907), 90, 87; *The Normal Advance* 13 (Sept. 1906): 15.
62. Brush, *In Honor*, 288; *The Normal Pennant* 3 (Feb. 1900): 11; *The Normal Index* 1 (Nov. 1885): 9, 3 (Nov. 1887): 28–30; *The Normalian* 1 (March 5, 1900): 8, 1 (June 20, 1900): 17–18; Arethusa Sorority, Program, "Anniversary exercises held at Normal Hall, Geneseo, by Arethusa Fraternity with Gamma Sigma Fraternity," 1900.
63. *The Normal Advance* 2 (Jan.–Feb. 1896): 60; Society records excerpted in Work Projects Administration, *The State Teachers College at Westfield*, 72; *The Normal Advance* 6 (Oct. 1899): 33; Dixie Club, Roll and minutes of meetings, Oct. 25, 1912; Ladies' Literary Society, Minutes, Oct. 8, 1880 and Feb. 4, 1881, UWOA, 127, 136; *The Normal Star* 2 (Feb. 23, 1912), SWTA, 1; *The Normal Advance* 18 (Jan. 1912): 128.
64. Park, *Cortland*, 41; Society records excerpted in Work Projects Administration, *The State Teachers College at Westfield*, 72; Agonian, Constitution and Alpha Chapter Secretaries' Reports, April 12, 1895 and Oct. 31, 1899, 173, 367; *The Pedagogue* (1905), 71; *The Normal Advance* 6 (Jan. 1900): 92; *The Pedagogue* (1907), 51; Rogers, *Oswego*, 88; Bugaighis, "Liberating Potential," 208–211.
65. *The Normal Student* 2 (Nov. 21, 1902), CSCA, 5; *The Normal Advance* 4 (Feb. 1898): 100–101, 5 (Dec. 1898): 67, 9 (Dec. 1903): 167; Cook and McHugh, *A History*, 153–154; *The Normal Advance* 3 (Sept.–Oct. 1896): 17; Agonian, Constitution and Alpha Chapter Secretaries' Reports, Dec. 7, 1895, 196; *The Normal Pennant* 1 (March 1898): 2; *The Normal Advance* 2 (Sept.–Oct. 1895): 16; Agonian, Constitution and Alpha Chapter Secretaries' Reports, Oct. 5, 1894, 145; Dialectical-Lafayette debate program, April 1905, in Programs, 1857–1938, file, UCUNA.
66. *The Normal Advance* 2 (May–June 1896): 105–106; *Senior Year Book: San Jose Normal* (1910), SJSUA, 36; Agonian, Constitution and Alpha Chapter Secretaries' Reports, April 12, 1901, 423–424; *The Normal Pennant* 4 (June 1901): 19–20; Brush, *In Honor*, 52; *The Pedagogue* (1909), 74; Fiegel, "A History," 50; Philalethean Society, Constitution and 2 volumes of Meeting Minutes, CASUNYG, Feb. 3, 1894, vol. 1, 186; *The Normal Advance* 2 (Sept.–Oct. 1895): 16, special edition (Nov. 1, 1912).

67. Brush, *In Honor*, 49; Dedman, *Cherishing*, 127–128, 124–125; Edwin H. Cates, *A Centennial History of St. Cloud State College* (Minneapolis, MN: Dillon Press, 1968), 78; Cook and McHugh, *A History*, 115; *A History of the State Normal School of Kansas*, 67.
68. Timbie quoted in Boyden, *Albert Gardner Boyden*, 146.

4 TEACHER EDUCATION: BREATHING "THE OZONE OF TEACHING"

1. Burton J. Bledstein, *The Culture of Professionalism: The Middle Class and the Development of Higher Education in America* (New York: W. W. Norton & Company, Inc., 1976), chapter 3; Merle L. Borrowman, *The Liberal and Technical in Teacher Education: A Historical Survey of American Thought* (New York: Teachers College Bureau of Publications, 1956), quotations on 229. In the same vein as Borrowman, Paul Mattingly argues that in the late nineteenth century normal schools were "unprofessedly mechanical," concentrating on "vocational training for specific skills." Paul H. Mattingly, *The Classless Profession: American Schoolmen in the Nineteenth Century* (New York: New York University Press, 1975), chapter 7, quotations on 166. Jurgen Herbst, *And Sadly Teach: Teacher Education and Professionalization in American Culture* (Madison, WI: The University of Wisconsin Press, 1989), 8.
2. Penina Migdal Glazer and Miriam Slater, *Unequal Colleagues: The Entrance of Women into the Professions, 1890–1940* (New Brunswick, NJ: Rutgers University Press, 1987), chapter 1, quotation on 3; Barbara J. Harris, *Beyond Her Sphere: Women and the Professions in American History* (Westport, CT: Greenwood Press, 1978), chapter 4. Nancy Hoffman explains that "the common image of teaching as 'women's work' resulted in its demotion to a second-rate profession. Because women themselves were viewed as subordinate to men, women's profession could not equal law, medicine, theology, or even the management of education." Nancy Hoffman, *Women's "True" Profession: Voices from the History of Teaching* (New York: The Feminist Press, 1981), 15. While Geraldine Jonçich Clifford points out that work with young children, low wages, teachers' youth, and low public regard for schooling have all contributed to the problematic image and status of teachers, she cites gendered views of women teachers as a key factor: the "perceptions that teachers were 'mothering,' or that women teachers were only marking time until marriage, had unfortunate effects for the image of professionalism that teaching was trying to cultivate." Geraldine Jonçich Clifford, "Man/Woman/Teacher: Gender, Family, and Career in American Educational History," in *American Teachers: Histories of a Profession at Work*, ed. Donald Warren (New York: MacMillan Publishing Company, 1989), 315–319. See also Michael W. Apple, "Teaching and 'Women's Work': A Comparative Historical and Ideological Analysis," *Teachers College Record* 86 (Spring 1985): 455–473; Linda Eisenmann, "Teacher Professionalism: A New Analytical Tool for the History of Teachers," *Harvard Educational Review* 61 (May 1991): 215–224.
3. Lough quoted in *The Normal Advance* 7 (January 1901), UWOA, 77.
4. Mark Ginsburg points out that class and gender ideology granted more status to "liberal" than "technical" approaches to teacher education. He continues, "The ideological struggle that transpired should be conceived of not only in terms of 'liberal' versus 'technical,' but also in relation to 'mental' versus 'manual'—terms

which have a clearer association with social class relations. The terms of the debate also connect gender relations, although not explicitly in terms of 'mental' versus 'emotional' labor." And it was their "association with less elite members of the population—working class, farmers, and women—which helped to locate the normal schools at the lower end of the liberal-technical hierarchical relations." Mark B. Ginsburg, "Teacher Education and Class and Gender Relations: A Critical Analysis of Historical Studies of Teacher Education," *Educational Foundations* 2 (Spring 1987): 6, 12–13. Andrew Gitlin asserts that normal schools' efforts to professionalize teaching focused on "experience and a more applied approach to education," as the institutions "viewed reflection on experience as a way of knowing" in addition to insisting "that teachers be instilled with a type of commitment, a sense of being called. . . . This professionalization project, with its emphasis on experience and commitment" Gitlin further argues, was unsuccessful in the face of prevailing gender ideology and conceptions of professionalism: "dominant groups in society, such as male industrialists who viewed science as a necessary part of increasing control, production, and efficiency, had little confidence in the experiential emphasis embodied in the normal school or the quality of candidates who entered this institution. Normal schools' functional approach to professionalization, as a consequence, did not garner their trust and thus did not act to uplift the authority, status, and work conditions of teachers." Andrew Gitlin, "Gender and Professionalization: An Institutional Analysis of Teacher Education and Unionism at the Turn of the Twentieth Century," *Teachers College Record* 97 (Summer 1996): 595–596, 598–599.

5. Albert Boyden quoted in Arthur C. Boyden, *The History of Bridgewater Normal School* (Bridgewater, MA: Bridgewater Normal Alumni Association, 1933), 78.
6. George C. Purington, *History of the State Normal School, Farmington, Maine* (Farmington, ME: Knowlton, McLeary & Co., 1889), 42; Elisabeth Ann Bowles, *A Good Beginning: The First Four Decades of the University of North Carolina at Greensboro* (Chapel Hill: The University of North Carolina Press, 1967), 19; *A History of the State Normal School of Kansas for the First Twenty-Five Years* (Emporia, KS, 1889), 54; Egbert R. Isbell, *A History of Eastern Michigan University, 1849–1965* (Ypsilanti, MI: Eastern Michigan University Press, 1971), 56–57.
7. *A History of the State Normal School of Kansas*, 54, 44, 56, 48; Irving H. Hart, *The First 75 Years* (Cedar Falls, IA: Iowa State Teachers College, 1951), 59–60, Seerley quoted on 59. Historian Andrew Gitlin states, "The importance of experience, in terms of knowledge production, was also evident in the normal schools' emphasis on hiring experienced teachers as opposed to more theoretically-minded, research-trained academics." Gitlin, "Gender and Professionalization," 596.
8. Dorothy Rogers, *Oswego: Fountainhead of Teacher Education: A Century in the Sheldon Tradition* (New York: Appleton-Century-Crofts, Inc., 1961), 154; Hart, *The First 75 Years*, 60; Erwin S. Selle, ed., *The Winona State Teachers College: Historical Notes, 1910–1935* (Winona, MN, 1935), 57.
9. Rogers, *Oswego*, 48, 51; *A History of the State Normal School of Kansas*, 48, 54; Ben Burks, "A Stepping Stone in Academic Careers: Virginia's Female Normal Schools and Their Instructors" (Paper presented at History of Education Society annual meeting, October 2000, San Antonio), 2, 10–11; Isbell, *A History*, 238, 396, 248, 56–57.
10. Isbell, *A History*, 56–57, 96, 115; David Sands Wright, *Fifty Years at the Teachers College: Historical and Personal Reminiscences* (Cedar Falls, IA: Iowa State

Teachers College, 1926), 105–107, 73–74, 12; Albert Reynolds Taylor, *Autobiography of Albert Reynolds Taylor* (Decatur, IL: Review Printing and Stationery Co., 1929), 76–78; *Semi-Centennial History of the Illinois State Normal University, 1857–1907* (Normal, IL: Illinois State Normal University, 1907), 98, 113, 115; M. Janette Bohi, *A History of Wisconsin State University Whitewater, 1868–1968* (Whitewater, WI: Whitewater State University Foundation, 1968), 72; Michael Francis Bannon, "A History of State Teachers College, Troy, Alabama" (Ed.D. diss., George Peabody College for Teachers, 1954), 41; Maxine Ollie Merlino, "A History of the California State Normal Schools—Their Origin, Growth, and Transformation into Teachers Colleges" (Ed.D. diss., University of Southern California, 1962), 180.

11. Walter H. Ryle, *Centennial History of the Northeast Missouri State Teachers College* (Kirksville, MO: Northeast Missouri State Teachers College, 1972), 106; Isbell, *A History*, 116, 238, 258, 396, Bellows quoted on 116; James O. Knauss, *The First Fifty Years: A History of Western Michigan College of Education* (Kalamazoo, MI: Western Michigan College of Education, 1953), 15; Wright, *Fifty Years*, 12, 129; Burks, "A Stepping Stone," 16.
12. *A History of the State Normal School of Kansas*, 55, 47, 56, 48.
13. *Annual Catalogue of the State Normal School, Florence, Alabama, 1876–77*, UCUNA, 11, *1883 and 1884*, 18; Hector Richard Carbone, "The History of the Rhode Island Institute of Instruction and the Rhode Island Normal School as Agencies and Institutions of Teacher Education, 1845–1920" (Ph.D. diss., University of Connecticut, 1971), 278; John C. Almack, "History of Oregon Normal Schools," *The Quarterly of the Oregon Historical Society* 21 (June 1920): 109; *Vermont State Normal Schools, 1898–99*, CSCA, no page numbers; *Catalogue of the California State Normal School, San Jose, 1898*, SJSUA, 11, *1904*, 14; *Annual Catalogue of the State Normal School at Oshkosh, Wis., for the School Year 1872–73*, UWOA, 23, *1879–80*, 29.
14. J. Orin Oliphant, *History of the State Normal School at Cheney, Washington* (Spokane: Inland-American Printing Company, 1924), 128–129; *Annual Catalogue of the State Normal College, Florence, Alabama 1889–90*, UCUNA, 21; *State Normal School, Geneseo, New York, 1901–1902*, CASUNYG, 9–10; Hart, *The First 75 Years*, 81, 87; Jean Talbot, *First State Normal School 1860, Winona State College, 1960* (Winona, MN: *Quarterly Bulletin of Winona State College* 55 [August 1959]—56 [August 1960]), 12; Samuel R. Mohler, *The First Seventy-Five Years: A History of Central Washington State College, 1891–1966* (Ellensburg, WA: Central Washington State College, 1967), 90; *State Normal School, Geneseo, New York, 1905–1906*, 7; Avis Leo Sebaly, "Michigan State Normal Schools and Teachers Colleges in Transition, With Special Reference to Western Michigan College of Education" (Ph.D. diss., University of Michigan, 1950), 199, 208.
15. Ypsilanti catalog excerpted in Daniel Putnam, *A History of the Michigan State Normal School at Ypsilanti, Michigan, 1849–1899* (Ypsilanti, MI: Michigan State Normal College, 1899), 78; *Annual Catalogue of the State Normal College, Florence, Alabama 1899–1900*, 29; *Catalogue of the California State Normal School, San Jose, 1900*, 31–32, *1904*, 15; *Announcement of the Southwest Texas State Normal School for the Annual Session 1904–1905* (Austin, TX: State Printers, 1904), SWTA, 24.
16. *Annual Catalogue of the State Normal School at Oshkosh, Wis., for the School Year 1872–73*, 22, *1879–80*, 28; *Catalogue and Circular of the California State*

Normal School, San Jose, 1878–79, SJSUA, 45, *1884–85*, 45; Purington, *History*, 195; Emporia catalog printed in *A History of the State Normal School of Kansas*, 47.

17. *Annual Catalogue of the State Normal School at Oshkosh, Wis., for the School Year 1897–98*, 52; Elizabeth Tyler Bugaighis, "Liberating Potential: Women and the Pennsylvania State Normal Schools, 1890–1930" (Ph.D. diss., The Pennsylvania State University, 2000), 301; Oliphant, *History*, 128; *Catalogue of the California State Normal School, San Jose, 1900*, 30; *Annual Catalogue of the State Normal College, Florence, Alabama 1899–1900*, 28.
18. *Annual Catalogue of the State Normal College, Florence, Alabama 1890–91*, 17; Carroll quoted in Herbert E. Fowler, *A Century of Teacher Education in Connecticut: The Story of the New Britain State Normal School and the Teachers College of Connecticut, 1849–1949* (New Britain, CT: The Teachers College of Connecticut at New Britain, 1949), 68; Ypsilanti catalog excerpted in Putnam, *A History*, 78; *Catalogue of the California State Normal School, San Jose, 1900*, 31; *Announcement of the Southwest Texas State Normal School for the Annual Session 1904–1905*, 24.
19. *Annual Catalogue of the State Normal School at Oshkosh, Wis., for the School Year 1872–73*, 22; Carey W. Brush, *In Honor and Good Faith: A History of the State University College at Oneonta, New York* (Oneonta, NY: The Faculty-Student Association of State University Teachers College at Oneonta, Inc., 1965), 14. At Ypsilanti by the late 1880s, "the Pestalozzian emphasis on the psychology of the child (with its obverse side of de-emphasis on the acquisition of facts, of the mere hearing of recitations by the teacher) had developed into a major emphasis on psychology in the professional curriculum." Isbell, *A History*, 53.
20. Albert Boyden quoted in Boyden, *The History*, 39; Albert Salisbury, *Historical Sketch of Normal Instruction in Wisconsin* (n.p., 1893), 56.
21. Bridgewater publication quoted in Boyden, *The History*, 38; C. O. Ruggles, *Historical Sketch and Notes: Winona State Normal School, 1860–1910* (Winona, MN: Jones & Kroeger Co., 1910), 126; Ypsilanti catalogs excerpted in Putnam, *A History*, 58, 62; Robert T. Brown, *The Rise and Fall of the People's Colleges: The Westfield Normal School, 1839–1914* (Westfield, MA: Institute for Massachusetts Studies, Westfield State College, 1988), 51; Charles A. Harper, *Development of the Teachers College in the United States, with Special Reference to the Illinois State Normal University* (Bloomington, IL: McKnight & McKnight, 1935), 127; *Catalogue and Circular of the California State Normal School, San Jose, 1878–79*, 46; *Catalogue of the State Normal School at Castleton, Vermont, 1881–1883*, CSCA, 13; Bannon, "A History," 21; Wright, *Fifty Years*, 109; James B. Bonder, "The Growth and Development of the State Teachers Colleges of Pennsylvania" (Ed.D. diss., Temple University, 1952), 334; Almack, "History," 109; Fredrick Chambers, "Historical Study of Arkansas Agricultural, Mechanical and Normal College, 1873–1943" (Ed.D. diss., Ball State University, 1970), 67; Emporia catalog printed in *A History of the State Normal School of Kansas*, 46; Boyden, *The History*, 38.
22. Ypsilanti catalog excerpted in Putnam, *A History*, 76–77; *Annual Catalogue of the State Normal College, Florence, Alabama 1898–99*, 21, *1899–1900*, 28; *Annual Catalogue of the State Normal School at Oshkosh, Wis., for the School Year 1897–98*, 55; *Catalogue of the California State Normal School, San Jose, 1900*, 28, 30.
23. Harper, *Development*, chapters 15 and 16, quotation on 235–236; Helen E. Marshall, *Grandest of Enterprises: Illinois State Normal University, 1857–1957* (Normal, IL: Illinois State Normal University, 1956), 188; Herbst, *And Sadly*

Teach, 145–146. See also Kathleen Ann Cruikshank, "The Rise and Fall of American Herbartianism: Dynamics of an Educational Reform Movement" (Ph.D. diss., University of Wisconsin-Madison, 1993).

24. Brush, *In Honor*, 22; W. Charles Lahey, *The Potsdam Tradition: A History and a Challenge* (New York: Appleton-Century-Crofts, 1966), 109; Rogers, *Oswego*, 12; Cook quoted in Harper, *Development*, 235. Rogers reported that at Oswego, "Herbartian influence was clearly evident in local slavery to lesson planning. . . . Herbart, as translated in McMurry's *Text on General Method*, was followed as slavishly in many classrooms as object teaching ever was." And a historian of Southern Illinois Normal University in Carbondale observed that "the five formal steps . . . tended greatly to formalization and exaggeration." Rogers, *Oswego*, 102; Eli G. Lentz, *Seventy-Five Years in Retrospect: Southern Illinois University, 1874–1949* (Carbondale, IL: University Editorial Board, Southern Illinois University, 1955), 35.
25. Boyden, *The History*, 90; Robert McGraw, "A Century of Service," in *The First 100 Years: Worcester State College*, ed. Herb Taylor (Worcester, MA: Worcester State College, Office of Community Services, 1974), section 2, no page numbers. See also Borrowman, *The Liberal and Technical*, 107–109.
26. Arthur Charles Forst, Jr., "From Normal School to State College: The Growth and Development of Eastern Connecticut State College From 1889 to 1959" (Ph.D. diss., University of Connecticut, 1980), 73; *Catalogue of the California State Normal School, San Jose, 1900*, 31; Oliphant, *History*, 128–131; Mohler, *The First Seventy-Five Years*, 18–19; Brown, *The Rise and Fall*, 106; Work Projects Administration in the State of Massachusetts, *The State Teachers College at Westfield* (Boston: State Department of Education, 1941), 65; *Vermont State Normal Schools, 1898–99*, 16; Isbell, *A History*, 55; Albert Reynolds Taylor, "History of Normal School Work in Kansas," *Kansas State Historical Quarterly* 6 (1900): 118; Harper, *Development*, 246.
27. Purington, *History*, 195; Ernest J. Hopkins and Alfred Thomas, Jr., *The Arizona State University Story* (Phoenix, AZ: Southwest Publishing Co. Inc., 1960), 89; I. F. Boughter, ed., *Fairmont State Normal School: A History* (Fairmont, WV: Fairmont State Normal School, 1929), 90; W. Wayne Dedman, *Cherishing This Heritage: The Centennial History of the State University College at Brockport, New York* (New York: Appleton-Century-Crofts, 1969), 113; Carbone, "The History," 338; Chambers, "Historical Study," 143; *Annual Catalogue of the State Normal School at Oshkosh, Wis., for the School Year 1879–80*, no page numbers, *1891–92*, 21–22; *Catalogue of the California State Normal School, San Jose, 1895*, 15; Bugaighis, "Liberating Potential," 301–302.
28. Emporia catalog printed in *A History of the State Normal School of Kansas*, 46–47; *Annual Catalogue of the State Normal College, Florence, Alabama 1899–1900*, 28; *Catalogue of the California State Normal School, San Jose, 1900*, 30.
29. *Catalogue and Circular of the California State Normal School, San Jose, 1878–79*, 45; *Catalogue of the California State Normal School, San Jose, 1900*, 30; Brush, *In Honor*, 20; Bowles, *A Good Beginning*, 20.
30. *Annual Catalogue of the State Normal School at Oshkosh, Wis., for the School Year 1897–98*, 55–56; Brown, *The Rise and Fall*, 49; Emporia catalog printed in *A History of the State Normal School of Kansas*, 46.
31. *Annual Catalogue of the State Normal College, Florence, Alabama 1898–99*, 20–21; *Announcement of the Southwest Texas State Normal School for the Annual Session 1904–1905*, 24.

32. *Catalogue of the California State Normal School, San Jose, 1900*, 30–31; *Annual Catalogue of the State Normal College, Florence, Alabama 1899–1900*, 28; Ypsilanti catalogs excerpted in Putnam, *A History*, 71, 77.
33. Estelle Greathead, *The Story of an Inspiring Past: Historical Sketch of the San Jose State Teachers College From 1862 to 1928 with an Alphabetical List of Matriculants and Record of Graduates by Classes* (San Jose, CA: San Jose State Teachers College, 1928), 42; Terry A. Barnhart, "Educating the Masses: The Normal-School Movement and the Origins of Eastern Illinois University, 1895–1899," *Journal of Illinois History* 4 (Autumn 2001): 215; Matthis quoted in *Historical Sketch of the State Normal School at San Jose, California, with a Catalogue of Its Graduates and a Record of Their Work for Twenty-Seven Years* (Sacramento, CA: J.D. Young, 1889), 98; *Annual Catalogue of the State Normal College, Florence, Alabama 1886–87*, 11; Tempe catalog quoted in Pamela Claire Hronek, "Women and Normal Schools: Tempe Normal, a Case Study, 1885–1925" (Ph.D. diss., Arizona State University, 1985), 125.
34. Maude B. Gerritson, "Training School, 1839–1939, Especially 1914–1939," in *First State Normal School in America: The State Teachers College at Framingham, Massachusetts* (Framingham, MA: The Alumnae Association of the State Teachers College at Framingham, Massachusetts, 1959), 56–57, student quoted on 56; George W. Neel, "A History of the State Teachers College at Edinboro, Pennsylvania" (Ed.D. diss., Rutgers University, 1950), 41–43; Troy catalog quoted in Bannon, "A History," 19; Edward M. Shackelford, *The First Fifty Years of the State Teachers College at Troy, Alabama, 1887–1937* (Montgomery, AL: The Paragon Press, 1937), 186–187; Susan Vaughn, "The History of State Teachers College, Florence, Alabama," *Bulletin of the State Teachers College, Florence, Alabama* 18 (Supplemental, 193?), UCUNA, 26; *Catalogue and Circular of the Branch Normal College of the Arkansas Industrial University for the Year Ending Friday, June 8th, 1900* (Little Rock, AR, 1900), 22; Benjamin Franklin Gilbert, *Pioneers for One Hundred Years: San Jose State College, 1857–1957* (San Jose, CA: San Jose State College, 1957), 94.
35. *Vermont State Normal Schools, 1898–99*, 7; Arthur Clarke Boyden, *Albert Gardner Boyden and the Bridgewater State Normal School: A Memorial Volume* (Bridgewater, MA: Arthur H. Willis, 1919), Albert Boyden quoted on 46, 64.
36. Brown, *The Rise and Fall*, 63, 90; Hronek, "Women," 125; I. N. Mitchell, ed., *Quarter Century of the Milwaukee State Normal School, 1886–1911* (Milwaukee, WI, 1911), 6, 9; William Frederick Hartman, "The History of Colorado State College of Education: The Normal School Period, 1890–1911" (Ph.D. diss., Colorado State College of Education, 1951), 135; Buffalo bulletin quoted in *New York State Teachers College At Buffalo: A History, 1871–1946* (Buffalo: New York State Teachers College at Buffalo, 1946), 46; Fowler, *A Century*, 67; Adelaide R. Pender, "At the New Britain Normal School, 1886–1888," Pender Collection, University Archives, Elihu Burritt Library, Central Connecticut State University, New Britain, Connecticut, 17.
37. Buffalo bulletin quoted in *New York State Teachers College At Buffalo*, 46; Rogers, *Oswego*, 24.
38. *Annual Catalogue of the State Normal College, Florence, Alabama 1898–1899*, 22; Wilson quoted in Carbone, "The History," 329–330; Brush, *In Honor*, 21.
39. Boyden quoted in Boyden, *Albert Gardner Boyden*, 67; Hartman, "The History," 137; *Annual Catalogue of the State Normal School at Oshkosh, Wis., for*

the School Year 1897–98, 52; Matthis quoted in *Historical Sketch of the State Normal School at San Jose*, 98.

40. Chambers, "Historical Study," 90; Hart, *The First 75 Years*, 105–108, Gilchrist quoted on 107; Dixon quoted in Boyden, *Albert Gardner Boyden*, 144–145.
41. *Annual Catalogue of the State Normal College, Florence, Alabama 1886–87*, 11, *1898–99*, 21; training-school report quoted in Carbone, "The History," 362; McGraw, "A Century," section 6, no page numbers; Edwards quoted in Sandra D. Harmon, " 'The Voice, Pen and Influence of Our Women Are Abroad in the Land': Women and the Illinois State Normal University, 1857–1899," in *Nineteenth-Century Women Learn to Write*, ed. Catherine Hobbs (Charlottesville: University of Virginia Press, 1995), 92–93.
42. Pender, "At the New Britain Normal School," 17–18; *Annual Catalogue of the State Normal College, Florence, Alabama 1898–99*, 21; *Catalogue of the California State Normal School, San Jose, 1900*, 33; evaluation form reproduced in Bannon, "A History," 83–84.
43. Harmon, " 'The Voice,' " 93; McGraw, "A Century," section 6, no page numbers; catalog quoted in Neel, "A History," 41–42; *Oshkosh State Normal School Bulletin* 7 (June 1910), UWOA, 75; legislative report excerpted in Thomas E. Finegan, *Teacher Training Agencies: A Historical Review of the Various Agencies of the State on New York Employed in Training and Preparing Teachers for the Public Schools of the State* (Albany: The University of the State of New York, 1917), 127.
44. State Superintendent of Education (Alabama), *Report, 1883* (Montgomery, AL, 1883), 19.
45. Cheers printed in Mohler, *The First Seventy-Five Years*, 109; *Echoes from the Geneseo Normal* 1 (June 1905), CASUNYG, no page numbers.
46. *The Normal Advance* 1 (Sept.–Oct. 1894): 8, 4 (Jan. 1898): 68–81, 8 (March 1902): 141–148, 8 (April 1902): 167–169, 9 (Nov. 1902): 36–38, 10 (Dec. 1903): 77–78; *The Normal Student* 1 (Nov. 26, 1901), CSCA, 6; *The Normal Pennant* 1 (Jan. 1898), SJSUA, 1–2, 1 (April 1898): 1–2; *The Normal Index* 1 (Nov. 1885), SJSUA, 15; *The Normal Student* 1 (Nov. 26, 1901): 9, 2 (Nov. 21, 1902): 7–9, 2 (May 20, 1903): 2–8.
47. *The Normal Gem* 1 (April 1, 1889), June Waite Collection, UCUNA, 3; Minutes of Dixie Club, 1899–1903, April 25, 1902, in Organizations, Files, UCUNA, 75. *The Pedagogue* (San Marcos, TX, 1905), SWTA, 65–66; *The Pedagogue* (1911), 102–103.
48. *The Normal Index* 4 (May 1889): 90–92, 6 (Sept. 25, 1890): 5–6, 1 (Feb. 1886): 45, 5 (Feb. 25, 1890): 57–59; Isbell, *A History*, 333; Rural Sociology Seminar records quoted in Knauss, *The First Fifty Years*, 18–19.
49. *The Normal Advance* 9 (Nov. 1902): 40; *The Normalian* 2 (May 13, 1901), CASUNYG; Frank A. Cooper, *The Plattsburgh Idea in Education, 1889–1964* (Plattsburgh, NY: Plattsburgh College Benevolent and Educational Association, Inc., 1964), 41; *The Normal Index* 6 (April 25, 1891): 94.
50. Lafayette Society: Records, 1880–1883, March 12, 1881, in Organizations, Files, UCUNA, 62; Philalethean Society, Constitution and 2 volumes of Meeting Minutes, Sept. 1890–Feb. 1905, March 18, 1892, CASUNYG, vol. 1, 87; Agonian, Constitution and Alpha Chapter Secretaries' Reports, April 22, 1892–Nov. 14, 1901, April 22, 1892, CASUNYG, 27; *The Normal Index* 4 (Feb. 25, 1889): 58.
51. Ladies' Literary Society, Minutes, Aug. 30, 1878–May 25, 1882, Nov. 27, 1879, UWOA, 84; *The Normal Advance* 19 (June 1904): 237–238; debate question

quoted in John W. Cook and James V. McHugh, *A History of the Illinois State Normal University* (Normal, IL: Illinois State Normal University, 1882), 145; Lafayette Society: Records, Feb. 17, 1882, 115; Philalethean Society, Constitution and 2 volumes of Meeting Minutes, Oct. 16, 1891, 63; *The Pedagogue* (1906), 120.

52. Lafayette Society: Records, April 24, 1880–May 1, 1880, 19–21; Lafayette Society: Records, Oct. 6, 1882–Oct. 13, 1882, 131–132; *The Normal Gem* 4 (Nov. 1891): 1; *The Normal Index* 4 (Feb. 25, 1889): 52–54; Agonian, Constitution and Alpha Chapter Secretaries' Reports, Feb. 23, 1894 and Feb. 10, 1896, 119, 207; *The Normal Advance* 10 (Oct. 1902): 34.
53. Agonian, Constitution and Alpha Chapter Secretaries' Reports, Feb. 1, 1897, 248; *The Normal Index* 5 (April 25, 1890): 82; Lafayette Society: Records, May 1, 1880, 21; Ladies' Literary Society, Minutes, April 4, 1879, 54, Nov. 7, 1879, 81; *The Normal Advance* 7 (Dec. 1900): 59–63.
54. *The Normal Advance* 3 (Nov.–Dec. 1896): 31; *The Normal Index* 5 (Feb. 25, 1890): 50–51; *Echoes from the Geneseo Normal* 2 (June 1906), no page numbers; *The Normal Advance* 14 (Jan. 1908): 97, 3 (Jan.–Feb. 1897): 61, 7 (Jan. 1901): 77; Dixie Club, in Organizations, Files, UCUNA, Jan. 31, 1913, no page numbers.
55. *The Normal Index* 4 (Feb. 25, 1889): 49; *The Normal Advance* 7 (March 1901): 124, 9 (Oct. 1902): 28, 1 (Nov.–Dec. 1894): 6–9, quotation on 8; *The Oneontan* quoted in Brush, *In Honor*, 70.
56. *The Normal Index* 1 (Jan. 1886): 41–41, 3 (Jan. 1888): 55–56; *The Normal Pennant* 2 (Sept. 1899): 4–6; Cooper, *The Plattsburgh Idea*, 42; *The Normal Advance* 1 (Jan.–Feb. 1895): 6; *The Normalian* 2 (Oct. 29, 1900): 4; *The Normal Advance* 8 (March 1902): 157.
57. *The Normal Index* 3 (Sept. 1887): 2; *The Crucible* quoted in Hartman, "The History," 143; *Normal Outlook* quoted in Mohler, *The First Seventy-Five Years*, 107; *The Normal Student* 1 (Feb. 6, 1902): 4, 1 (June 26, 1902): 6; *The Pedagogue* (1904), 6; *The Normal Advance* 7 (Jan. 1901): 77; *Normal Crescent* (Oshkosh, WI: The Phoenix Society, April 25, 1885–May 8, 1886), Aug. 29, 1885, UWOA, 47.
58. *The Pedagogue* (1905), 105.

5 "Noble" Men and "Not Necessarily Bloomer Women": The Public Sphere, Gender Attitudes, and Life Choices

1. Social capital is the benefits of one's network of connections to other people and involvement in social groups, or, in the words of Bourdieu, "the aggregate of the actual or potential resources which are linked to . . . membership in a group—which provides each of its members with the backing of the collectively-owned capital, a 'credential' which entitles them to credit, in the various senses of the word." One's social capital "depends on the size of the network of connections he can effectively mobilize and on the volume of the capital (economic, cultural or symbolic) possessed in his own right by each of those to whom he is connected." Pierre Bourdieu, "The Forms of Capital," in *Handbook of Theory and Research for the Sociology of Education*, ed. John G. Richardson (New York: Greenwood Press, 1986), 248–249. See also David Swartz, *Culture & Power: The Sociology of Pierre Bourdieu* (Chicago: The University of Chicago

Press, 1997), chapter 4; James Coleman, "Social Capital and the Creation of Human Capital," *American Journal of Sociology* 94 (1988, Issue Supplement): S95–S120; Alejandro Portes, "Social Capital: Its Origins and Applications in Modern Sociology," *Annual Review of Sociology* 24 (1998): 1–24; Robert D. Putnam, *Bowling Alone: The Collapse and Revival of American Community* (New York: Simon & Schuster, 2000), 18–21.

2. Some work in women's history emphasizes that separate-spheres ideology had little impact on the less-privileged classes and races. See, e.g., Gerda Lerner, *The Majority Finds Its Past: Placing Women in History* (New York: Oxford University Press, 1979), 15–30; Paula Giddings, *When and Where I Enter: The Impact of Black Women on Race and Sex in America* (New York: Bantam Books, 1984), chapter 3. For a thorough description of separate-spheres ideology in middle-class society and on college campuses, see Lynn D. Gordon, *Gender and Higher Education in the Progressive Era* (New Haven, CT: Yale University Press, 1990), chapter 1. On "conventional coeds," see Helen Lefkowitz Horowitz, *Campus Life: Undergraduate Cultures from the End of the Eighteenth Century to the Present* (Chicago: The University of Chicago Press, 1987), chapter 9; Barbara Miller Solomon, *In the Company of Educated Women: A History of Women and Higher Education in America* (New Haven, CT: Yale University Press, 1985), chapter 7.
3. Rose Swart discussed "The New Woman" in an all-school talk on November 27, 1895. *The Normal Advance* 2 (Nov.–Dec. 1895), UWOA, 34.
4. Clarence G. Fassett, " 'The Noble Nineteen' or the Boys of S.N.S.," *The Normal Pennant* 7 (October 1903), SJSUA, 16.
5. M. Janette Bohi, *A History of Wisconsin State University Whitewater, 1868–1968* (Whitewater, WI: Whitewater State University Foundation, 1968), 45; Egbert R. Isbell, *A History of Eastern Michigan University, 1849–1965* (Ypsilanti, MI: Eastern Michigan University Press, 1971), 309–310; Edwin H. Cates, *A Centennial History of St. Cloud State College* (Minneapolis, MN: Dillon Press, 1968), 40; Bridgewater students quoted in *As We Were . . . As We Are: Bridgewater State College, 1840–1976* (Bridgewater, MA: Alumni Association, Bridgewater State College, 1976), 51.
6. Fairmont's rules in the mid 1870s and 1892 quoted in I. F. Boughter, ed., *Fairmont State Normal School: A History* (Fairmont, WV: Fairmont State Normal School, 1929), 64, 77–78; Dixon quoted in Arthur Clarke Boyden, *Albert Gardner Boyden and the Bridgewater State Normal School: A Memorial Volume* (Bridgewater, MA: Arthur H. Willis, 1919), 144; *Annual Catalogue of the State Normal School at Oshkosh, Wis., for the School Year 1871–72*, UWOA, 19, *1879–80*, 47–48, *1885–86*, 44–45; *Catalogue and Circular of the California State Normal School, San Jose, 1884–85*, SJSUA, 57–58.
7. Janette Bohi, "Whitewater, A Century of Progress (1868–1968)," in *History of the Wisconsin State Universities*, ed. Walker D. Wyman (River Falls, WI: River Falls State University Press, 1968), 63; Dixon quoted in Boyden, *Albert Gardner Boyden*, 144; Clark quoted in Albert Salisbury, *Historical Sketches of the First Quarter-Century of the State Normal School at Whitewater, Wisconsin* (Madison: Tracy, Gibbs & Co., 1893), 51; Kutztown catalogs quoted in Mark K. Fritz, "The State Normal Schools: Teaching Teachers and Others," *Pennsylvania Heritage* 11 (Fall 1985): 5; John E. Hubley, *Hilltop Heritage: Shippensburg State's First Hundred Years* (Shippensburg, PA, 1971), 46; Elizabeth Tyler Bugaighis, "Liberating Potential: Women and the Pennsylvania

State Normal Schools, 1890–1930" (Ph.D. diss., The Pennsylvania State University, 2000), 158–159.

8. Burton J. Bledstein, *The Culture of Professionalism: The Middle Class and the Development of Higher Education in America* (New York: W. W. Norton & Company, Inc., 1976), chapter 2, statistics on 46; Stuart M. Blumin, *The Emergence of the Middle Class: Social Experience in the American City, 1760–1900* (New York: Cambridge University Press, 1989); Paul DiMaggio, "Cultural Entrepreneurship in Nineteenth-Century Boston: The Creation of an Organizational Base for High Culture in America," *Media, Culture and Society* 4 (1982): 33–50; Paul DiMaggio, "Cultural Entrepreneurship in Nineteenth-Century Boston, Part II: The Classification and Framing of American Art," *Media, Culture and Society* 4 (1982): 303–322; Alan Trachtenberg, *The Incorporation of America: Culture & Society in the Gilded Age* (New York: Hill and Wang, 1982), chapter 4; Putnam, *Bowling Alone*, chapter 23.
9. Margaret Gibbons Wilson, *The American Woman in Transition: The Urban Influence, 1870–1920* (Westport, CT: Greenwood Press, 1979), chapters 2, 5, 7.
10. United States Bureau of the Census, *Census of Population* (Washington, DC: U.S. Government Printing Office, 1870–1910); Edward Noyes, "Oshkosh: From Normal School to State University (1871–1969)," in *History of the Wisconsin State Universities*, 95–96; *Oshkosh State Teachers College: The First Seventy-Five Years* (Oshkosh, WI: Oshkosh State Teachers College, 1946), 6; *The Quiver* (Oshkosh, WI, 1946), UWOA, no page numbers; *The First Half Century of the Oshkosh Normal School* (Oshkosh, WI: The Faculty of State Normal School, 1921), UWOA, 7–8; Benjamin Franklin Gilbert, *Pioneers for One Hundred Years: San Jose State College 1857–1957* (San Jose, CA: San Jose State College, 1957), 49; S. E. Rothery, "Some Educational Institutions: Pilgrimages About San Jose," *The Overland Monthly* 30 (July 1897): 74–77.
11. United States Bureau of the Census, *Census of Population*; W. Wayne Dedman, *Cherishing This Heritage: The Centennial History of the State University College at Brockport, New York* (New York: Appleton-Century-Crofts, 1969), chapter 8; Robert T. Brown, *The Rise and Fall of the People's Colleges: The Westfield Normal School, 1839–1914* (Westfield, MA: Institute for Massachusetts Studies, Westfield State College, 1988), 90; Helen E. Marshall, *Grandest of Enterprises: Illinois State Normal University, 1857–1957* (Normal, IL: Illinois State Normal University, 1956), 116; Ronald C. Brown, *Beacon on the Hill: Southwest Texas State University, 1903–1978* (San Marcos, TX: Southwest Texas State University), 6; *Announcement of the Southwest Texas State Normal School for the Annual Session 1903–1904* (Austin, TX: State Printers, 1903), SWTA, 8; *Bulletin of the State Normal College, Florence, Alabama* 1 (March 1912), UCUNA, 8–9.
12. United States Bureau of the Census, *Census of Population*; Robert A. Laflar, *The First 100 Years: Centennial History of the University of Arkansas* (Fayetteville, AR: University of Arkansas Foundation, Inc., 1972), 273–274; Tom Baskett, Jr., ed., *Persistence of the Spirit: The Black Experience in Arkansas* (Little Rock, AR: Arkansas Endowment for the Humanities, 1986), 27, 32; D. B. Gaines, *Racial Possibilities as Indicated by the Negroes of Arkansas* (Little Rock, AR: Philander Smith College, 1898), 131, 137.
13. United States Bureau of the Census, *Census of Population*; John H. Black, *Oral History: John Black Remembers the Early 1900s in Geneseo* (sound recording, 1980), CASUNYG; Rochester, NY *Express*, March 25, 1870, quoted in Rosalind R. Fisher, *". . . the stone strength of the past . . .": Centennial History of State*

University College of Arts and Science at Geneseo, New York (Geneseo, NY, 1971), 39–42; *The History of Rutland County, Vermont: Civil, Ecclesiastical, Biographical and Military* (White River Junction, VT: White River Paper Company, 1882), 513; *Vermont State Normal Schools, 1902–1903*, CSCA, 33, *1908–1909*, 29.

14. United States Bureau of the Census, *Census of Population*; Ernest J. Hopkins and Alfred Thomas, Jr., *The Arizona State University Story* (Phoenix, AZ: Southwest Publishing Co., Inc., 1960), viii, 128; Melvin Frank Fiegel, "A History of Southwestern State College, 1903–1953" (Ed.D. diss., Oklahoma State University, 1968), 14–15; James H. Thomas and Jeffrey A. Hurt, "Southwestern Normal School: The Founding of an Institution," *The Chronicles of Oklahoma* 54 (Winter 1976–1977): 465.

15. Shippensburg rule quoted in Hubley, *Hilltop Heritage*, 50; *Catalogue of the State Normal School at Castleton, Vermont, 1872–73*, CSCA, 15, *1883–1885*, 15; Dedman, *Cherishing*, 122, 185; Fisher, *". . . the stone strength,"* 78; Fredrick Chambers, "Historical Study of Arkansas Agricultural, Mechanical and Normal College, 1873–1943" (Ed.D. diss., Ball State University, 1970), 92–93, 99–100, 123–124; State Superintendent of Public Instruction (Arkansas), *Biennial Report, 1905–1906* (Little Rock, AR, 1906), 70; Dudley S. Brainard, *History of St. Cloud State Teachers College* (St. Cloud, MN, 1953), 12, 31; Elisabeth Ann Bowles, *A Good Beginning: The First Four Decades of the University of North Carolina at Greensboro* (Chapel Hill, NC: The University of North Carolina Press, 1967), 15.

16. Carey W. Brush, *In Honor and Good Faith: A History of the State University College at Oneonta, New York* (Oneonta, NY: The Faculty-Student Association of State University Teachers College at Oneonta, Inc., 1965), 47; Boughter, *Fairmont State Normal*, 63–64; *Annual Catalogue of the State Normal College, Florence, Alabama, 1898–99*, UCUNA, 41; Fiegel, "A History," 24; Adelaide R. Pender, "At the New Britain Normal School, 1886–1888," Pender Collection, University Archives, Elihu Burritt Library, Central Connecticut State University, New Britain, Connecticut, 9, 12.

17. Cecil Dryden, *Light for an Empire: The Story of Eastern Washington State College* (Cheney, WA: Eastern Washington State College, 1965), 191; Samuel R. Mohler, *The First Seventy-Five Years: A History of Central Washington State College, 1891–1966* (Ellensburg, WA: Central Washington State College, 1967), 45; Dorothy Rogers, *Oswego: Fountainhead of Teacher Education: A Century in the Sheldon Tradition* (New York: Appleton-Century-Crofts, Inc., 1961), 63–64, 120.

18. For example, official regulations at San Marcos in 1905 included the following: ". . . Home study periods will be from 3:30 P.M. to 5:30 P.M. and 7 P.M. till 10 P.M. each Monday, Tuesday, Wednesday, Thursday and Friday. . . . The hours designated above as study periods are to be spent by all students in their own rooms, or in the library, or in the laboratory, in the prosecution of school work. . . . During the after-noons and evenings of Saturdays and Sundays, students may, within proper bounds, make and receive social visits, but such visits may not extend later than 10 P.M. and the proprietors of boarding homes are directed to see that their parlors are in all cases vacated by students not later than this hour. . . ." "Regulations," Sept. 1, 1905, T. G. Harris, Correspondence, 1903–1915, box 2, series 2, SWTA.

19. Principal Burr quoted in Arthur Charles Forst, Jr., "From Normal School to State College: The Growth and Development of Eastern Connecticut State

College From 1889 to 1959" (Ph.D. diss., University of Connecticut, 1980), 100; Fisher quoted in State Superintendent of Public Instruction (Arkansas), *Biennial Report, 1905–1906*, 70.

20. Sabbath rules quoted in George Frank Sammis, "A History of the Maine Normal Schools" (Ph.D. diss., University of Connecticut, 1970), 127; Hubley, *Hilltop Heritage*, 51; Lowell H. Harrison, "Gordon Wilson's Normal Education: Western Kentucky State Normal School, 1908–1913," *Register of the Kentucky Historical Society* 86 (Winter 1988): 33; Edith Tammany, "W. A. Brodie Class Banquet," *The Normalian* 1 (June 20, 1900), CASUNYG, 1; Thomas Rothrock, "Joseph Carter Corbin and Negro Education in the University of Arkansas," *The Arkansas Historical Quarterly* 30 (Winter 1971): 292; Harrison, "Gordon Wilson's Normal Education," 43.
21. "The Church Life of the Students," *The Normalian* 2 (June 10, 1901): 8; Tammany, "W. A. Brodie Class Banquet," 1; *The Normal Index* 2 (April 1887), SJSUA, 117; Harrison, "Gordon Wilson's Normal Education," 33; *The Normal School Bulletin* 2 (July 1913), SWTA, 8; *Bulletin of the State Normal College, Florence, Alabama* 1 (June 1912): 9.
22. Elizabeth L. Wheeler, "Isaac Fisher: The Frustrations of a Negro Educator at Branch Normal College, 1902–1911," *The Arkansas Historical Quarterly* 41 (Spring 1982): 14; *The Normal Star* 1 (April 21, 1911), SWTA; *The Normal Student* 1 (Feb. 6, 1902): 12, 1 (March 27, 1902): 13, 2 (Nov. 21, 1902): 5; Chico instructor quoted in Maxine Ollie Merlino, "A History of the California State Normal Schools—Their Origin, Growth, and Transformation into Teachers Colleges" (Ed.D. diss., University of Southern California, 1962), 102; William Frederick Hartman, "The History of Colorado State College of Education: The Normal School Period, 1890–1911" (Ph.D. diss., Colorado State College of Education, 1951), 164; Forst, "From Normal School," 123.
23. *The Normal Index* 6 (May 25, 1891): 98; *The Normalian* 2 (Oct. 29, 1900); Work Projects Administration in the State of Massachusetts, *The State Teachers College at Westfield* (Boston: State Department of Education, 1941), 66; Fisher, *". . . the stone strength,"* 84; Wheeler, "Isaac Fisher," 33; Harrison, "Gordon Wilson's Normal Education," 40; *A History of the State Normal School of Kansas for the First Twenty-Five Years* (Emporia, Kansas, 1889), 31–32.
24. Putnam, *Bowling Alone*, chapter 23, quotation on 383.
25. Rothery, "Some Educational Institutions," 75.
26. Bessie L. Park, *Cortland—Our Alma Mater: A History of Cortland Normal School and the State University of New York Teachers College at Cortland, 1869–1959* (Cortland, NY, 1960), 36; Fiegel, "A History," 37; Bohi, *A History*, 63, committee quoted on 63; Charles H. Coleman, "Eastern Illinois State College: Fifty Years of Public Service," *Eastern Illinois State College Bulletin* 189 (January 1, 1950): 135–136, 139, Lord quoted on 136; *New York State Teachers College At Buffalo: A History, 1871–1946* (Buffalo: New York State Teachers College at Buffalo, 1946), 123; Harrison, "Gordon Wilson's Normal Education," 31.
27. *Report of the Proceedings Commemorating the One-Hundredth Anniversary of the Establishment of a Chartered School, Known at Different Periods as the Rutland County Grammar School, Castleton Seminary and State Normal School, in Castleton, Vermont, 1787–1887*, 1888, CSCA; Fisher, *". . . the stone strength,"* 143; *Oshkosh State Teachers College*, 57; Boyden, *Albert Gardner Boyden*, 61–62, *Boston Herald* quoted on 61; Boughter, *Fairmont State Normal*, 40; Black quoted in Rogers, *Oswego*, 72.

28. Class Day Programs (Oshkosh), 1888–1891, UWOA; Commencement Programs, Castleton State Normal School, 1874–1940, CSCA; *The Normal Student* 1 (June 20, 1902), CSCA, 3; "Graduate Recalls Etiquette of Early Days," in "Geneseo State Normal School: Special Dedication Edition," *Livingston County Leader* 54 (May 5, 1939), CASUNYG, 7; Agonian Sorority, Programs, "Anniversary exercises held at Normal Hall, Geneseo, by Agonian Fraternity with Philalethean Fraternity," 1900–1905, CASUNYG; Arethusa Sorority, Programs, "Anniversary exercises held at Normal Hall, Geneseo, by Arethusa Fraternity with Gamma Sigma Fraternity," 1900–1903, CASUNYG; Clionian Fraternity, Programs, "Anniversary exercises held at Normal Hall, Geneseo, by Clionian Fraternity with Delphic Fraternity," 1897, 1900, 1902, CASUNYG; Commencement Programs (Florence), 1870s–1930s, UCUNA; Wheeler, "Isaac Fisher," 14; Commencement programs in "Keepers of the Spirit: The L. A. Davis, Sr. Historical Collection," Exhibit, Isaac S. Hathaway—John M. Howard Fine Arts Center, University of Arkansas at Pine Bluff, Pine Bluff, AR, May 1995.
29. Work Projects Administration, *The State Teachers College*, 50; *The Cortland Standard* quoted in Park, *Cortland*, 43–44; *The Normal Advance* 5 (Oct. 1898), 32, 6 (Nov. 1899): 55, 7 (Jan. 1901), 76–77; *The Normal Index* 1 (April 1886): 37.
30. *The Normal Index* 2 (May 1887): 131, 4 (Oct. 2, 1888): 2; John Edward Marryman, "Indiana University of Pennsylvania: From Private Normal School to Public University" (Ph.D. diss., University of Pittsburgh, 1972), 237; *The Normal Advance* 1 (May–June 1895): 21, 5 (May 1899): 169, 6 (June 1900): 205, 8 (June 1902): 230, 9 (May 1903): 164–166, 13 (June 1907): 249; Program, Philalethean-Agonian Fifth Annual Picnic, in Maude Wiard, Scrapbook and Diary (class of 1908), CASUNYG; *Announcement of the Southwest Texas State Normal School for the Annual Session 1908–1909* (Austin, TX: State Printers, 1908), SWTA, 45; Mary Francis McKinney, May C. Hansen, and Gladys Tyng, "Student Life," in *Three Decades of Progress: Eastern Kentucky State Teachers College*, ed. Jonathan T. Dorris (Richmond, KY: Eastern Kentucky State Teachers College, 1936), 150; Hopkins and Thomas, *The Arizona State University*, 177.
31. Victoria Ann Smith, "A Social History of Marshall University During the Period as the State Normal School, 1867–1900," *West Virginia History* 25 (October 1963): 36; Cates, *A Centennial History*, 57–59; Dignam quoted in Work Projects Administration, *The State Teachers College*, 53; Eugene Coleman Ney, "Slippery Rock State Normal School, 1890–1916: The Maltby Era" (Ph.D. diss., University of Pittsburgh, 1998), 140–141, Maltby quoted on 140–141; *Semi-Centennial History of the Illinois State Normal University, 1857–1907* (Normal, IL: Illinois State Normal University, 1907), 167; Cook quoted in Sandra D. Harmon, " 'The Voice, Pen and Influence of Our Women Are Abroad in the Land': Women and the Illinois State Normal University, 1857–1899," in *Nineteenth-Century Women Learn to Write*, ed. Catherine Hobbs (Charlottesville: University of Virginia Press, 1995), 89; Wilson quoted in Harrison, "Gordon Wilson's Normal Education," 29.
32. Marshall, *Grandest*, 148–149; *A History of the State Normal School of Kansas*, 71–72; Students' Christian Union quoted in Park, *Cortland*, 97; Hubley, *Hilltop Heritage*, 70–71; Mohler, *The First Seventy-Five Years*, 110–111; Walter H. Ryle, *Centennial History of the Northeast Missouri State Teachers College* (Kirksville, MO: Northeast Missouri State Teachers College, 1972), 576–578.

33. *New York State Teachers College At Buffalo*, 143; Rogers, *Oswego*, 126; Bowles, *A Good Beginning*, 134; Pamela Dean, "Covert Curriculum: Class, Gender, and Student Culture at a New South Woman's College, 1892–1910" (Ph.D. diss., University of North Carolina at Chapel Hill, 1994), 161–162; Marshall, *Grandest*, 225; Brown, *Beacon*, 15–16; *The Normal Index* 6 (March 25, 1891): 73; Agonian, Constitution and Alpha Chapter Secretaries' Reports, Sept. 22, 1898, CASUNYG, 310; *The Normalian* 2 (Nov. 12, 1900): 4; *The Normal Advance* 5 (Dec. 1898): 71.
34. *The Normal Index* 1 (April 1886): 72–73; *The Normal Advance* 13 (March 1907): 153; *The Normal Student* 1 (May 22, 1902): 7; *The Normal Advance* 5 (June 1899): 194; *Florence Normal College Quarterly* 1 (May 1909), UCUNA, 39; *The Normal Advance* 8 (Feb. 1902): 135; Agonian, Constitution and Alpha Chapter Secretaries' Reports, April 22, 1892–Nov. 14, 1901; Philalethean Society, Constitution and 2 volumes of Meeting Minutes, Sept. 1890–Feb. 1905, CASUNYG; Agonian Sorority, Programs, "Anniversary exercises," 1900–1905; Arethusa Sorority, Programs, "Anniversary exercises," 1900–1903; Clionian Fraternity, Programs, "Anniversary exercises," 1897, 1900, 1902; Chambers, "Historical Study," 188–189; Wheeler, "Isaac Fisher," 21.
35. *The Lamron: 50th Anniversary, Geneseo Normal School* (Geneseo, NY: Students of the State Normal School, 1921), CASUNYG, 67; Ryle, *Centennial History*, 722; *Oshkosh State Teachers College*, 60; *Students' Hand Book, 1906–1907* (Oshkosh, WI: Students' Christian Association, 1906), UWOA, 23–24; Dedman, *Cherishing*, 128–129; David Sands Wright, *Fifty Years at the Teachers College: Historical and Personal Reminiscences* (Cedar Falls, IA: Iowa State Teachers College, 1926), 203–205; Marshall, *Grandest*, 225–226; Brown, *Beacon*, 15; *The Pedagogue* (1905), SWTA, 80–83; *The Normal Pennant* 12 (June 1909): 57.
36. *Catalogue and Circular of the California State Normal School, San Jose, 1888*, 41; Pender, "At the New Britain Normal School," 11; catalog cited in Hopkins and Thomas, *The Arizona State University*, 131; *A History of the State Normal School of Kansas*, 34; *Annual Catalogue of the State Normal College, Florence, Alabama, 1890–91*, 23; student quoted in Rogers, *Oswego*, 113; Park, *Cortland*, 78–80.
37. Smith, "A Social History," 37; *Catalogue of the State Normal School at Castleton, Vermont, 1883–1885*, 15; Wright, *Fifty Years*, 176–178; Hopkins and Thomas, *The Arizona State University*, 173; Frank T. Reuter, *West Liberty State College: The First 125 Years* (West Liberty, WV: West Liberty State College, 1963), 30; "Normal Index Department," *The Pacific Coast Teacher: A Monthly Magazine Devoted to the Educational Interests of the Pacific Coast* 1 (April 1892), SJSUA, 273.
38. Ronald A. Smith, "Athletics in the Wisconsin State University System: 1867–1913," *Wisconsin Magazine of History* 55 (Autumn 1971): 3–6; Dedman, *Cherishing*, 132.
39. Dedman, *Cherishing*, 133; Fisher, *". . . the stone strength,"* 145–147; Smith, "Athletics," 9–13, 17; Bohi, *A History*, 106, newspaper quoted on 110; Hopkins and Thomas, *The Arizona State University*, 135; newspaper quoted in Hartman, "The History," 174; Van Oot quoted in Rogers, *Oswego*, 127–128. On men's athletics at colleges and universities, see Frederick Rudolph, *The American College and University: A History* (Athens: University of Georgia Press, [1962] 1990), chapter 18.
40. Poem by "E. I." (full name not listed), in *The Normal Advance* 3 (Jan.–Feb. 1897): 54; *The Normal Pennant* 1 (Nov. 1898): 10, 3 (Oct. 1900): 9;

The Normalian 3 (Feb. 1902): 12; Brown, *Beacon*, 15; *The Pedagogue* (1904), 55–57; *The Pedagogue* (1906); local newspaper and *Normal Outlook* quoted in Mohler, *The First Seventy-Five Years*, 97.

41. Lee Graver, *A History of the First Pennsylvania State Normal School, Now the State Teachers College at Millersville* (Millersville, PA: State Teachers College, 1955), 237; *The Normal Pennant* 3 (Feb. 1900): 10; Ryle, *Centennial History*, 676; newspaper quoted in Cates, *A Centennial History*, 114–115; *The Normal Pennant* 7 (March 1904): no page numbers, 8 (June 1905): 92; *Echoes from the Geneseo Normal* 1 (June 1905), CASUNYG, no page numbers; *Florence Normal College Quarterly* 2 (May 1910): 32.
42. Ryle, *Centennial History*, 676; *The Normalian* 3 (March 1902): 10; *The Normal Advance* 3 (March–April 1897): 78; *The Quiver* (1898), 113; *The Normal Advance* 6 (Dec. 1899): 71.
43. Park, *Cortland*, 27, 82; Marshall, *Grandest*, 183; *The Normal Pennant* 1 (March 1898): 2, 7 (June 1904): 48, 8 (June 1905): 92; *Echoes from the Geneseo Normal* (1905), no page numbers; San Diego catalogs quoted in Merlino, "A History," 167; *The Normal Advance* 13 (Dec. 1906): 73–74, 17 (Oct. 1910): 49–50; Hartman, "The History," 173; *The Normal Star* 2 (Nov. 4, 1911): 2.
44. *The Normal Advance* 3 (May–June 1897): 95, 5 (May 1899); *The Normal Pennant* 12 (June 1909): 57; *The Pedagogue* (1911); *The Quiver* (1897), 92; *The Normal Advance* 4 (April 1898): 144, 7 (Oct. 1900): 29, 2 (May–June 1896): 106; *Echoes from the Geneseo Normal* (1906), 11–12; *The Normal Pennant* 4 (June 1901): 12, 6 (June 1903): 73, 8 (June 1905): 95.
45. *The Normalian* 1 (June 20, 1900), 3.
46. Class Day Programs (Oshkosh), 1888–1897; *The Normal Advance* (various issues); *The First Half Century of the Oshkosh Normal School*, 25; *Students' Hand Book, 1906–1907*, 35–37; PerLee quoted in Park, *Cortland*, 104; *The Normal Pennant* 1 (May 1898): 4; Estelle Greathead, *The Story of an Inspiring Past: Historical Sketch of the San Jose State Teachers College From 1862 to 1928 with an Alphabetical List of Matriculants and Record of Graduates by Classes* (San Jose, CA: San Jose State Teachers College, 1928), 69.
47. *Annual Catalogue of the State Normal School, Florence, Alabama, 1883 and 1884*, UCUNA, 14.
48. David B. Tyack and Elisabeth Hansot, *Learning Together: A History of Coeducation in American Public Schools* (New Haven, CT: Yale University Press), 1991, chapter 6, quotation on 147; Michael S. Kimmel, "The Contemporary 'Crisis' of Masculinity in Historical Perspective," in *The Making of Masculinities: The New Men's Studies*, ed. Harry Brod (Boston: Allen & Unwin, 1987), 137–146; Gail Bederman, *Manliness & Civilization: A Cultural History of Gender and Race in the United States, 1880–1917* (Chicago: The University of Chicago Press, 1995). See also Carroll Smith-Rosenberg, *Disorderly Conduct: Visions of Gender in Victorian America* (New York: Oxford University Press, 1985).
49. Anne Firor Scott, "The Ever Widening Circle: The Diffusion of Feminist Values from the Troy Female Seminary, 1822–1872," *History of Education Quarterly* 19 (Spring 1979): 4–5; *The Normal Advance* 2 (Nov.–Dec. 1895): 34; Nancy F. Cott, *The Grounding of Modern Feminism* (New Haven, CT: Yale University Press, 1987).
50. Lee quoted in Rogers, *Oswego*, 49–50; Marshall, *Grandest*, 185; Robin O. Harris, " 'To Illustrate the Genius of Southern Womanhood': Julia Flisch and Her Campaign for the Higher Education of Georgia Women," *The Georgia*

Historical Quarterly 80 (Fall 1996): 519; Isbell, *A History*, 238–239, 256; Bugaighis, "Liberating Potential," 152–153.

51. Agonian, Constitution and Alpha Chapter Secretaries' Reports, May 24, 1901, 429; *The Normal Advance* 6 (Nov. 1899): 53; *Annual Catalogue of the State Normal School, Florence, Alabama, 1884–85*, 23, Chambers, "Historical Study," 122; Commencement Programs, Castleton State Normal School, 1878, 1881, 1887, 1891; Agonian, Constitution and Alpha Chapter Secretaries' Reports, April 30, 30, 1892, 29, Sept. 18, 1894, 141, Sept. 13, 1897, 268; debate topics quoted in Harmon, " 'The Voice,' " 92.
52. Ladies' Literary Society, Minutes, Sept. 6, 1878, UWOA, 4, Oct. 25, 1878, 17; *The Normal Advance* 5 (Feb. 1898): 111–115; *The Normal Star* 2 (Oct. 13, 1911): 3, 2 (Feb. 23, 1912): 1; *The Normal Advance* 4 (Dec. 1897): 55, 8 (Dec. 1901): 87, 15 (May 1909): 155, 18 (June 1912): 274; *Catalogue of the California State Normal School, San Jose, 1906*, SJSUA, 7; *The Livingston Republican*, Feb. 17, 1876, quoted in Fisher, *". . . the stone strength,"* 83.
53. *The Normalian* 2 (Nov. 12, 1900): 3; *The Normal Index* 4 (April 25, 1889): 76–77; article by "M. G." (probably Maud Gardner, associate editor), "Normal Index Department" 1 (Feb. 1892): 196.
54. *The Normal Pennant* 1 (May 1898): 8; *The Pedagogue* (1905), 17, 20; *The Normal Star* 2 (March 8, 1912): 1.
55. *The Normal Index* 1 (Feb. 1886): 49; *The Normal Gem* 2 (Nov. 1891), UCUNA, 1; *The Normalian* 1 (March 5, 1900): 6.
56. *The Normal Index* 4 (Feb. 25, 1889): 49; *The Normal Pennant* 2 (May 1899): 6; *The Normal Advance* 1 (May–June 1895): 6–9; *The Quiver* (1899), 41; I. N. Mitchell, ed., *Quarter Century of the Milwaukee State Normal School, 1886–1911* (Milwaukee, 1911), 20.
57. Nellie Patton, Paper and Speech, 1897, UCUNA.
58. *Normal Crescent* (Oshkosh, WI: The Phoenix Society), June 15, 1885, UWOA, 40–42; Class Day Program (Oshkosh), 1896, 50–56; *The Normal Student* 1 (June 26, 1902): 13–15, 2 (June 25, 1903): 11–14; *The Pedagogue* (1904), 76.
59. *The Normal Index* 1 (October 1885): 3; "The Normal Index Department" 1 (June 1892): 374.
60. *The Normal Index* 6 (Oct. 25, 1890): 24; *The Normal Advance* 2 (Sept.–Oct. 1895): 15, 4 (Oct. 1897): 15.
61. Florence principal quoted in United States Commissioner of Education, *Report*, 1887–1888 (Washington, DC: U.S. Government Printing Office, 1888), 408–409; *The Normal Index* 3 (September 1887): 12; *Senior Year Book: San Jose Normal* (San Jose, CA: Students of San Jose Normal School, 1910), SJSUA, 35; Childs and McNaughton quoted in Gilbert, *Pioneers*, 93, 107–108; Smith, "Athletics," 3.
62. PerLee quoted in Park, *Cortland*, 93; *The Normal Student* 2 (February 19, 1903): 11.
63. *The Normal Index* 1 (March 1886): 62, 2 (October 1886): 33, 6 (Feb. 25, 1891): 71; *The Normal Student* 1 (Nov. 26, 1901): 9; *The Pedagogue* (1904), 10; *The Normal Index* 1 (Feb. 1888): 69; *The Normal Pennant* 1 (Jan. 1899): 7.
64. *The Normal Student* 1 (Nov. 26, 1901): 7; *The Normal Pennant* 7 (Oct. 1903): 16.
65. *The Normal Student* 1 (June 26, 1902): 10; *The Normal Advance* 2 (Sept.–Oct. 1895): 16; Lafayette Society, Records, 1880–1883, in Organizations, Files, UCUNA, April 24, 1880, 19, May 1, 1880, 21, Oct. 6, 1882, 131, Oct. 13,

1882, 132; Philalethean Society, Constitution and 2 volumes of Meeting Minutes, Jan. 19, 1899, vol. 2, 77; *The Normal Advance* 5 (March 1899): 121; *Purple and Gold* (Florence, AL: Students of State Normal College, 1913), UCUNA, no page numbers.

66. Christopher Jencks and David Riesman, *The Academic Revolution* (Garden City, NY: Doubleday & Company, Inc., 1968), 69.
67. The majority of students who attended state normal schools did not actually finish the course of studies. Yet, the institutions generally tracked the activities of students who graduated. Thus, it would be difficult to analyze the life paths of nongraduates, who probably enjoyed a certain amount of class mobility, but perhaps not as much as actual graduates.
68. Boyden, *Albert Gardner Boyden*, 74; percentages of male San Jose graduates who taught calculated from information included in "Record of Graduates By Classes," in Greathead, *The Story*, 290–506; percentages of male Oshkosh graduates who worked in the schools and average tenure on the job calculated from listing of individual graduates' "Years experience since graduating" in "Graduates' Directory," *Oshkosh State Normal School Bulletin* 4 (Oct. 1906), UWOA, 8–53. Because some Oshkosh graduates, especially of the later years, were still teaching or working as administrators in 1906, the average "years experience" were probably slightly higher than what is reported.
69. George C. Purington, *History of the State Normal School, Farmington, Maine* (Farmington, ME: Knowlton, McLeary & Co., 1889), 80–81; "50th Anniversary, class of 1884, Geneseo Normal School," Folder: Education: Normal School, LCHS, no page numbers.
70. "50th Anniversary, class of 1884"; Purington, *History*, 78; "Register of the Alumni," 50, 53; *Esto Lux* (Florence, AL: Alumni of State Normal College, 1898), UCUNA, no page numbers; *A History of the State Normal School of Kansas*, 66, 69, 77; percentages of male San Jose graduates who worked in administration calculated from information included in "Record of Graduates By Classes"; *Directory of Graduates, Oshkosh State Normal School* (Oshkosh, WI: State Normal School Alumni Association, 1912), UWOA, 16.
71. *Directory of Graduates*, 24, 17, 80, 85, 47; Brush, *In Honor*, 72, 82; Ryle, *Centennial History*, 114; Work Projects Administration, *The State Teachers College*, 54; "Register of the Alumni," 58; "Some of Florence Alumni," *Florence Times*, May 28, 1930, in Alumni: Clippings, file, UCUNA.
72. *A History of the State Normal School of Kansas*, 161–169; *General Catalogue of the State Normal School, Westfield, Mass., 1839–1889* (Boston: Wright & Potter Printing Company, 1890), 94; Purington, *History*, 113, 102–103; *Directory of Graduates*, 107.
73. Park, *Cortland*, 90; Marshall, *Grandest*, 197; Purington, *History*, 78, 184; *Directory of Graduates*, 80, 85, 47; Work Projects Administration, *The State Teachers College*, 54; *General Catalogue of the State Normal School, Westfield, Mass.*, 121, 123; Brush, *In Honor*, 76–79, 82.
74. *A History of the State Normal School of Kansas*, 68, 71; Brush, *In Honor*, 76–77; "50th Anniversary, class of 1884"; *Directory of Graduates*, 89, 105, 63; Jerry G. Nye, *Southwestern Oklahoma State University: The First 100 Years* (Weatherford, OK: Southwestern Oklahoma State University, 2001), 22.
75. Percentages of male San Jose graduates who studied law calculated from information included in "Record of Graduates By Classes"; percentages of male Oshkosh graduates who worked as attorneys calculated from listings of current

occupations in "Graduates' Directory"; *Directory of Graduates*, 32, 44; Brush, *In Honor*, 82; Purington, *History*, 91, 128; *A History of the State Normal School of Kansas*, 161–169, 69; "Register of the Alumni," 51, 69; "Some of Florence Alumni"; *Catalogue and Circular of the Branch Normal College of the Arkansas Industrial University for the Year Ending June 8th, 1900* (Little Rock, AR, 1900), 4, 7; *Annual Catalogue of the Arkansas State College for Negroes, 1928–1929* (Little Rock, AR, 1929), 64.

76. Percentages of male San Jose graduates who studied medicine calculated from information included in "Record of Graduates By Classes"; percentages of male Oshkosh graduates who worked as medical doctors calculated from listings of current occupations in "Graduates' Directory"; *Directory of Graduates*, 95, 113; Brown, *The Rise and Fall*, 76; *A History of the State Normal School of Kansas*, 79; "50th Anniversary, class of 1884"; "Register of the Alumni"; *Catalogue and Circular of the Branch Normal College of the Arkansas Industrial University for the Year Ending June 8th, 1900*, 4; *Annual Catalogue of the Arkansas State College for Negroes, 1928–1929*, 62–66.
77. Percentages of male San Jose graduates who worked in various fields calculated from information included in "Record of Graduates By Classes," Matthews profiled on 362, Cathbertson profiled on 365; *A History of the State Normal School of Kansas*, 161–169, Kirker profiled on 66; *Directory of Graduates*, 24; "50th Anniversary, class of 1884"; *Esto Lux*; "Register of the Alumni," 60.
78. Work Projects Administration, *The State Teachers College*, 53–54; "Some of Florence Alumni"; "50th Anniversary, class of 1884"; Brush, *In Honor*, 78; J. S. Nasmith, "An Open Letter From J. Nasmith," *Platteville Witness* LXIII (April 13, 1932), 2.
79. California Board of Trustees of the State Normal Schools, *Report, Years Ending June 30, 1883 and June 30, 1884* (Sacramento, CA: J.D. Young, 1884), 30; Frank E. Welles, "Historical Address: Review of the First Twenty Years of the Geneseo State Normal School," June 15, 1891, CASUNYG; Commissioners of Higher Education through the Commission on Coordination, "The Kansas State Normal Schools," in *The Organization, Government and Results in the State Educational Institutions of Kansas* (Presented before the Association of Higher Education, 1913), 24–25; percentages of female San Jose graduates who taught calculated from information included in "Record of Graduates By Classes"; Harmon, " 'The Voice,' " 95; Boyden, *Albert Gardner Boyden*, 74; percentages of female Oshkosh graduates who worked in the schools and average teaching tenures calculated from listing of individual graduates' "Years experience since graduating" (even-numbered years only) in "Graduates' Directory."
80. Geraldine Jonçich Clifford, " 'Daughters into Teachers' ": Education and Demographic Influences on the Transformation of Teaching into 'Women's Work' in America," in *Women Who Taught: Perspectives on the History of Women and Teaching*, ed. Alison Prentice and Marjorie R. Theobald (Toronto: University of Toronto Press, 1991), 119; Purington, *History*, 184; *Biographical Directory and Condensed History of the State Normal School, Mankato, Minn, 1870–1890* (Mankato, MN: Mankato State Normal School Alumni Association, 1891), 19–39; "50th Anniversary, class of 1884."
81. *Crucible* quoted in Kathleen Underwood, "The Pace of Their Own Lives: Teacher Training and the Life Course of Western Women," *Pacific Historical Review* 55 (November 1986): 526; marriage rates for Florence alumnae calculated using information in "Register of the Alumni"; marriage rate for Castleton

alumnae calculated using information in *General Catalogue of the Graduates of Castleton Normal School, 1867–1937* (1937), CSCA; Harmon, " 'The Voice,' " 96; marriage rates for Oshkosh alumnae calculated using information (even-numbered years only) in "Graduates' Directory."

82. Richardson and Bishop quoted in Pamela Claire Hronek, "Women and Normal Schools: Tempe Normal, a Case Study, 1885–1925" (Ph.D. diss., Arizona State University, 1985), 106; average teaching tenure for married Oshkosh alumnae calculated using information (even-numbered years only) in "Graduates' Directory"; Bugaighis, "Liberating Potential," 248; Reuter, *West Liberty*, 24; Solomon, *In the Company*, 118; Madelyn Holmes and Beverly J. Weiss, *Lives of Public Schoolteachers: Scenes From American Educational History* (New York: Garland Publishing, Inc., 1995), 232. On marriage rates among the general population and college graduates, as well as a discussion of college women's career paths, see Solomon, *In the Company*, chapter 8; Gordon, *Gender and Higher Education*, 30–33. Underwood summarizes: "Graduates [of Greeley] clearly did not race to the altar. . . . This is in sharp contrast to the patterns of their peers." Colorado State Normal School graduates "were older than their peers . . . when they married, and when they bore children." Underwood, "The Pace of Their Own Lives," 526–527, 524.
83. Reuter, *West Liberty*, 24–25; *General Catalogue of the State Normal School, Westfield, Mass.*, 111; "50th Anniversary, class of 1884"; *Directory of Graduates*, 27; *Esto Lux*; Bugaighis, "Liberating Potential," 246.
84. Purington, *History*, 80, 111, 115, 121; Salisbury, *Historical Sketches of the First Quarter-Century*, 153; "Record of Graduates By Classes," Mitchess and Packham on 298; Harmon, " 'The Voice,' " 96; Bugaighis, "Liberating Potential," 247, 255–257.
85. *Catalogue and Circular of the Branch Normal College of the Arkansas Industrial University for the Year Ending Friday, June 8th, 1900*, 4; Louie G. Ramsdell, "First Hundred Years of the First State Normal School in America: The State Teachers College at Framingham, Massachusetts—1839–1939," in *First State Normal School in America: The State Teachers College at Framingham, Massachusetts* (Framingham, MA: The Alumnae Association of the State Teachers College at Framingham, Massachusetts, 1959), 11; "Register of the Alumni"; *Esto Lux.* Even in the South, married women teachers were more common than social norms and official regulations would suggest; see, for example, Sylvia Hunt, "To Wed and to Teach: The Myth of the Single Teacher," in *Women and Texas History: Selected Essays*, ed. Fane Downs and Nancy Baker Jones (Austin: Texas State Historical Association, 1993), 127–142.
86. "50th Anniversary, class of 1884"; Brush, *In Honor*, 111, 81; Rogers, *Oswego*, 157–158, 104.
87. *Catalogue and Circular of the Branch Normal College of the Arkansas Industrial University for the Year Ending Friday, June 8th, 1900*, 4–7; *Annual Catalogue of the Arkansas State College for Negroes, 1928–1929*, 64–67; "Register of the Alumni"; Harmon, " 'The Voice,' " 96; *Biographical Directory*, 24; *Directory of Graduates*, 49; "Record of Graduates By Classes," 296.
88. Brush, *In Honor*, 72; *Florence Normal College Quarterly* 2 (Jan. 1911): 21; Hollis quoted in Rogers, *Oswego*, 83; *The Normal School Bulletin* (San Marcos) 11 (April 1922): 151–174; "Record of Graduates By Classes"; Greathead, *The Story*, 93; Deward Homan Reed, *The History of Teachers Colleges in New Mexico* (Nashville, TN: George Peabody College for Teachers, 1948), 68–69. It was

actually more common for women to serve as school superintendents in the early twentieth century than in the mid twentieth century. See Jackie M. Blount, *Destined to Rule the Schools: Women and the Superintendency, 1873–1995* (Albany: State University of New York Press, 1998).

89. *The Normalian* 2 (November 12, 1900), 6; Purington, *History*, 184; Harmon, " 'The Voice,' " 96.
90. *Albany Journal* quoted in Rogers, *Oswego*, 83; letter from Edna E. Bacon, February 11, 1906, in Rufus Henry Halsey, Personal Papers and Correspondence, UWOA; Oswego statistic in Rogers, *Oswego*, 83; Ramsdell, "First Hundred Years," 11; Purington, *History*, 135; letter from Daisy W. Chapin, Jan. 17, 1907, in Halsey, Personal Papers; Purington, *History*, 143; *Catalogue and Circular of the Branch Normal College of the Arkansas Industrial University for the Year Ending Friday, June 8th, 1900*, 4–7; Rogers, *Oswego*, 88; C. O. Ruggles, *Historical Sketch and Notes: Winona State Normal School, 1860–1910* (Winona, MN: Jones & Kroeger Co., 1910), 182–184. Many women in the nineteenth century used teaching as a means of broadening their geographical horizons; see, e.g., Mary Hurlbut Cordier, *Schoolwomen of the Prairies and Plains: Personal Narratives from Iowa, Kansas, and Nebraska, 1860s–1920s* (Albuquerque: The University of New Mexico Press, 1992); Jacqueline Jones, *Soldiers of Light and Love: Northern Teachers and Georgia Blacks, 1865–1873* (Chapel Hill: The University of North Carolnia Press, 1980).
91. Brown, *The Rise and Fall*, 76; Purington, *History*, 114; Pender, "At the New Britain Normal School," title page; *Directory of Graduates*, 93; "Record of Graduates By Classes," 360; Bugaighis, "Liberating Potential," 253, 255; "Record of Graduates By Classes"; *Annual Catalogue of the Arkansas State College for Negroes, 1928–1929*, 64–66; "Register of the Alumni," 62; Ramsdell, "First Hundred Years," 11; Bowles, *A Good Beginning*, 161.
92. "Record of Graduates By Classes," Bowers on 295, Bertola on 322; Brush, *In Honor*, 82; Bowles, *A Good Beginning*, 162; Salisbury, *Historical Sketches of the First Quarter-Century*, 156, 159; Brown, *The Rise and Fall*, 76; "And the Croix de Guerre Too!" *Geneseo Alumni News* 3 (May 1963), 6.
93. Holmes and Weiss, *Lives of Women Public School Teachers*, 2.

EPILOGUE: "LOTS OF PEP! LOTS OF STEAM!"

1. Jeffrey Selingo, "Facing New Missions and Rivals, State Colleges Seek a Makeover," *The Chronicle of Higher Education* (November 17, 2000): A40. See also Jeffrey Selingo, "Mission Creep?" *The Chronicle of Higher Education* (May 31, 2002): A19; Kit Lively, "What's in a Name? Just Ask Colleges that Want to Be Called Universities," *The Chronicle of Higher Education* (June 13, 1997): A33–A34. For a scholarly review of similar issues, see Christopher C. Morphew, " 'A Rose by Any Other Name': Which Colleges Became Universities," *The Review of Higher Education* 25 (Winter 2002): 207–223. Scholarship on teacher education also makes reference to this loss of identity; see, e.g., Geraldine Jonçich Clifford and James W. Guthrie, *Ed School: A Brief for Professional Education* (Chicago: University of Chicago Press, 1988); John I. Goodlad, Roger Soder, and Kenneth Sirotnik, eds., *Places Where Teachers Are Taught* (San Francisco: Jossey-Bass, 1990).
2. David Riesman, *Constraint and Variety in American Education* (Lincoln: University of Nebraska Press, [1956] 1958), chapter 1.

3. See Appendix for more details on when individual normal schools became teachers colleges.
4. Chico yearbook quoted in Maxine Ollie Merlino, "A History of the California State Normal Schools—Their Origin, Growth, and Transformation into Teachers Colleges" (Ed.D. diss., University of Southern California, 1962), 337.
5. Carey W. Brush, *In Honor and Good Faith: A History of the State University College at Oneonta, New York* (Oneonta, NY: The Faculty-Student Association of the State University Teachers College at Oneonta, Inc., 1965), 89; Robert T. Brown, *The Rise and Fall of the People's Colleges: The Westfield Normal School, 1839–1914* (Westfield, MA: Institute for Massachusetts Studies, Westfield State College, 1988), 121, 1914 curriculum printed on 142–156; Shackelford quoted in Michael Francis Bannon, "A History of State Teachers College, Troy, Alabama" (Ed.D. diss., George Peabody College for Teachers, 1954), 80.
6. Brown, *The Rise and Fall*, 116–118, Snedden quoted on 117; Estelle Greathead, *The Story of an Inspiring Past: Historical Sketch of the San Jose State Teachers College From 1862 to 1928 with an Alphabetical List of Matriculants and Record of Graduates by Classes* (San Jose, CA: San Jose State Teachers College, 1928), 47–49; John Marvin Smith, "The History and Growth of Southwest Texas State Teachers College," (M.A. thesis, University of Texas-Austin, 1930), 90–91; Dorothy Rogers, *Oswego: Fountainhead of Teacher Education: A Century in the Sheldon Tradition* (New York: Appleton-Century-Crofts, Inc., 1961), 145–146.
7. Rogers, *Oswego*, 98, 149; John Hugh Reynolds and David Yancey Thomas, *History of the University of Arkansas* (Fayetteville: University of Arkansas, 1910), 309–310; *Bulletin of the State Normal College, Florence, Alabama* 6 (June 1917), UCUNA, 24–26; Susan Vaughn, "The History of State Teachers College, Florence, Alabama," *Bulletin of the State Teachers College, Florence, Alabama* 18 (Supplemental, 193?), UCUNA: 23; *Bulletin of the State Normal College, Florence, Alabama* 5 (June 1916): 23.
8. Charles A. Harper, *Development of the Teachers College in the United States, with Special Reference to the Illinois State Normal University* (Bloomington, IL: McKnight & McKnight, 1935), 339–340; *Oshkosh State Normal School Bulletin* 8 (June 1911), UWOA: 20–23, 29, 11 (June 1914): 7–8; *The Normal School Bulletin* 1 (July 1912), SWTA: 15–21, 2 (July 1913): 14, 9 (July 1920): 33–40; other Oshkosh and San Marcos bulletins.
9. *Bulletin of the State Teachers College, Oshkosh, Wisconsin* 18 (July 1, 1922), UWOA: 46; *Oshkosh State Normal School Bulletin* 15 (Sept. 1, 1918): 64–65.
10. Mary Moffett Ledger, *The Social Background and Activities of Teachers College Students* (New York: Teachers College Bureau of Publications, 1929), 66; *The Normal Pennant* 6 (June 1903), SJSUA, 82, 12 (June 1909): 41, Greathead, *The Story*, 74–77; *Senior Year Book: San Jose Normal* (San Jose, CA: Students of San Jose Normal School, 1915).
11. *The Normal Star* 2 (Feb. 23, 1912), SWTA, 1; Alethean Society, Constitution and Minutes, Jan. 31, 1917, UWOA: 24; *The Normalian* (1917), CASUNYG, 75; Brush, *In Honor*, 116; Lee Graver, *A History of the First Pennsylvania State Normal School, Now the State Teachers College at Millersville* (Millersville, PA: State Teachers College, 1955), 232–233; Walter H. Ryle, *Centennial History of the Northeast Missouri State Teachers College* (Kirksville, MO: Northeast Missouri State Teachers College, 1972), 596–597; Helen E. Marshall, *Grandest of Enterprises: Illinois State Normal University, 1857–1957* (Normal, IL: Illinois State Normal University, 1956), 261, 272–274.

12. *The Quiver* (Oshkosh, WI, 1920), UWOA, 122; "Some Reminiscences of Irma L. Bruce at SWT," in Ronald C. Brown, "Oral History: San Marcos Veterans," boxes 153–154, series 7, SWTA, 9; *Bulletin of the State Teachers College, Florence, Alabama: Extra-Curricular Activities* 17 (Nov. 1929), UCUNA, no page numbers; David Sands Wright, *Fifty Years at the Teachers College: Historical and Personal Reminiscences* (Cedar Falls, IA: Iowa State Teachers College, 1926), 213–214; Brush, *In Honor*, 187–188.
13. George Frank Sammis, "A History of the Maine Normal Schools" (Ph.D. diss., University of Connecticut, 1970), 103; Edward M. Shackelford, *The First Fifty Years of the State Teachers College at Troy, Alabama, 1887–1937* (Montgomery, AL: The Paragon Press, 1937), 139; Mary Clough Cain, *The Historical Development of State Normal Schools for White Teachers in Maryland* (New York: Teachers College Bureau of Publications, 1941), 136; Cecil Dryden, *Light for an Empire: The Story of Eastern Washington State College* (Cheney, WA: Eastern Washington State College, 1965), 127; *The First Half Century of the Oshkosh Normal School* (Oshkosh, WI: The Faculty of State Normal School, 1921), UWOA, 43; *Bulletin of the State Normal College, Florence, Alabama* 2 (June 1913): 20–21, 5 (June 1916): 18, 6 (June 1918): 9–10; George Kimball Plochman, *The Ordeal of Southern Illinois University* (Carbondale, IL: Southern Illinois University Press, 1957), 15.
14. Egbert R. Isbell, *A History of Eastern Michigan University, 1849–1965* (Ypsilanti, MI: Eastern Michigan University Press, 1971), 357; Ronald A. Smith, "Athletics in the Wisconsin State University System: 1867–1913," *Wisconsin Magazine of History* 55 (Autumn 1971): 17–18; Samual R. Mohler, *The First Seventy-Five Years: A History of Central Washington State College, 1891–1966* (Ellensburg, WA: Central Washington State College, 1967), 98; Jerry G. Nye, *Southwestern Oklahoma State University: The First 100 Years* (Weatherford, OK: Southwestern Oklahoma State University, 2001), 55.
15. Smith, "Athletics," 20, 22; M. Janette Bohi, *A History of Wisconsin State University Whitewater, 1868–1968* (Whitewater, WI: Whitewater State University Foundation, 1968), 110–111; Ryle, *Centennial History*, 676–677; *The Normal Advance* 18 (November 1911), UWOA, 66–67; *La Torre of 1922, Being the Book of the Pioneer Class of the State Teachers College at San Jose* (San Jose, CA: Students of the State Teachers College, 1922), SJSUA, 29; *The Normal Advance* 17 (Nov. 1910): 70.
16. Wright, *Fifty Years*, 213.
17. Cheer printed in Mohler, *The First Seventy-Five Years*, 109.

Index

CPSIA information can be obtained at www.ICGtesting.com
Printed in the USA
LVOW120715160213

320404LV00005B/564/P